STATISTICAL ANALYSIS
A Computer
Oriented
Approach

A. A. AFIFI

UNIVERSITY OF CALIFORNIA, LOS ANGELES

S. P. AZEN

UNIVERSITY OF SOUTHERN CALIFORNIA, LOS ANGELES

ACADEMIC PRESS New York and London

ACADEMIC PRESS, INC.
111 Fifth Avenue, New York, New York 10003

United Kingdom Edition published by
ACADEMIC PRESS, INC. (LONDON) LTD.
24/28 Oval Road, London NW1

LIBRARY OF CONGRESS CATALOG CARD NUMBER: 70-182654

PRINTED IN THE UNITED STATES OF AMERICA

To our

MOTHERS: *Nazira Afifi — Shirley Azen*

WIVES: *Beverly Afifi — Colleen Azen*

CHILDREN: *Lolita, Osama, and Mostafa Afifi
and prospective Azens*

CONTENTS

1
Introduction to Data Analysis

2
Preliminary Data Analysis

3
Regression and Correlation Analysis

4
The Analysis of Variance

5
Multivariate Statistical Methods

Appendix I
Review of Fundamental Concepts

Appendix II
Statistical Tables

PREFACE

When a prospective reader picks up a statistics book he usually asks the questions: (1) What is the level of this book? (2) What are its contents? (3) Why is it different from the many available statistics books? and (4) What are its uses? The answers to these questions are as follows:

1. *Level of this book.* This book is written for readers with one elementary course in the fundamentals of statistical inference and no previous training in computer applications or programming. Appendix I serves as a review of the principles of statistical inference and Chapter 1 introduces the reader to the computer terminology and techniques used throughout the book. The minimum required level of mathematics is college algebra. Although we discuss concepts requiring mathematical sophistication beyond college algebra, such concepts are immediately "translated" into their operational meaning and usage. In addition, the book includes starred sections in which greater detail is given for the mathematically advanced reader.

2. *Content of the book.* This book contains both elementary and advanced topics. The reader will find a review of the probabilistic foundations of statistics and the standard statistical inference procedures. In addition, regression and correlation analysis, as well as the analysis of variance and multivariate analysis are explained. To cover such a wide range of material, we have eliminated mathematical derivations and computational formulas and concentrated on the essentials—the applications and interpretations of the statistical tools.

3. *Uniqueness of the book.* (a) It assumes that computations are performed on a digital computer. This enables us to eliminate the boring computational details usually found in standard texts. It also enables us to discuss techniques such as stepwise regression and stepwise discriminant analysis, which heretofore have been accessible only on an advanced mathematical level.

(b)   Many complicated topics are explained in words as well as in equations. Examples from actual research assist in motivating the concepts.

(c)   We show how simple programs can be used for complicated analysis. For example, we show how simple linear regression may be performed using a packaged descriptive program.

(d)   We demonstrate how a packaged program can be used for data analysis, for example, transforming variables to induce normality, examining residuals to verify the assumptions of the model, and so forth.

(e)   We illustrate original uses for packaged programs. For example, we demonstrate how a factorial analysis of variance program may be used to analyze a Latin square design. We also show how a descriptive program may be used to test for the linearity of the regression model.

(f)   Interspersed throughout the text are comments which highlight important supplementary information.

4. *Uses of book.* This book will serve as a *reference* book on statistics for researchers, particularly those with access to packaged computer programs. Since manuals accompanying these packaged programs usually describe only the mechanics of the programs, that is, how to set up the input in order to obtain the prescribed output, this book serves as a supplement to such manuals.

As a *textbook* this book could be used for a variety of courses. The following diagrams give four different courses at various levels:

Course 1   Elementary Applied Statistical Analysis
                         (1 semester, undergraduate)

```
┌──────────────┐
│ Appendix I   │
│ Chapter 1    │
│ Chapter 2    │
└──────────────┘
```

Course 2   Applied Statistical Analysis
                         (1 year, first-year graduate)

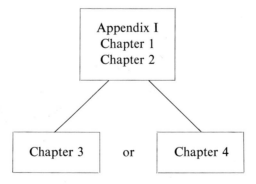

```
┌──────────────┐
│ Appendix I   │
│ Chapter 1    │
│ Chapter 2    │
└──────────────┘
      /        \
┌───────────┐     ┌───────────┐
│ Chapter 3 │  or │ Chapter 4 │
└───────────┘     └───────────┘
```

Course 3   Applied Multivariate Analysis
(1 semester, second year graduate)

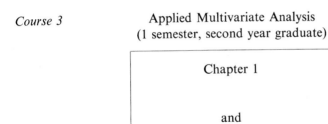

Chapter 1
and
Sections 3.2–3.4	Chapter 5

Course 4   Intensive Course in Data Analysis
(1 week, 8 hours per day)

Chapter 1			
and			
Sections 2.4–2.5	Sections 3.1, 3.3	Section 4.6	Sections 5.3–5.5

The numbering system of the book is as follows: Sections follow sequentially within a chapter; subsections, equations, comments, figures and tables follow sequentially within a section. Thus,

Section i.j means section j of Chapter i
Section i.j.k means subsection k of Section i.j
Equation i.j.k means equation k of Section i.j
Table i.j.k means table k of Section i.j
Figure i.j.k means figure k of Section i.j
Comments i.j.k means comments k of Section i.j

In the text

Comment i.j.k.h means comment h within Comments i.j.k.

Finally, as mentioned earlier, Comments identified by * are accessible only to the more mathematically advanced reader and may be skipped without loss of meaning.

ACKNOWLEDGMENTS

We wish to give our special thanks to our students Takamura Ashikaga, Anthony Auriemma, Stuart Beal, Charles Breverman, Ikbal Fahmy, Thomas Farver, Ronald Huss, Vinnie Liu, George Meier, Susan Sacks, and Girma Wolde-Tsadik for their comments, criticisms, and assistance in performing many of the computations in the book. We also thank Shirley Azen and Colleen Gillen Azen for their editorial and technical assistance.

We are indebted to Professor Richard Bellman of the University of Southern California for his interest, encouragement, and advice in the inception and completion of this book. Thanks also to Virginia Zoitl and Leona Povandra whose administrative talents were indispensible in "getting it together."

Our appreciation goes to the marvelous typists who miraculously translated our scrawls into real words—Ann Eiseman, Betty Horvath, Kay Isleib, Georgia Lum, Jean Roth, Kathy Scofield, and Shari Wilcox.

Most of the data used in examples in the book was obtained from Afifi's association with the Shock Research Unit of the University of Southern California. Much of the insight into the data discussed in the book was chiefly due to our discussion and work with Doctors Max H. Weil and Herbert Shubin of the Shock Unit. To them we owe a special measure of gratitude.

In addition, we would like to thank Norm Palley and David Erbeck of the same unit for valuable discussion of the data processing parts of the book, and to Professor Virginia Clark of the University of California, Los Angeles, for other valuable discussions.

The data for Example 1.4.2 and for many problems in the text was obtained from the Los Angeles County Heart Study with the kind permission of Dr. John

Chapman and Mrs. Anne H. Coulson of the University of California, Los Angeles.

We are indebted to the Literary Executor of the late Sir Ronald A. Fisher, F.R.S., to Dr. Frank Yates, F.R.S., and to Oliver and Boyd, Edinburgh, for permission to reprint Table III from their book "Statistical Tables for Biological, Agricultural and Medical Research."

We also wish to thank the staff and reviewers of Academic Press for their assistance, editorial remarks, etc.

Computing assistance was obtained from the Health Science's Computing Facility, UCLA, sponsored by NIH Special Research Resources Grant RR-3.

The work was done with partial support from National Institutes of Health Grant No. GM 16197-03, the United States Public Health Service research grants HE05570 and GM16462 from the National Heart Institute, and grant HS00238 from the National Center for Health Services Research and Development.

A PARTIAL LIST
OF COMMONLY USED NOTATIONS

The following is a list of the symbols introduced in Appendix I.

Symbol	Description	Section
$b_n(i, p)$	binomial distribution	I.2.1
cdf	cumulative distribution function	I.1.4
$E(X)$	expected value of a random variable X	I.1.5
$f(x)$	density of a continuous random variable X	I.1.4
$F(x)$	cumulative distribution function of a random variable X	I.1.4
$F(v_1, v_2)$	F distribution with v_1 and v_2 degrees of freedom	I.2.8
$F_{q/100}(v_1, v_2)$	qth percentile of the F distribution with v_1 and v_2 degrees of freedom	I.2.8
Gesund	practicing physician	I.1.1
H_0	null hypothesis	I.5
H_1	alternative hypothesis	I.5
$N(0, 1)$	standard normal distribution	I.2.5
$N(\mu, \sigma^2)$	normal distribution	I.2.5
$N(\mathbf{\mu}, \mathbf{\Sigma})$	multivariate normal distribution	I.6.3
P	P value	I.5.2
$p(x)$	probability function of a discrete random variable X	I.1.4
$\Pr(E)$	probability of the event E	I.1.3

Symbol	Description	Section
s	sample standard deviation	I.4.2
s^2	sample variance	I.4.2
$t(v)$	Student's t distribution with v degrees of freedom	I.2.7
$t_{q/100}(v)$	qth percentile of the Student's t distribution with v degrees of freedom	I.2.7
$U(a,b)$	uniform distribution on $[a,b]$	I.2.3
$V(X)$	variance of a random variable X	I.1.5
w	individual or experimental unit in population	I.1.1
W	population or universe	I.1.1
x	realization, observation or measurement	I.1.2
X	random variable	I.1.2
$X^{p \times 1}$	random vector	I.6.1
$X^{p \times m}$	random matrix	I.6.1
$\bar{x}$	sample mean	I.4.2
$\bar{\mathbf{x}}$	sample mean vector	I.6.2
$z_{q/100}$	qth percentile of $N(0,1)$ distribution	I.2.6
α	significance level	I.5
$1-\alpha$	confidence level	I.5
β	probability of type II error	I.5
μ	population mean	I.1.5
$\boldsymbol{\mu}$	population mean vector	I.6.3
v	degrees of freedom	I.2.6
π	power	I.5
σ	population standard deviation	I.1.5
σ^2	population variance	I.1.5
σ_{ij}	covariance between X_i and X_j	I.6.2
Σ	covariance matrix	I.6.3
θ	parameter	I.4.2
$\hat{\theta}$	estimate of parameter θ	I.4.2
$\Phi(z)$	cdf for $N(0,1)$	I.2.5
$\phi(z)$	frequency curve for $N(0,1)$	I.2.5
$\chi^2(v)$	chi square distribution with v degrees of freedom	I.2.6
$\chi^2_{q/100}(v)$	qth percentile of the chi square distribution with v degrees of freedom	I.2.6

1 INTRODUCTION TO DATA ANALYSIS

As stated in the preface there are two main objectives of this book. The first is to present in a practical manner the fundamental techniques of classical statistical analysis—both univariate and multivariate. The second main objective is to illustrate the best usage of packaged statistical programs. By this we mean (a) how to select the best program to perform a desired analysis, (b) how to interpret the various options that are available in a packaged program, (c) how to translate the output from a typical program and (d) how to use simple programs for complicated analyses.

In this chapter we treat some preliminary definitions and concepts not usually found in a typical statistics book or course. Thus, in Section 1.1 we define the kinds of data and types of measurement which arise in applications. We also comment upon elementary tools for statistical computations.

In Sections 1.2 and 1.3 we define general terminology relating to computer usage. Thus, in Section 1.2 we outline the commonly used computer *hardware* and in Section 1.3 we deal with *software* as it pertains to packaged programs. In this section we also give a list of commonly used packaged programs.

In Section 1.4 we take up the preparation of data for a packaged program and discuss the *coding sheet* and the *format statements*. In this section two sets of data are given which are used throughout the book in examples and/or exercises. Section 1.5 is a discussion of the properties of a good packaged statistical program, and finally, Section 1.6 is an outline of other important uses of the computer as a statistical tool.

1

1.1 Data, Measurements, and Computational Tools

The term *data* is used extensively in scientific investigations. In general, it refers to factual material used as a basis for discussion or decision making and in statistics, it refers to the information available for analysis and interpretation. Indeed, some statisticians view statistical analysis as data analysis (Tukey, 1962). In this book *observations*, which are realizations of underlying random variables, will constitute the data for a given problem. Hence, the words "data," "observations," and "realizations" will be used interchangeably.

In this section we discuss the types of data which arise in scientific investigations. Data is the result of making *measurements* on individuals or experimental units from the population under investigation. By measurement we mean the assignment of *numerals* to these experimental units according to some rule. Numerals may be either letters representing *classes* or *categories* in the population or they may be numbers. The numbers themselves may also represent categories of the population, in which case the rules of arithmetic are meaningless, or they may represent quantities which follow the rules of arithmetic. For example, if 1 represents the class of males and 2 the class of females, then $1 + 2$ has no meaning in this context. On the other hand, if 1 is the number of dollars earned on one day and 2 is the number of dollars earned on the next day, then $1 + 2 = 3$ has meaning in this situation; that is, it is the total number of dollars earned in the two days.

The scales of measurement that can be made on an individual are quite varied. For example, for any individual in the population of adults in the United States, we may measure (a) his sex, (b) his socio-economic status, (c) his temperature, and (d) his height. It is apparent that these four scales are quite different in nature since in (a) we can only say that one sex is *different* from the other; in (b) we can say that one status is different and *greater* than another; in (c) we can say that one temperature is different, greater, and so much *more* than another; while in (d) we can say that one height is different, greater, so much more, and so many *times* another. These four examples suggest the four scales of measurement developed by S. S. Stevens (Churchman and Ratoosh, 1959, Ch. 2) called the *nominal scale, ordinal scale, interval scale,* and *ratio scale.* We now briefly define and discuss each scale.

1. *The nominal scale*   This scale is used only to categorize individuals in the population. For each category a numeral is assigned so that two different categories will be identified by distinct numerals. For example, if individuals are classified by sex, then we may assign letters M and F, words MALE and FEMALE, or numbers 1 and 2 to the two categories.

The structure of the nominal scale is undistorted under any one-to-one substitution of the numerals. Thus, we may substitute 1 for M and 2 for F in the above example. Alternatively, we may substitute 2 for M and 1 for F or 100 for M and 1000 for F, etc.

We repeat that arithmetic operations are not meaningful for the nominal scale. Hence, the *median* and *mean* are not meaningful measures of central tendency. The appropriate statistic in this case is the *mode* since it remains unchanged under any one-to-one substitution. Thus, if there are more males than females, then the mode describes the category "male" regardless of whether we call it M, 1, 2, or 1000.

2. *The ordinal scale* In addition to categorizing individuals in the population, this scale orders the categories. For each category we assign a distinct numeral so that the order of the numerals corresponds to the order of the categories. Thus, if we assign numbers to the categories, then the categories are in numerical order; if we assign letters to the categories, then they are in alphabetical order; if we assign words to the categories, then the order is according to the meaning of the words. For example, we may wish to classify individuals into one of three socio-economic classes—low, average, high. If we choose to order these categories from low to high, then we may assign the numbers $1 = $ low, $2 = $ average, $3 = $ high; or the letters $X = $ low, $Y = $ average, $Z = $ high; or call the categories LOW, AVERAGE, HIGH. Alternatively, we could have ordered the categories from high to low; thus, $1 = $ high, $2 = $ average, $3 = $ low, and so forth. Although the numbers and letters in this example are sequential, any ordered numbers or letters will do, for example, $1 = $ low, $10 = $ average, $100 = $ high, or $A = $ low, $P = $ average, $Z = $ high, and so forth.

The structure of the ordinal scale is undistorted under any one-to-one substitution which preserves the order. For example, the transformation $1 \to 2$, $2 \to 3$, $3 \to$ any number larger than 3 is a permissible substitution, while $1 \to 2$, $2 \to 3$, $3 \to 1$ is not permissible.

Again, arithmetic operations are not meaningful for this scale so that an appropriate statistic for central tendency is one not dependent on the value of the numeral. Rather, appropriate statistics depend on the order of the numerals. Thus, the median as well as the mode are meaningful measures of central tendency.

3. *The interval scale* This scale not only categorizes and orders individuals, but also quantifies the comparison between categories. Thus, we can determine how much *more* one category is than another. To make such comparisons we need a *unit of measurement* and an *arbitrary zero point*. For example, the temperature in degrees Fahrenheit of an individual is an interval scale where 0°F is the origin and 1°F is the unit of measurement. Thus, an individual with a temperature of 100.6°F has a temperature of 2°F above normal (98.6°F).

The structure of the interval scale is undistorted under linear transformations of the form $x' = ax + b, a > 0$. The effect of this transformation is to shift the origin b units and to multiply the unit of measurement by a. As an example of such a transformation consider $x' = \frac{5}{9}(x - 32) = 0.55x - 17.8$, where x is temperature in °F. Such a transformation changes the Fahrenheit scale to the Centigrade scale.

For this scale, arithmetic operations are meaningful so that the mean as well as the median and mode are appropriate measures of central tendency.

4. *The ratio scale*   This scale is the same as the interval scale except that there is an *absolute zero point*. For this scale we can determine how many *times* as great one measurement is over another. For example, the height in inches of an individual is a ratio scale where 0 in. is a fixed zero point and 1 in. is the unit of measurement. Thus, an individual 72 in. tall is twice as tall as an individual who measures 36 in.

The structure of the ratio scale is undistorted under scalar transformations of the form $x' = cx, c > 0$. Thus, if $y = 2x$, then $y' = 2x'$. In both cases one measurement is twice the other. An example of such a transformation is $x' = \frac{1}{12}x$, which transforms inches into feet. All statistics appropriate for the interval scale are also appropriate for the ratio scale.

In choosing statistical inference techniques, the investigator should use those techniques which utilize the properties of his measurement scale. Thus, in the nominal scale only statistical methodology for nonordered categories is appropriate. This includes the χ^2 test for the multinominal distribution, the χ^2 test for association, and inferences concerning the binominal distribution. Some of these techniques are given in Sections 2.1 and 2.5. For the ordinal scale methods based on ranks are appropriate. These methods are part of the general area known as *nonparametric statistics* and are not discussed here (see, for example, Brownlee, 1965; Gibbons, 1971; Noether, 1967; Siegel, 1956; Walsh, 1965). Statistical techniques appropriate for the interval scale are also proper for the ratio scale and include all available statistical methodology.

The reader should note that in addition to the above classification of measurement scales, observations may also be classified as either discrete or continuous. Realizations of continuous random variables are called *continuous observations*, and realizations of discrete random variables are called *discrete observations*. Nominal and ordinal data are necessarily discrete, while interval and ratio data may be either discrete or continuous. For example, temperature (°F or °C) is a continuous measurement on an interval scale, while the number of phone calls received by an exchange in one hour is a discrete measurement on a ratio scale. The emphasis of this book is mostly on the analysis of continuous data measured on the interval or ratio scale.

In analyzing data, many tools are available to facilitate computations. For example, the *slide rule* is used for some simple computations ($\cdot$, $\div$, $\sqrt{x}$, $\log x$, e^x), but, of course, accuracy is sacrificed for computational ease.

One of the most useful inventions is the *desk calculator* which makes fairly sophisticated statistical analyses possible. The desk calculator is a mechanical tool which performs, $+$, $-$, $\cdot$, $\div$, and sometimes $\sqrt{x}$. Most elementary statistics texts assume that a desk calculator is available to the reader and, as a consequence, direct much effort toward expressing statistical formulas in a form suitable for the desk calculator.

Desk calculators are inefficient for large-scale multivariate statistical analyses. The revolutionary advance in computational facility came with the advent of the *digital computer*. Analyses heretofore infeasible, now become common practice. Indeed, the invention of the high-speed computer made possible several significant theoretical advances in scientific theory. We discuss the digital computer in Section 1.2.

An intermediary between the desk calculator and the high-speed computer is the *desk electronic computer*, which possesses the desirable properties of portability and low cost as well as the ability to perform fairly sophisticated arithmetic.

Finally, two aids for statistical analyses are *mathematical and statistical tables* such as Burrington (1970); "Handbook of Mathematical Tables" (1952), (Chemical Rubber Publishing Co.); Fisher and Yates (1963); Hald (1952); and Pearson and Hartley (1958).

1.2 Components of a Computer Center—the Hardware

In this section we discuss the three types of components of a computer center—the *central processor*, the *peripheral equipment*, and the *ancillary equipment*. The discussion is in no way meant to be complete, but rather it is intended to familiarize the reader with the basic terminology that is commonly used, in order to enable him to efficiently use packaged statistical programs.

The *central processor* is the component of the computer system which carries out the processing assignments of the user. Hence, the processor is that component which is commonly referred to as the *computer*. It consists of a *memory*, which stores information as a series of electronic pulses, and *logical components*, which designate different types of activity within the processor. Each type of activity is called an *instruction* to the processor; instructions may be computational, manipulative, data generating, decision making, or concerned with the input–output of information. Computational instructions primarily perform addition—subtraction is complementary addition, multiplication is successive addition, and division is successive subtraction. Functions such as log, square root, cos, and so forth, are expressible as sequences of these arithmetic operations. Manipulative instructions transfer information within the processor; input–output instructions are the means of transmitting information to or from the input–output devices; data-generating instructions generate and store symbols; and decision-making instructions compare two pieces of information and decide if they are equal.

The commonly used *media* for the input of information from user to processor are the *punched card*, *magnetic tape*, and *magnetic disc*. The commonly used media for transmitting output from processor to user include these as well as the *listing sheet* and *digital graph*. The punch card is designed so that up to 80 symbols (or characters) may be coded on each card. The magnetic tape, which resembles that used for tape recordings, may contain as many as 15,000,000 characters of information. The magnetic disc, which resembles a recording disc and contains even larger amounts of information, has an advantage over the tape in that information can be located and retrieved very fast. These last two media—the tape and the disc—are also used by the processor as *memory extensions*, that is, they store information or serve as "scratch pads" by storing intermediate results. The listing sheet displays tables of numbers and text as well as crude graphs; the digital graph displays refined graphs.

The *input–output devices* are peripheral equipment which enable the user to communicate with the central processor using the above media. Input information on punched cards is transmitted to the processor by the *card reader* which reads approximately 1000 cards per minute. The *card punch* is used to punch cards of output information transmitted by the processor. The *magnetic tape* unit may be used for both reading an input tape or writing on an output tape (at a rate of about 150 inches per second). The *magnetic disc unit* also reads or writes discs (at a rate of about 100,000 characters per second). The *high-speed line printer* produces the listing sheets, and the *digital graph plotter* makes graphs on paper by a recording pen controlled by the processor.

Another type of peripheral equipment, called a remote *terminal*, permits the user to *interact* directly with the computer system. One kind of remote terminal is the *typewriter* (or the *teletype*) by which the user types instructions to the processor, and the processor, in turn, displays results on the typewriter (or teletype). Another kind of device is the *cathode-ray* terminal which displays output on a cathode-ray tube (CRT). Input may be either by typewriter or by an electric marker called a *light pen*. One advantage of using the terminal is that the user may make decisions based on intermediate output. In the next section we discuss a packaged program which permits this kind of interaction.

The *ancillary equipment* of the computer center includes those components which enable the user to quickly perform certain mechanical operations on cards. Thus, the *key punch* punches holes in any of the 80 columns and 12 rows of a punch card. A single hole in a column represents a digit from 0–9 (a *numeric punch*) or a special symbol such as − or +, while combinations of holes in a column represent either a letter or a special symbol such as *, /, ·, $, and so forth.

Characters which are numeric, alphabetic or special symbols are called *alphanumeric*. Other ancillary devices include the *verifier*, which is used to check the punched cards; the *lister*, which prints information from punched cards onto paper; the *reproducer*, which produces a duplicate or modified copy of a set of punched cards; and the *interpreter*, which decodes each card and prints the information on the card itself. Finally, the *sorter* selects cards, reading one column on each pass, and places them in numbered bins according to the values of the digits punched in that column.

1.3 The Software

Software refers to the *programs*, that is, the sets of instructions to the central processor, which perform certain tasks. *Systems programs* reside permanently in memory and enable the user to process *applications programs*. In this book software refers to applications programs. These programs are usually written in a *problem-oriented language*, that is, one which is removed from the machine syntax (*machine language*) and closer to the user's language. Commonly used problem-oriented languages are CØBØL, FØRTRAN, and PL/1. Since most packaged programs are written in FØRTRAN, we will emphasize this language in this book.

Depending on the complexity of the task (or *job*), a program may or may not consist of a set of smaller subprograms called *subroutines*. If a program does contain subroutines, then each one performs a part of the total task, and all of them are under the control of a supervising program called the *main routine*. An advantage of using a subroutine is that it may be used (or *called*) more than once, thus saving space in the memory of the central processor. Another advantage is that a subroutine may be used with more than one program, thus saving time and effort in writing (*programming*) and correcting (*debugging*) that part of the task.

Programs may be *general* in that they can be used for a variety of tasks specified by different users. For example, a program which calculates the mean of 5 observations can be generalized to one which calculates the mean of n observations, where n is an input specified by the user. Another program, which plots a histogram with 5 class intervals, may be generalized to one which permits the number of class intervals to be specified by the user.

To be useful, a general program needs to be *documented* so that any user can apply it to his specific task. Documentation may consist of a set of explanatory cards (called *comment cards*) inserted in the deck of cards of the original program (the *source deck*), or it may consist of a write-up or a manual. For some general tasks, a set of programs and an accompanying manual are available to users at different computer centers. These programs are said to be *packaged*. In particular, if the package of programs performs statistical tasks, then these programs are called *packaged statistical programs*—the concern of this book.

If a given packaged program is available at a computer center, then it usually is stored on tape or on disc and is "on call" by the central processor. The user has access to the program through the use of certain *control statements* which retrieve the program and put it in the memory of the processor. The control statements may be specified on cards or by a peripheral device such as a typewriter or a light pen. Whichever the case, the manual accompanying the packaged programs describes the methods of accessing and using the program. (A word of caution: since many installations modify the method of accession of the program, a person knowledgeable about a given package should be consulted by the novice.)

We now discuss three popular packaged statistical programs in an effort to explain some of the above ideas.

EXAMPLE 1.3.1   One of the most popular packages of statistical programs is the Biomedical Computer Programs (BMD) written in FØRTRAN and developed by W. J. Dixon at the Medical Center of the University of California at Los Angeles. These programs are arranged in six classes: descriptive, multivariate analysis, regression analysis, special programs, time series analysis, and variance analysis. For each program the manual accompanying the package describes the type of program, the input and output of the program, limitations on the size of the input parameters, and special features of the program. A recent addition is the BMDX package which extends the capabilities of the BMD series.

To use a given program the user must prepare an input deck of cards. The deck consists of:

a. System control cards—these cards inform the central processor of the requirements of the task. For example, on the IBM 360, the JØB card specifies the job number and user, the EXEC card specifies the BMD program being used, and the DD card identifies and describes each data input or output device.

b. Program control cards—these cards inform the central processor of the details of the BMD program, for example, the parameter values, the optional computations and output, and so forth.

c. Special cards—these cards perform special operations, for example, labelling the variables, transforming the variables (called *transgeneration* cards), defining the arrangement of input and output (called *format* cards), notifying the processor of the end of a job, and so forth.

d. Data—the data may be card input, but large quantities of data are usually on tape.

EXAMPLE 1.3.2   Another popular package is the IBM Scientific Subroutine Package (SSP). It is a collection of over 250 FØRTRAN subroutines classified either as statistical or mathematical. The subroutines are purely computational in nature and do not contain any references to input–output devices. Furthermore, the subroutines do not contain fixed maximum dimensions for the variables named in their calling sequences. Thus, the user must write a FØRTRAN main program which defines the dimensions of the variables (called a DIMENSIØN statement) as well as necessary input–output statements for the solution of his problem. The subroutines in the package are entered by means of the FØRTRAN CALL statement. Details of these programs are explained in the manual accompanying the package.

EXAMPLE 1.3.3   A third package is the Service Bureau Corporation STATPACK. The purpose of these programs is to permit the user to interact with the computer through a remotely located terminal. The program allows the user to input his parameters and then to make decisions at different times during the statistical analysis. The user can apply more than one program to the same data. The program works on the 2741 terminal and types 33 and 35 teletypes. Details of this set of programs are given in the manual accompanying the package.

We now give a partial list of available packaged programs and their developers. Packaged programs are usually obtainable from the developers for a fee. (Caution: some packaged programs may use subroutines not available at the purchaser's installation.)

PACKAGED PROGRAM	DEVELOPER
(1) BMD—Biomedical Computer Programs	W. J. Dixon, Health Sciences Computing Facility, Univ. of California, Los Angeles, California
(2) SSP—Scientific Subroutine Package	IBM, 112 East Post Rd., White Plains, New York
(3) STATPACK	Service Bureau Corporation, 1350 Avenue of the Americas, New York, New York
(4) SPSS—Statistical Package for the Social Sciences	Norman H. Nie, National Opinion Research Center, Univ. of Chicago, Chicago, Illinois
(5) OSIRIS	John Sonquist, Inst. for Social Research, Univ. of Michigan, Ann Arbor, Michigan

PACKAGED PROGRAM	DEVELOPER
(6) Data Text	Arthur Couch and David Armor, Harvard Univ., Cambridge, Massachusetts
(7) SSIPP—Social Sciences Instructional Programming Project	Michael A. Hall, Beloit College, Beloit, Wisconsin
(8) Regression Analysis for Economists	William Raduchel, Harvard Univ., Cambridge, Massachusetts
(9) ATSTAT–TSTAT	Rand/System Development Corp., Santa Monica, California
(10) SPAN	System Development Corp., Santa Monica, California
(11) PSTAT	Princeton Univ., Princeton, New Jersey
(12) PERSUB	Lackland Air Force Base
(13) BASIS	Burroughs Corp.,
(14) SSUPAC	Univ. of Illinois, Urbana, Illinois
(15) TSR	Univ. of Michigan, Ann Arbor, Michigan
(16) MSA	Harvard Univ., Cambridge, Massachusetts

COMMENTS 1.3.1   1.   The results of a packaged statistical program should not make decisions for the researcher. The program relieves the researcher from the burdens of the computations, but the interpretation of the results depends on the sophistication and thought of the analyst.

2.   Some disadvantages in using packaged programs are:

(a)   The researcher must adjust to the notation and requirements of a packaged program. Manuals explaining the use of packaged programs are often not helpful in interpreting the statistical output.

(b)   The user of a statistical packaged program is restricted to the numerical methods used by the program. These methods may not be the most desirable or efficient for the researcher's problem.

(c)   A packaged program may not print all the information desired by the researcher. For example, some programs calculate a point estimate but not a confidence interval for a parameter.

(d)   Packaged programs are written only for standard statistical techniques. The researcher wishing to perform a nonstandard analysis may have to write his own program. In fact, he should not use a packaged program if it is not the most appropriate for his problem.

1.4 Preparation of Data for Packaged Programs

In this section we discuss the preparation of data for computer usage. Much of the data arising from scientific investigations is collected and recorded manually, for example, laboratory sheets, questionnaires, special forms, and so forth. The reasons are that it is easier and cheaper than automating the collection of data, and also, in many situations the investigator must make intermediate decisions concerning the measurement. On the other hand, in sophisticated data processing situations the volume of data justifies the adoption of particular automated procedures entailing the use of specialized hardware and software. We discuss such a situation in Example 1.4.1.

If the data has been collected and recorded manually, it is necessary to prepare the data in some logical and concise manner which is compatible with the input medium. Since discs and tapes are sophisticated forms of input, we discuss here only the preparation of data for the most commonly used medium of input—the punched card. Before the data is punched on cards, it must be transcribed or *coded* onto a special form called a *coding sheet*. The coding sheet specifies the assignment of the 80 columns of the card to the variables in the study. Although the assignment is arbitrary, some general recommendations appropriate for FØRTRAN programs are as follows:

1. Each individual (or experimental unit) should have a unique identifying numeral to be distinguished from the other individuals in the sample. The identification numeral is usually coded in the first or the last columns of the form. In particular, columns 73–80 are often used for this purpose.

One common method of identification, the assignment of successive integers to the individuals, has the advantage that the cards can be sorted on the identification number in case they are out of sequence. The number of columns used for this integer is determined by the known or anticipated size of the sample. Thus, if the sample size is 493, then at least three columns are used. Here, and in all data coding, integers must be *right justified*. Thus, if columns 78–80 are used for the ID number, then the integer 2 is coded in column 80 and not in either of the other two columns. Similarly, the integer 32 should be coded in columns 79 and 80. Sometimes leading zeroes are coded to avoid mistakes, so that, for example, 2 and 32 are coded as 002 and 032.

Other identification numerals are the unique 9-digit social security number or a composite of letters and numbers representing attributes such as sex, race, first and last letters of the first name, date of birth, and so forth.

2. Each observation should be coded to the accuracy with which the measurement was made—it is generally inadvisable to round off or truncate the original measurement. For example, if temperature is recorded to one decimal place, it is

not recommended to code it as an integer. Thus the detail which went into the measuring technique is preserved, at least for the initial stages of analysis. The rounding or truncating of the observations can always be made at a later stage.

3.   For any variable a sufficient number of columns should be allowed to accommodate all of the observations in the sample. The minimum number of columns is determined by the observation with the most significant places. Thus, the integer observations 386, 7232, and 24 need at least 4 columns. An integer observation, requiring a smaller number of columns, is always right justified and may be preceded by leading zeroes. Thus, the above observations may be coded as

Col.	1	2	3	4		Col.	1	2	3	4
		3	8	6	or as		0	3	8	6
	7	2	3	2			7	2	3	2
			2	4			0	0	2	4

If some of the observations are negative, an additional column must be reserved for the minus sign. Thus if the third observation above is -24, then we may code these observations as

Col.	1	2	3	4	5
			3	8	6
		7	2	3	2
	—			2	4

Although it is not necessary, a plus sign may precede positive observations.

4.   If the values of the variable include decimal fractions, then a column may or may not be used for the decimal point. If we do code the decimal point, then it need not appear in the same column for each observation. Thus, for example, the observations 723.2, 38.6, and 0.24 may be coded as

Col.	1	2	3	4	5	6
		7	2	3	.	2	0
			3	8	.	6	0
				0	.	2	4

in which case the decimal always appears in column 4, or as

Col.	1	2	3	4	5
		7	2	3	.	2
			3	8	.	6
		0	.	2	4	0

in which case the decimal appears in different columns. Note that the observations need not be right justified if the decimal is punched. If we do not code the decimal point, then decimal placement is handled during processing by the *format statement* (discussed later). In this case the assumed decimal point *must* appear in the same place and all observations are right justified. Thus, the above observations would be coded as

Col.	1	2	3	4	5
	7	2	3	2	0
		3	8	6	0
			0	2	4

where the assumed decimal is between columns 3 and 4. It would be incorrect to code it as

Col.	1	2	3	4	5
	7	2	3	2	0
		3	8	6	0
		0	2	4	0

5. A *missing value* (that is, an observation which was not obtained or was lost) may be coded as a blank or it may be assigned a *special value*, that is, a number which cannot possibly occur. For example, -10.0 (feet) is a special value for the height of an individual, 999 (years) is a special value for the age of an individual, and 9 is a special value for an ordinal measurement which is scaled from 1 to 7.

A disadvantage of using blanks is that some computers do not distinguish them from zeroes. Thus, if 0 is a possible value of a variable, then a blank would incorrectly be read as a 0. (Some computers read blanks as -0 making it possible to distinguish between the two symbols.)

A disadvantage of using special values is that the coder may forget to use the assigned special value and code a dash or a blank instead.

6. It is advisable to assign numbers rather than letters to measurements made on the nominal or ordinal scale. It is best to avoid using 0 as one of the codes since it may be confused with a blank.

7. It is not advisable to change measurements made on the interval or the ratio scale to the ordinal scale. For example, measurements of age in years should not be coded as $1 =$ under 21, $2 = 21$ or older but less than 35, and $3 = 35$ or older. This would unnecessarily sacrifice some of the information in the original measurement. It can be done either by the processor or by the investigator at a later stage in the analysis.

8. Sometimes two or more variables may be combined into a single variable without loss of information. For example,

$$X_1 = \begin{cases} 0 & \text{if no children in household} \\ 1 & \text{if children in household,} \end{cases}$$

$$X_2 = \text{age of oldest child in household,}$$

could be combined into

$$Y_1 = \text{age of oldest child in household,}$$

where $Y_1 = 0$ implies there are no children in the household. Combining variables in this way eliminates missing data for X_2 and also saves column space.

9.   The order of the variables may be arranged to form meaningful clusters. For example, the items on an application for admission to a university may be arranged in the following groups:

(a)   identification number;

(b)   personal data—height, weight, marital status, and so forth;

(c)   educational data—achievement test scores, IQ score, grade-point average, and so forth;

(d)   work experience—kind of employment, number of years at last job, and so forth.

10.   More than one card may be used to code all the measurements for each subject. The investigator is not restricted to the 80 columns on a card. When more than one card is used, it is advisable to code each type of card on a separate coding sheet. It is also recommended to code the identification number on each card and also to reserve columns for the card sequence number.

EXAMPLE 1.4.1   At the Shock Research Unit at the University of Southern California, Los Angeles, California, data on many physiological variables are collected successively in time on each patient. Specialized automated data collection procedures were developed for collecting this data and recording it on tape as well as on listing sheets (Stewart *et al.*, 1968; Palley *et al.*, 1970). From the wealth of data that was collected, a special set was extracted and coded for keypunching to be used in examples and exercises in this book.

In this set, initial measurements (that is, measurements upon admission) and final measurements on the same variables (that is, measurements just before death or discharge) were collected on 113 critically ill patients. The coding sheet for these variables is presented in Fig. 1.4.1. Note that the first four columns are reserved for the patient identification number and the last column is coded as 1 for initial measurements and 2 for final measurements. Details of the coding of the variables are given in Table 1.4.1 and the data is given in Table 1.4.2. This data will be called *Data Set A*.

EXAMPLE 1.4.2   As another example we present a subset of the data from an epidemiological heart disease study on Los Angeles County employees. The details of the coding of the variables are given in Table 1.4.3 and the data is given in Table 1.4.4. This data will be called *Data Set B*.

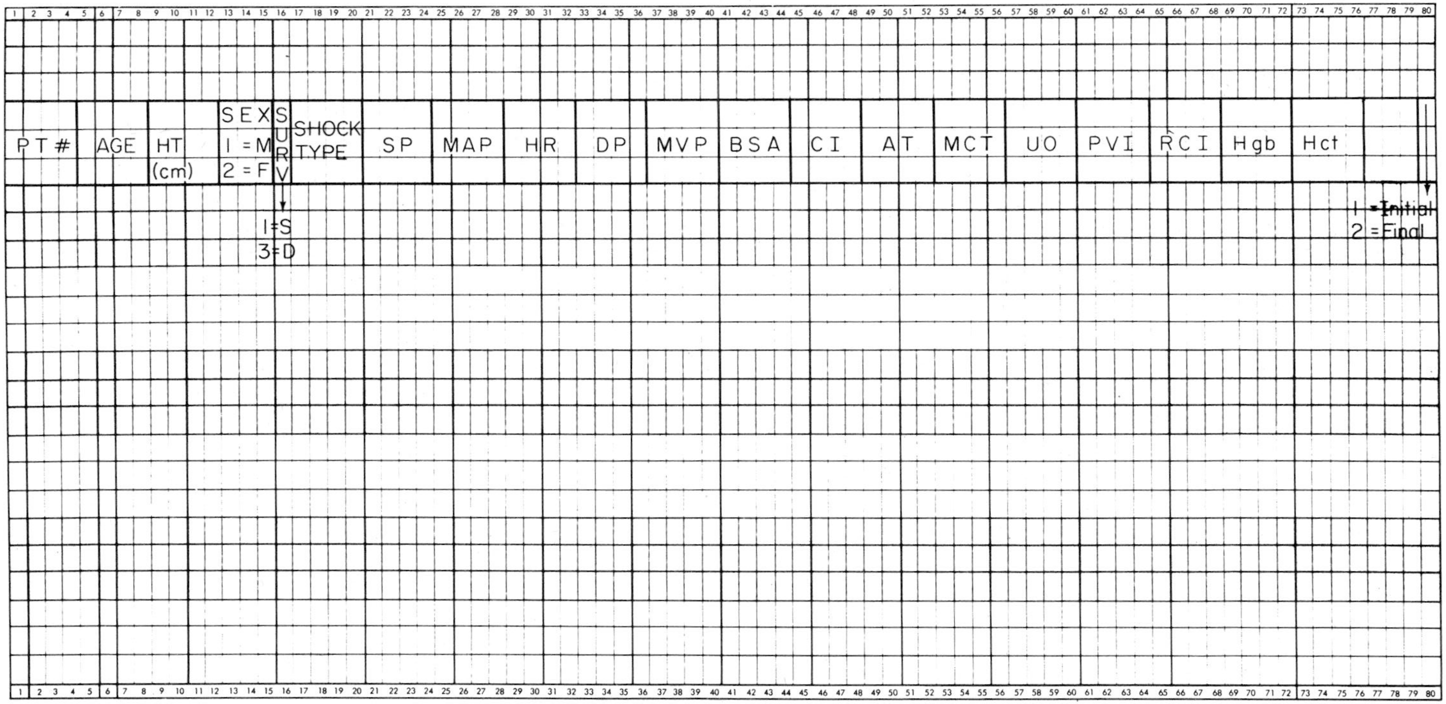

FIG. 1.4.1. *Coding sheet for Example 1.4.1.*

TABLE 1.4.1. *Details of the Variables in Example 1.4.1 (Data Set A)*

Column	Variable	Units	Scale	Comments
1–4	ID	none	Nominal	Patients numbered sequentially
5–8	Age	yr	Ratio	Age at last birthday
9–12	Height	cm	Ratio	
13–15	Sex	none	Nominal	1 = Male, 2 = Female
16	Survival	none	Nominal	1 = Survived, 3 = Died
17–20	Shock type	none	Nominal	2 = Nonshock
				3 = Hypovolemic shock
				4 = Cardiogenic shock
				5 = Bacterial shock
				6 = Neurogenic shock
				7 = Other
21–24	Systolic pressure	mm Hg	Ratio	Recorded to nearest integer
25–28	Mean arterial pressure	mm Hg	Ratio	Recorded to nearest integer
29–32	Heart rate	beats/min	Ratio	Discrete variable
33–36	Diastolic pressure	mm Hg	Ratio	Recorded to nearest integer
37–40	Mean central venous pressure	mm Hg	Ratio	Decimal between cols. 39, 40
41–44	Body surface area	m²	Ratio	Decimal between cols. 42, 43
45–48	Cardiac index	liters/min/m²	Ratio	Decimal between cols. 46, 47
49–52	Appearance time	sec	Ratio	Decimal between cols. 51, 52
53–56	Mean circulation time	sec	Ratio	Decimal between cols. 55, 56
57–60	Urinary output	ml/hr	Ratio	Recorded to nearest integer
61–64	Plasma volume index	ml/kg	Ratio	Decimal between cols. 63, 64
65–68	Red cell index	ml/kg	Ratio	Decimal between cols. 67, 68
69–72	Hemoglobin	gm	Ratio	Decimal between cols. 71, 72
73–76	Hematocrit	percent	Ratio	Decimal between cols. 75, 76
77–79	Blank			
80	Card sequence	none	Ordinal	1 = Initial, 2 = Final

TABLE 1.4.2. *Data for Example 1.4.1* (*Data Set A*)

Patient ID	Age	Height	Sex	Survival	Shock Type	SP	MAP	HR	DP	MVP	BSA	CI	AT	MCT	UO	PVI	RCI	Hgb	Hct	1=Init 2=Final
517	68	165	11		2	114	88	95	73	17	141	66	115	225	110	562	206	113	340	1
517	68	165	11		2	131	98	81	76	48	141	241	89	183	180	667	292	100	335	2
537	37	171	11		2	149	115	76	97	36	182	355	82	156	40	507	234	127	390	1
537	37	171	11		2	144	106	104	86	30	182	519	63	138	50	507	234	107	325	2
546	50	175	11		2	146	101	76	74	80	169	405	56	125	0	644	239	134	410	1
546	50	175	11		2	125	85	77	61	46	171	383	72	150	40	644	239	101	330	2
563	53	157	21		2	107	83	183	70	198	174	95	64	380	0	294	278	155	460	1
563	53	157	21		2	127	92	97	73	105	179	305	92	178	625	459	175	131	310	2
562	75	177	11		2	141	65	100	82	41	175	190	126	297	42	471	294	137	420	1
562	75	177	11		2	173	115	75	92	115	175	222	145	251	37	471	294	127	360	2
629	66	178	13		2	114	59	102	44	138	189	348	90	168	0	495	206	93	280	1
629	66	178	13		2	72	46	100	35	128	190	228	69	147	0	440	206	91	250	2
634	52	185	13		2	112	67	73	49	150	200	380	82	151	0	525	152	92	280	1
634	52	185	13		2	89	44	57	30	124	202	253	90	170	4	525	145	89	260	2
583	68	169	21		2	95	65	97	53	131	174	140	149	446	0	458	260	124	400	1
583	68	169	21		2	124	76	87	56	82	173	137	146	411	381	532	199	132	355	2
585	73	155	21		2	154	97	73	67	55	167	365	104	167	0	430	281	130	390	1
585	73	155	21		2	160	108	85	74	69	167	365	89	164	150	430	281	116	350	2
594	53	168	21		2	138	101	110	70	31	151	330	51	113	205	632	191	103	300	1
594	53	168	21		2	155	110	103	78	87	151	411	48	100	152	663	156	95	290	2
630	64	152	21		2	146	109	114	88	50	152	233	134	234	0	425	164	110	330	1
630	64	152	21		2	129	105	106	91	34	152	191	172	292	226	443	140	109	375	2
642	65	178	11		2	124	80	130	64	77	184	291	60	173	0	475	230	100	310	1
642	65	178	11		2	127	87	107	60	88	184	471	76	150	65	651	197	103	280	2
639	74	165	11		2	105	74	97	53	95	174	369	96	190	0	620	158	78	250	1
639	74	165	11		2	118	78	85	54	123	174	268	111	188	30	594	169	100	295	2
651	31	170	11		2	131	82	129	70	57	176	183	48	173	0	223	370	123	460	1
651	31	170	11		2	164	104	112	82	94	176	344	70	131	43	396	178	118	410	2
649	29	170	11		2	146	100	54	74	68	181	135	81	152	0	386	210	133	410	1
649	29	170	11		2	120	93	101	79	4	181	260	79	162	30	393	143	134	360	2
648	56	155	21		2	91	72	81	55	136	129	410	20	122	405	701	162	79	240	1
648	56	155	21		2	106	61	87	40	55	130	296	65	154	44	679	189	112	265	2
667	22	165	21		2	113	84	101	65	40	164	277	51	193	510	393	195	122	365	1
667	22	165	21		2	117	86	137	67	33	161	312	14	71	75	467	159	118	300	2
665	24	165	21		2	114	84	87	67	27	148	260	80	162	377	486	259	133	420	1
665	24	165	21		2	123	81	149	65	5	149	406	41	105	200	587	202	102	320	2
664	46	163	21		2	123	72	111	56	32	162	332	44	116	12	433	148	101	315	1
664	46	163	21		2	164	101	114	76	48	162	424	39	112	97	489	190	113	340	2

TABLE 1.4.2 (continued)

Patient ID	Age	Height	Sex	Survival	Shock Type	SP	MAP	HR	DP	MVP	BSA	CI	AT	MCT	UO	PVI	RCI	Hgb	Hct	1=Init 2=Final
685	40	183	1	1	2	108	73	28	59	95	195	234	147	278	0	715	247	100	340	1
685	40	183	1	1	2	109	75	77	60	93	195	280	147	267	0	715	247	100	340	2
684	77	168	1	1	2	74	53	93	42	97	183	300	95	194	15	668	178	105	270	1
684	77	168	1	1	2	107	61	97	44	79	184	327	97	178	58	617	176	125	370	2
679	50	178	1	1	2	166	105	140	78	26	195	421	22	98	318	482	151	83	250	1
679	50	178	1	1	2	99	96	122	92	38	195	378	29	110	350	493	151	92	271	2
715	76	152	2	3	2	116	88	122	70	83	144	188	144	342	23	498	171	96	290	1
715	76	152	2	3	2	109	78	84	58	64	144	168	158	337	1	529	237	109	315	2
687	67	161	1	1	2	102	61	74	41	75	167	155	111	209	60	397	212	136	380	1
687	67	161	1	1	2	146	83	77	58	20	167	226	137	210	95	393	212	129	330	2
689	37	169	1	1	2	97	63	96	45	56	178	617	51	114	200	645	116	66	200	1
689	37	169	1	1	2	83	56	92	42	34	178	594	69	122	200	652	161	66	200	2
698	55	168	1	1	2	137	84	25	60	77	165	551	50	98	450	621	336	133	400	1
698	55	168	1	1	2	117	60	125	48	82	165	542	51	102	42	845	336	133	420	2
700	60	154	2	1	2	148	73	96	44	186	167	34	168	327	160	590	175	98	300	1
700	60	154	2	1	2	145	75	102	39	151	167	185	134	260	850	590	175	102	290	2
705	44	161	2	1	2	132	96	86	76	87	142	254	119	237	0	656	467	143	450	1
705	44	161	2	1	2	141	103	85	81	94	142	247	76	225	0	656	467	143	450	2
713	54	170	1	1	2	150	104	66	77	31	178	426	91	165	0	463	252	116	350	1
713	54	170	1	1	2	162	112	65	82	19	178	328	126	232	0	463	252	116	350	2
716	65	170	1	1	2	141	75	55	44	152	159	294	191	344	3	867	215	77	250	1
716	65	170	1	1	2	130	75	81	45	79	159	702	115	237	52	781	205	91	225	2
721	54	178	1	1	2	120	89	95	72	4	172	166	133	228	375	593	158	83	265	1
721	54	178	1	1	2	146	112	97	88	18	172	270	122	216	400	559	261	118	345	2
722	57	164	2	1	2	171	117	92	80	302	186	443	32	81	0	494	169	93	280	1
722	57	164	2	1	2	165	114	36	77	319	186	389	34	101	0	494	169	92	290	2
732	22	179	1	1	2	153	110	126	88	90	185	758	47	93	190	744	162	103	310	1
732	22	179	1	1	2	138	105	119	84	81	185	794	25	96	176	601	199	99	370	2
742	52	182	1	1	2	159	119	106	99	28	204	405	77	137	30	398	228	136	410	1
742	52	182	1	1	2	163	117	221	100	95	204	175	68	385	20	468	228	124	385	2
543	52	152	2	3	3	82	52	106	38	189	155	589	28	97	0	663	124	71	300	1
543	52	152	2	3	3	77	35	101	26	124	164	334	50	132	1	745	146	63	200	2
541	59	169	2	3	3	99	58	140	45	82	158	472	60	124	0	479	194	85	280	1
541	59	169	2	3	3	147	97	115	72	115	158	386	60	117	23	500	275	79	260	2
560	70	173	1	3	3	80	49	82	38	67	185	249	130	232	0	460	182	112	335	1
560	70	173	1	3	3	63	40	49	32	147	185	178	170	325	1	550	179	122	230	2
573	79	152	2	3	3	68	49	175	40	143	158	124	59	296	0	333	154	110	330	1
573	79	152	2	3	3	65	55	98	49	142	165	108	191	390	3	353	164	135	410	2

TABLE 1.4.2 (continued)

Patient ID	Age	Height	Sex	Survival	Shock Type	SP	MAP	HR	DP	MVP	BSA	CI	AT	MCT	UO	PVI	RCI	Hgb	Hct	1=Init 2=Final
593	61	149	1	3	3	75	47	135	35	26	137	559	53	100	0	755	215	97	290	1
593	61	149	1	3	3	72	25	70	21	1	142	135	157	341	37	696	146	111	275	2
588	56	168	1	3	3	105	52	73	37	114	166	418	63	122	0	479	108	74	235	1
588	56	168	1	3	3	42	24	59	17	93	166	203	94	196	0	479	108	71	245	2
596	89	171	1	3	3	26	15	103	10	184	170	313	157	256	0	756	179	75	230	1
596	89	171	1	3	3	60	38	96	27	180	177	158	210	401	0	673	323	113	340	2
584	69	168	1	1	3	96	72	107	61	58	184	143	110	237	0	210	176	152	480	1
584	69	168	1	1	3	168	114	131	84	92	189	365	58	118	109	399	198	144	400	2
650	56	155	2	3	3	106	79	98	63	61	131	225	128	253	0	835	235	76	235	1
650	56	155	2	3	3	65	31	106	21	64	131	302	46	133	0	730	235	69	250	2
625	53	165	2	1	3	122	86	81	72	64	178	139	208	441	0	324	179	135	415	1
625	53	165	2	1	3	127	89	86	74	106	181	102	26	271	88	444	167	141	375	2
613	61	163	2	1	3	74	45	116	36	35	192	172	53	129	0	349	151	100	300	1
613	61	163	2	1	3	168	103	81	72	129	192	340	42	101	2	349	203	102	250	2
692	26	177	1	3	3	150	116	91	96	60	184	286	59	139	41	377	231	106	270	1
692	26	177	1	3	3	93	59	81	45	149	184	352	30	114	0	545	158	90	260	2
672	80	166	1	1	3	63	49	110	43	57	180	92	261	498	57	321	236	160	470	1
672	80	166	1	1	3	140	98	123	76	46	179	330	123	212	44	567	159	137	265	2
719	66	151	2	3	3	67	54	96	46	94	133	83	85	205	1	300	119	99	300	1
719	66	151	2	3	3	132	89	81	65	46	133	131	80	191	10	465	105	98	170	2
693	18	166	1	1	3	136	107	100	97	20	177	107	96	192	383	207	234	112	300	1
693	18	166	1	1	3	133	89	128	65	25	177	688	32	85	42	593	168	104	310	2
695	47	165	2	1	3	85	65	99	56	26	166	171	75	191	53	403	133	111	325	1
695	47	165	2	1	3	131	95	93	76	59	166	238	21	176	300	460	133	83	255	2
734	50	173	1	1	3	129	86	154	72	75	186	366	78	133	5	483	210	123	365	1
734	50	173	1	1	3	146	97	144	78	111	186	399	57	117	270	483	210	123	365	2
444	75	140	2	3	4	62	51	97	43	130	130	60	150	590	5	335	208	147	430	1
444	75	140	2	3	4	70	48	78	37	110	130	120	180	510	10	415	208	109	320	2
340	70	160	2	3	4	62	38	53	29	100	187	90	190	390	0	394	241	131	400	1
340	70	160	2	3	4	129	74	72	53	190	187	120	130	300	15	394	241	112	365	2
529	60	165	2	3	4	145	99	110	75	220	190	156	184	393	10	335	200	125	425	1
529	60	165	2	3	4	182	103	106	72	210	190	217	159	370	15	335	200	125	450	2
426	47	176	1	1	4	80	64	84	55	10	180	110	120	280	80	373	272	146	490	1
426	47	176	1	1	4	87	68	77	52	40	180	410	100	170	75	508	217	99	320	2
412	56	173	1	1	4	83	66	110	60	10	182	126	221	407	110	362	240	166	500	1
412	56	173	1	1	4	102	75	108	63	90	182	281	100	206	50	564	266	154	330	2
518	71	164	2	1	4	102	74	112	65	19	169	133	153	313	80	321	141	130	403	1
518	71	164	2	1	4	121	79	84	56	35	169	256	85	184	90	398	141	94	290	2

TABLE 1.4.2 (continued)

Patient ID	Age	Height	Sex	Survival	Shock Type	SP	MAP	HR	DP	MVP	BSA	CI	AT	MCT	UO	PVI	RCI	Hgb	Hct	1=Init 2=Final
575	69	150	23		4	82	59	126	48	80	155	141	124	290	0	333	169	120	370	1
575	69	150	23		4	63	52	135	47	129	151	155	128	252	11	333	169	103	350	2
568	60	155	21		4	151	92	119	74	21	133	172	76	208	0	428	164	105	325	1
568	60	155	21		4	152	88	113	64	16	133	361	116	208	125	474	164	71	210	2
655	90	147	23		4	137	94	101	72	61	144	133	181	324	0	272	210	137	415	1
655	90	147	23		4	92	59	80	45	33	144	131	143	300	20	396	210	96	310	2
592	62	168	11		4	98	71	104	59	112	183	218	70	253	0	531	278	180	540	1
592	62	166	11		4	113	78	73	59	142	185	311	121	229	25	805	141	90	260	2
598	63	177	11		4	115	97	78	85	181	180	115	148	361	0	395	254	152	470	1
598	63	177	11		4	103	72	62	56	118	180	254	93	191	235	699	219	148	290	2
660	47	155	23		4	103	81	78	66	180	164	141	110	374	0	262	178	122	430	1
660	47	155	23		4	62	43	83	33	121	165	290	36	146	0	401	162	106	310	2
638	38	163	21		4	144	82	105	63	175	183	169	120	258	0	440	229	137	420	1
638	38	163	21		4	119	75	118	56	80	158	326	67	168	17	450	229	119	360	2
686	70	164	13		4	85	51	76	36	111	180	356	79	161	0	569	212	119	360	1
686	70	164	13		4	74	45	67	32	124	180	249	92	180	0	729	212	119	360	2
707	58	145	23		4	94	50	152	36	86	109	444	77	139	1	780	269	119	320	1
707	58	145	23		4	48	26	59	19	95	109	205	181	328	1	870	269	59	185	2
659	57	177	11		4	153	124	104	108	58	186	141	190	403	0	325	238	170	510	1
659	57	177	11		4	116	86	119	70	70	184	303	107	192	20	465	161	155	445	2
696	60	170	11		4	131	87	94	67	44	175	142	202	386	28	553	259	119	360	1
696	60	170	11		4	169	91	97	69	166	164	164	232	452	22	553	259	126	310	2
730	50	168	13		4	52	33	85	25	60	173	216	158	358	1	712	219	97	310	1
730	50	168	13		4	38	26	74	20	52	173	308	124	246	1	715	219	84	240	2
758	58	175	13		4	102	82	103	69	146	181	169	170	312	4	445	258	156	470	1
758	58	175	13		4	59	48	184	41	137	181	95	147	386	1	492	247	116	310	2
743	42	169	11		4	67	51	217	45	113	191	162	179	347	3	378	256	146	440	1
743	42	163	11		4	91	60	93	45	40	191	271	76	154	110	420	219	123	370	2
515	61	173	11		5	128	91	107	71	115	163	230	74	193	140	747	186	79	240	1
515	61	173	11		5	134	91	97	67	52	163	288	82	171	73	703	186	98	300	2
528	69	161	11		5	91	71	135	61	141	169	254	120	258	5	489	271	130	390	1
528	69	161	11		5	141	94	69	70	79	169	294	90	207	500	688	187	96	280	2
526	78	160	11		5	90	60	113	46	86	163	330	100	194	21	653	168	91	270	1
526	78	160	11		5	147	91	95	63	22	163	282	107	186	88	829	168	100	300	2
549	69	168	11		5	118	83	73	62	84	179	258	97	166	0	413	149	100	300	1
549	69	168	11		5	164	109	90	74	17	179	428	52	104	100	602	177	113	340	2
555	43	160	11		5	101	81	145	70	2	148	346	82	167	0	637	289	114	370	1
555	43	160	11		5	109	85	217	70	18	148	475	51	128	37	742	289	124	300	2

Table 1.4.2 (continued)

Patient ID	Age	Height	Sex	Survival	Shock Type	SP	MAP	HR	DP	MVP	BSA	CI	AT	MCT	UO	PVI	RCI	Hgb	Hct	1=Init 2=Final
658	37	160	2	3	5	100	59	120	43	67	179	349	55	133	0	389	117	105	330	1
658	37	160	2	3	5	44	25	120	17	64	179	296	46	124	0	573	171	87	310	2
702	61	168	1	3	5	116	64	105	48	128	164	354	141	300	320	561	185	96	290	1
702	61	168	1	3	5	90	42	53	31	130	164	240	144	338	20	561	185	89	310	2
657	48	154	2	1	5	86	63	86	51	76	154	207	61	143	0	490	107	73	210	1
657	48	154	2	1	5	126	84	80	60	46	154	352	47	108	77	518	108	92	280	2
725	17	150	2	3	5	72	60	169	51	166	142	362	45	116	7	547	186	103	310	1
725	17	150	2	3	5	77	41	87	29	122	142	401	40	110	1	473	160	89	270	2
699	55	187	1	1	5	80	62	135	50	161	225	469	38	137	1	687	133	67	200	1
699	55	187	1	1	5	162	117	94	88	140	225	428	82	169	42	560	187	105	295	2
729	62	155	2	3	5	56	30	95	22	100	167	259	97	171	1	384	124	83	265	1
729	62	155	2	3	5	67	36	101	29	152	172	225	84	173	1	337	156	79	220	2
733	50	163	2	3	5	88	52	109	42	144	145	183	78	197	1	504	189	110	330	1
733	50	163	2	3	5	62	23	57	16	116	146	155	110	232	1	640	183	72	215	2
720	82	177	1	1	5	98	62	126	43	44	188	497	48	114	15	712	244	91	270	1
720	82	177	1	1	5	164	112	91	78	34	188	273	57	171	17	587	182	93	295	2
741	61	155	2	3	5	103	75	134	62	142	161	196	160	303	2	542	209	116	350	1
741	61	155	2	3	5	57	32	71	22	122	163	177	59	226	3	419	110	96	290	2
723	58	161	1	1	5	82	61	82	48	152	153	763	145	275	1	1066	280	80	245	1
723	58	161	1	1	5	106	67	54	50	155	153	273	168	346	4	1066	280	102	260	2
731	78	160	1	1	5	96	65	130	50	23	171	535	74	133	1	585	242	105	315	1
731	78	160	1	1	5	134	95	95	69	99	173	347	102	186	1	556	185	104	340	2
530	53	173	1	3	6	106	83	79	69	167	208	200	85	174	50	355	221	122	380	1
530	53	173	1	3	6	123	88	129	69	83	208	542	56	128	75	461	221	88	270	2
545	42	157	2	3	6	90	66	83	56	91	149	120	42	115	0	362	187	143	430	1
545	42	157	2	3	6	52	33	92	24	162	149	228	13	206	5	443	133	62	190	2
522	30	163	2	1	6	97	71	93	55	151	158	286	60	144	10	557	151	99	300	1
522	30	163	2	1	6	124	82	95	57	7	158	411	39	95	55	565	134	100	300	2
540	30	160	2	1	6	126	105	101	90	115	151	121	54	191	0	308	172	139	420	1
540	30	160	2	1	6	125	88	92	67	97	151	169	41	107	115	385	161	116	350	2
554	28	169	2	1	6	88	69	59	57	90	166	187	97	211	100	459	201	112	340	1
554	28	169	2	1	6	132	88	111	65	3	163	401	40	91	525	436	165	116	285	2
620	45	170	2	3	6	71	56	176	50	31	166	110	78	232	0	348	290	157	470	1
620	45	170	2	3	6	53	34	84	26	98	167	176	101	229	0	348	290	136	220	2
662	37	160	2	3	6	45	37	89	32	64	154	137	90	184	0	436	137	93	280	1
662	37	160	2	3	6	78	42	105	28	34	162	300	43	96	5	345	289	112	340	2
676	24	169	1	3	6	91	75	83	66	115	164	227	102	202	110	552	355	141	420	1
676	24	169	1	3	6	42	22	56	17	46	164	150	70	197	0	586	166	98	290	2

TABLE 1.4.2 (continued)

Patient ID	Age	Height	Sex	Survival	Shock Type	SP	MAP	HR	DP	MVP	BSA	CI	AT	MCT	UO	PVI	RCI	Hgb	Hct	1=Init 2=Final
631	55	178	11		6	107	81	90	62	15	177	291	67	151	0	559	201	121	380	1
631	55	178	11		5	85	64	83	49	60	177	264	49	139	21	530	201	131	350	2
691	68	160	13		6	155	103	104	76	69	169	237	132	233	0	319	187	104	310	1
691	68	160	13		6	94	60	84	44	91	172	230	100	184	2	415	149	104	310	2
646	34	168	11		6	110	79	122	62	112	172	290	65	132	0	450	169	111	335	1
646	34	168	11		6	106	78	109	62	16	172	362	60	131	42	480	196	110	340	2
653	28	161	21		6	48	32	134	26	129	147	134	94	212	160	442	233	133	410	1
653	28	161	21		6	133	86	133	66	69	145	631	46	97	38	545	170	85	270	2
710	21	170	23		6	125	104	130	93	114	176	17	104	368	61	357	205	158	480	1
710	21	170	23		6	79	52	60	41	94	176	208	41	133	1	550	187	101	270	2
697	59	154	21		6	83	59	85	44	108	150	314	109	184	12	552	160	86	240	1
697	59	154	21		6	88	63	70	45	73	157	305	69	132	161	546	174	83	250	2
706	63	161	21		6	78	53	97	41	4	165	198	141	234	370	344	120	106	320	1
706	63	161	21		6	111	80	71	60	106	165	264	126	214	23	583	168	93	280	2
744	16	170	21		6	104	73	126	53	45	167	421	44	119	277	479	205	104	310	1
744	16	170	21		6	124	92	108	74	46	167	434	44	98	126	566	218	106	320	2
535	62	150	21		7	88	50	87	35	80	166	90	86	181	22	437	858	67	205	1
535	62	150	21		7	105	63	95	44	115	166	195	104	239	43	437	858	105	295	2
539	66	70	11		7	149	94	111	73	89	194	353	89	166	33	318	192	143	430	1
539	66	170	11		7	169	113	153	86	69	194	381	78	142	20	420	300	98	300	2
602	74	142	23		7	115	85	118	65	46	143	248	81	157	0	420	226	119	360	1
602	74	142	23		7	107	69	113	52	63	165	221	75	140	12	473	269	86	310	2
617	50	161	23		7	132	89	140	75	131	176	121	85	369	0	404	433	110	330	1
617	50	161	23		7	69	49	112	42	124	176	127	130	314	12	475	433	121	225	2
704	68	165	13		7	90	68	112	57	124	168	98	209	480	10	417	196	136	390	1
704	68	165	13		7	45	32	101	26	105	168	107	139	342	16	478	159	87	410	
712	75	165	23		7	68	44	102	35	14	147	193	100	258	1	531	292	111	350	1
712	75	165	23		7	126	89	89	76	81	166	66	227	546	1	401	159	104	310	2
740	71	160	23		7	59	45	112	40	44	170	83	195	463	1	351	153	155	475	1
740	71	160	23		7	43	33	72	27	50	170	185	101	219	1	637	152	124	260	2
718	42	168	21		7	153	104	127	88	55	224	204	72	168	30	407	108	90	270	1
718	42	168	21		7	126	83	77	67	98	224	193	129	252	48	410	168	111	295	2
527	40	163	13		7	112	61	136	48	257	170	120	107	345	1	709	162	139	420	1
527	40	163	13		7	89	44	148	32	144	170	350	29	106	1	709	162	91	285	2
724	52	163	13		7	55	37	58	29	42	168	332	94	168	1	576	261	104	310	1
724	52	163	13		7	60	28	25	22	172	169	120	208	426	1	702	235	67	200	2

TABLE 1.4.3. *Details of the Variables in Example 1.4.2 (Data Set B)*

Column	Variable	Units	Scale	Comments
2–5	ID	none	Nominal	Patients numbered sequentially
7–8	Age in 1950	yr	Ratio	Age at last birthday (in 1950)
11	Examining M.D. in 1950	none	Nominal	Coded from 1–4
12–14	Systolic blood pressure in 1950	mm Hg	Ratio	Recorded to nearest integer
15–17	Diastolic blood pressure in 1950	mm Hg	Ratio	Recorded to nearest integer
19–20	Height in 1950	in.	Ratio	Recorded to nearest inch
21–23	Weight in 1950	lb	Ratio	Recorded to nearest pound
24–26	Serum cholesterol in 1950	mg %	Ratio	Recorded to nearest integer
29	Socio-economic status	none	Ordinal	1 = high, ..., 5 = low
32	Clinical status	none	Nominal	0 = Other heart disease (h.d.) 1 = Coronary h.d. 2 = Coronary and hypertensive h.d. 3 = Hypertensive h.d. 4 = Hypertensive and rheumatic h.d. 5 = Rheumatic h.d. 6 = Possible/potential h.d. 7 = Hypertension without h.d. 8 = Normal
41	Examining M.D. in 1962	none	Nominal	Coded from 1–4 (different from col. 11)
42–44	Systolic blood pressure in 1962	mm Hg	Ratio	Recorded to nearest integer
45–47	Diastolic blood pressure in 1962	mm Hg	Ratio	Recorded to nearest integer
48–50	Serum cholesterol in 1962	mg %	Ratio	Recorded to nearest integer
51–53	Weight in 1962	lb	Ratio	Recorded to nearest integer
73–74	Ischemic h.d. diagnosis	none	Nominal	0 = Not known 1–3 = Myocardial infarction 4–7 = Angina pectoris 8–9 = Other
76–77	Year of death (up to 1968)	none	Interval	0 = Not dead, otherwise year of death recorded

TABLE 1.4.4. *Data for Example 1.4.2 (Data Set B)*

Case	Age	50 DR	SYST	DIAST	Height	Weight	SER-CH	SE	CS	62 DR	SYST	DIAST	SER-CH	Weight	DIAG	DTH
1	42	1	110	65	64	147	291	2	8	4	120	78	271	146	2	68
2	53	1	130	72	69	167	278	1	6	2	122	68	250	165	9	67
3	53	2	120	90	70	222	342	4	8	1	132	90	304	223	2	64
4	48	4	120	80	72	229	239	4	8	2	118	68	209	227	3	66
5	53	3	118	74	66	134	243	3	8	5	118	56	261	138	2	66
6	58	2	122	72	69	135	210	3	8	4	130	72	245	136	2	64
7	48	4	130	90	67	165	219	3	8	4	138	86	275	166	2	63
8	60	1	124	80	74	235	203	3	8	1	160	90	271	226	3	65
9	59	4	160	100	72	206	269	5	8	3	150	100	291	198	3	67
10	40	3	120	80	69	148	185	3	8	3	110	64	241	152	2	66
11	56	3	115	80	64	147	260	3	8	4	140	80	326	152	2	68
12	58	3	140	90	63	121	312	5	8	1	120	75	234	114	2	63
13	64	2	135	85	64	189	185	1	8	4	140	78	153	168	3	66
14	57	2	110	78	70	173	282	3	8	2	144	74	236	171	2	66
15	32	1	112	70	69	171	254	2	8	4	142	96	249	179	3	64
16	53	1	140	90	65	150	303	2	8	1	205	85	302	153	2	65
17	48	1	130	80	64	147	271	4	8	3	165	85	251	163	3	64
18	47	2	115	84	67	211	304	1	8	1	155	80	278	149	9	68
19	47	2	130	80	67	147	334	1	8	3	138	85	303	147	2	0
20	28	1	120	86	70	189	328	3	8	2	128	88	300	194	6	0
21	37	3	95	55	69	190	226	3	8	3	155	105	311	191	2	0
22	54	1	141	100	65	171	363	3	7	2	180	100	276	154	2	65
23	38	1	130	90	67	170	399	2	8	2	132	86	353	167	2	0
24	52	2	125	90	65	141	199	2	8	2	152	100	234	135	6	0
25	46	1	110	70	67	159	271	3	8	3	152	88	299	164	6	65
26	51	4	120	80	70	139	261	3	8	1	130	95	285	173	6	0
27	49	1	120	80	68	194	263	3	8	1	178	76	230	196	9	0
28	46	4	110	70	66	160	242	3	8	3	130	90	254	175	6	0
29	26	1	110	80	70	206	260	3	8	3	130	76	325	231	2	0
30	35	2	120	80	72	191	321	3	8	1	130	80	334	169	6	0
31	45	2	108	80	70	155	258	5	8	4	138	88	259	182	2	0
32	57	1	130	80	69	184	167	3	8	3	155	90	237	173	2	0
33	24	3	104	75	70	157	185	5	8	2	120	80	236	166	7	0
34	64	1	144	95	66	191	244	1	8	2	198	110	227	187	7	0
35	34	3	142	102	71	176	314	1	7	1	145	100	233	176	9	0
36	30	3	110	80	71	198	234	2	8	1	100	65	227	187	2	0
37	52	4	145	90	66	183	289	3	7	3	150	88	299	143	6	0
38	56	2	125	75	65	122	329	5	8	1	140	80	253	130	2	66
39	44	1	125	90	65	156	439	1	8	1	130	80	342	152	2	0
40	45	3	130	90	73	143	243	4	8	4	158	78	249	146	6	0
41	29	2	140	95	64	148	419	5	7	1	130	85	254	139	5	0
42	42	2	108	80	67	145	285	2	8	1	115	70	249	146	6	0
43	46	3	134	90	73	198	271	3	8	1	125	80	219	205	4	0
44	45	4	150	104	71	187	278	1	7	1	210	110	368	189	2	0
45	51	3	120	90	66	163	226	3	8	1	130	80	271	167	1	0
46	44	2	120	90	72	211	188	3	8	2	138	88	240	196	2	68
47	32	2	108	78	66	151	235	4	8	2	120	70	226	155	6	0
48	34	2	130	90	61	120	317	2	5	2	122	68	248	118	6	0
49	30	2	120	84	68	170	258	3	8	1	165	110	359	181	4	0
50	40	1	112	80	69	167	334	3	8	2	100	70	306	154	6	0

Table 1.4.4 (continued)

Case	Age	50 DR	SYST	DIAST	Height	Weight	SER-CH	SE	CS	62 DR	SYST	DIAST	SER-CH	Weight	DIAG	DTH
51	34	3	124	88	66	195	345	3	8	1	130	85	309	183	4	0
52	43	3	118	72	71	149	224	3	8	2	120	70	209	147	2	0
53	39	2	164	110	66	245	220	4	7	1	145	95	258	194	2	0
54	45	2	110	80	69	170	347	3	8	3	150	90	296	186	2	0
55	41	2	115	80	68	145	339	5	8	1	160	80	254	149	1	0
56	57	1	130	90	68	188	353	3	8	1	160	80	230	172	7	0
57	53	3	110	80	67	150	235	3	8	4	120	76	221	170	9	0
58	55	2	125	90	65	163	235	2	8	2	170	94	255	139	6	0
59	57	2	210	110	67	165	220	1	4	2	150	94	178	174	6	0
60	38	3	115	90	70	187	385	3	8	2	142	108	334	202	9	0
61	45	4	110	80	67	209	240	3	8	2	156	108	251	205	1	0
62	33	1	130	90	68	200	188	3	8	4	125	70	210	191	7	0
63	61	2	160	100	68	160	241	2	7	3	170	110	235	163	6	0
64	36	3	100	75	72	164	241	3	8	4	125	72	271	175	2	0
65	37	3	130	88	67	178	295	4	8	3	170	90	367	194	1	0
66	51	3	125	85	73	198	283	4	8	4	175	85	309	211	9	0
67	46	1	110	80	69	178	277	3	8	4	150	88	311	186	2	0
68	51	3	138	100	72	208	296	3	7	1	135	95	286	199	8	0
69	60	3	130	84	69	122	243	2	8	4	166	102	291	113	0	65
70	57	2	110	80	71	224	158	1	8	3	135	80	202	224	0	68
71	63	4	130	80	67	143	243	3	8	2	166	90	273	121	0	64
72	63	3	115	65	69	196	278	2	8	1	150	65	248	192	0	66
73	68	3	120	80	63	109	215	1	8	2	136	76	251	108	0	68
74	57	3	145	85	66	140	308	1	8	3	150	80	247	134	0	63
75	64	1	150	90	70	147	226	2	8	4	140	85	192	145	0	63
76	63	2	115	75	67	180	303	3	8	2	122	80	289	126	0	65
77	62	1	120	80	68	174	535	2	8	5	146	76	268	156	0	66
78	55	1	140	82	69	145	199	3	8	2	176	106	218	155	0	65
79	50	3	150	90	71	170	326	1	7	1	195	98	204	178	0	64
80	39	1	114	72	65	156	187	5	8	3	168	110	199	170	0	66
81	50	3	150	115	71	220	283	3	7	1	164	120	275	239	0	67
82	42	3	105	78	67	166	195	3	8	2	112	80	218	168	0	67
83	53	4	100	80	71	199	209	2	8	1	110	65	220	185	0	63
84	56	1	150	90	72	233	284	1	8	5	134	74	170	225	0	68
85	62	4	166	90	66	130	258	3	7	3	150	70	185	126	0	64
86	61	3	138	80	63	158	285	3	8	1	120	70	225	113	0	64
87	57	2	110	68	71	166	300	3	8	1	105	60	262	143	0	68
88	43	1	120	85	70	134	220	3	8	4	140	85	354	141	0	63
89	65	1	170	105	67	183	214	5	3	4	190	100	148	161	0	65
90	49	2	120	90	69	139	273	3	8	3	160	104	308	163	0	64
91	47	1	110	70	70	130	203	3	8	1	125	75	198	127	0	66
92	59	1	110	80	70	167	220	3	8	1	125	75	275	152	0	67
93	53	4	120	90	62	166	253	3	8	4	134	78	316	168	0	67
94	43	1	120	80	72	171	198	3	8	3	152	86	190	178	0	65
95	52	3	140	86	67	128	300	1	8	1	185	95	260	144	0	64
96	65	4	125	85	67	164	228	5	8	1	130	70	240	152	0	64
97	50	4	110	70	67	162	239	3	8	3	70	70	233	155	0	68
98	33	4	106	80	67	151	191	2	8	4	128	70	207	142	0	66
99	49	4	120	80	63	142	283	3	8	5	116	88	311	160	0	67
100	62	2	100	65	69	141	224	3	8	2	170	78	242	125	0	65

TABLE 1.4.4 (continued)

Case	Age	50 DR	SYST	DIAST	Height	Weight	SER-CH	SE	CS	62 DR	SYST	DIAST	SER-CH	Weight	DIAG	DTH
101	45	4	130	90	70	200	220	3	8	2	130	90	203	168	0	67
102	53	3	125	88	67	167	226	3	8	3	160	100	212	165	0	67
103	30	2	160	85	71	155	187	3	8	1	130	95	184	156	0	63
104	25	1	110	80	74	190	235	3	8	2	116	90	280	210	0	63
105	26	3	118	80	65	120	328	2	8	3	132	90	354	137	0	63
106	69	1	160	90	67	185	314	4	8	3	170	85	191	159	0	66
107	60	2	140	80	68	170	356	1	6	5	192	74	250	150	0	0
108	45	3	130	100	72	168	252	5	3	1	150	100	234	176	0	0
109	55	1	110	80	75	198	358	3	8	3	110	80	264	177	0	0
110	63	3	190	100	66	187	207	5	3	2	148	70	210	188	0	0
111	52	1	170	100	65	164	218	5	7	1	110	70	261	130	0	0
112	53	1	200	140	67	197	210	3	3	1	215	100	139	164	0	64
113	55	3	118	82	69	124	265	3	8	5	132	90	284	124	0	66
114	48	3	120	85	68	161	267	2	8	1	110	75	258	162	0	0
115	50	3	105	70	65	161	325	4	8	3	125	75	186	114	0	0
116	44	1	130	80	69	202	246	3	8	1	130	60	176	165	0	64
117	49	4	120	80	69	189	295	3	8	1	140	75	305	203	0	0
118	61	1	150	90	69	142	247	3	6	2	150	90	218	174	0	67
119	42	4	120	85	67	192	250	5	8	1	162	98	268	207	0	0
120	47	1	110	80	71	228	250	5	8	2	128	88	249	207	0	0
121	40	2	100	70	68	169	260	2	6	1	112	70	269	151	0	0
122	33	3	125	88	67	149	220	3	8	1	135	80	225	166	0	0
123	29	3	130	90	70	173	280	3	8	1	145	90	308	180	0	0
124	43	3	120	80	71	164	260	3	8	1	125	85	271	176	0	0
125	51	4	130	90	69	193	290	2	8	3	120	80	254	178	0	0
126	57	3	160	90	63	144	280	4	7	3	170	80	321	152	0	0
127	30	3	115	86	70	172	210	2	8	1	108	80	201	187	0	0
128	44	2	120	90	67	178	260	2	8	3	130	96	242	173	0	0
129	44	2	120	90	72	196	240	3	8	4	150	84	240	193	0	0
130	38	2	106	80	67	181	210	5	8	3	110	80	208	179	0	0
131	35	1	124	90	68	189	320	3	8	4	138	88	338	201	0	0
132	34	3	126	85	64	165	310	2	8	1	130	90	296	167	0	0
133	30	1	104	70	69	161	300	3	8	1	135	85	290	178	0	0
134	51	3	140	92	69	170	310	1	8	3	172	100	282	163	0	0
135	51	4	120	80	70	200	260	2	8	2	144	80	283	169	0	0
136	57	3	108	76	66	161	200	2	8	3	106	75	150	161	0	0
137	46	2	115	90	69	189	300	3	8	2	110	60	169	150	0	0
138	50	2	132	88	69	220	220	3	8	3	182	100	204	246	0	0
139	25	4	120	84	72	180	220	3	8	3	130	75	222	179	0	0
140	55	2	120	92	65	154	310	2	8	2	126	72	265	138	0	0
141	47	2	130	75	67	145	260	3	8	4	138	76	289	163	0	0
142	49	1	130	80	64	162	230	5	8	1	195	105	220	194	0	0
143	47	3	130	95	64	163	280	4	8	1	130	70	190	152	0	0
144	34	3	105	65	65	137	220	5	8	2	118	82	306	152	0	0
145	47	2	120	80	68	152	220	3	8	1	145	76	270	160	0	0
146	46	1	120	80	68	152	270	2	8	3	150	100	326	166	0	0
147	45	2	120	82	71	171	240	1	8	4	140	84	238	169	0	0
148	37	1	150	105	69	205	220	3	7	3	180	110	214	187	0	0
149	36	3	120	90	67	188	220	5	8	4	125	82	189	182	0	0
150	42	1	110	70	71	162	190	3	8	3	155	90	174	187	0	0

TABLE 1.4.4 (continued)

Case	Age	50 DR	SYST	DIAST	Height	Weight	SER-CH	SE	CS	62 DR	SYST	DIAST	SER-CH	Weight	DIAG	DTH
151	57	3	130	90	71	181	260	2	8	3	190	104	288	185	0	0
152	43	2	120	80	69	201	300	1	8	4	138	82	204	198	0	0
153	58	3	110	78	69	175	300	3	8	4	148	80	229	198	0	0
154	34	1	110	70	72	157	220	3	8	1	135	80	243	158	0	0
155	29	3	148	98	69	203	260	5	8	2	150	110	245	194	0	0
156	43	2	110	75	67	143	190	3	8	3	124	90	204	139	0	0
157	48	3	122	94	73	198	250	1	8	3	170	120	276	211	0	0
158	41	3	120	80	64	147	220	4	8	1	170	110	333	160	0	0
159	54	3	120	78	69	137	210	3	8	4	160	80	329	143	0	0
160	47	1	110	80	66	187	210	3	8	1	120	85	279	188	0	0
161	35	1	110	80	70	154	250	3	8	1	110	75	256	140	0	0
162	48	2	142	90	67	163	220	4	8	1	180	80	223	167	0	0
163	49	1	112	78	64	149	240	3	8	4	125	72	253	153	0	0
164	25	3	115	78	73	180	160	3	8	1	120	80	216	200	0	0
165	48	4	150	110	66	189	220	1	3	3	180	105	224	178	0	0
166	42	1	120	90	75	207	210	3	8	3	140	94	219	224	0	0
167	34	3	152	102	68	185	300	3	7	3	195	125	280	180	0	0
168	39	3	130	92	67	144	210	3	8	2	128	88	204	151	0	0
169	46	3	108	75	71	140	260	3	8	3	120	75	228	148	0	0
170	30	2	106	80	70	173	260	4	8	5	118	80	254	186	0	0
171	64	4	104	74	63	146	300	3	8	1	130	80	249	143	0	0
172	47	3	104	80	73	177	390	3	8	1	110	70	339	189	0	0
173	36	3	135	80	69	155	290	3	8	2	120	84	238	157	0	0
174	30	4	120	80	70	130	260	3	8	4	135	80	260	157	0	0
175	43	3	130	92	70	198	230	3	8	3	140	100	235	204	0	0
176	34	1	110	80	66	155	250	3	8	1	180	100	278	121	0	0
177	26	3	110	75	66	136	230	3	8	3	118	76	183	134	0	0
178	22	4	120	90	69	192	240	5	8	1	150	105	269	201	0	0
179	26	2	88	60	70	178	300	2	8	2	112	76	308	206	0	0
180	54	2	100	64	71	173	300	4	8	1	125	75	259	202	0	0
181	53	3	138	90	66	215	310	4	8	2	148	88	336	202	0	0
182	48	2	120	75	70	203	220	3	8	1	130	70	273	213	0	0
183	42	3	115	75	71	204	220	2	8	3	140	80	225	224	0	0
184	42	3	122	78	66	123	190	2	8	2	112	66	196	127	0	0
185	60	4	128	80	70	141	230	3	8	3	140	80	224	122	0	0
186	35	2	100	78	68	141	220	3	8	2	144	90	239	138	0	0
187	52	2	120	70	68	135	250	3	8	1	135	80	300	137	0	0
188	44	2	150	95	61	127	260	3	3	1	155	95	312	130	0	0
189	41	2	142	96	66	145	300	2	8	4	148	88	315	143	0	0
190	28	2	115	65	66	150	220	2	8	3	124	86	276	149	0	0
191	36	3	118	84	63	153	300	2	8	2	124	80	252	149	0	0
192	54	3	148	90	71	140	220	1	8	1	230	105	265	143	0	0
193	49	2	110	75	70	141	150	3	8	4	170	98	221	147	0	0
194	49	4	105	75	69	144	250	3	8	1	110	70	254	144	0	0
195	61	3	122	78	64	133	180	3	8	2	170	72	229	142	0	0
196	50	1	115	80	66	148	300	2	8	1	115	65	273	152	0	0
197	23	1	110	70	69	137	120	3	8	2	112	76	198	153	0	0
198	20	3	130	80	66	150	210	5	0	1	130	85	274	158	0	0
199	46	3	140	84	66	138	130	4	6	2	148	88	160	157	0	0
200	36	1	100	70	70	157	260	3	8	3	120	86	251	152	0	0

After the column assignment and coding has been completed, the data is ready to be keypunched on cards. When this task has been accomplished, the next problem is to communicate the layout of the data cards to the computer. This is achieved by the *format card* which specifies to the processor (a) which columns to skip, (b) which columns to read as one variable, (c) the location of the decimal point, if applicable, and (d) the number of cards for each individual.

Since most of the packaged statistical programs are written in FØRTRAN, we discuss only format statements written in this language. The general form of a FØRTRAN format statement is

Col. 7

FØRMAT $(A, B, C, ...)$

where $A, B, C, ...$, may be any of the following instructions:

1. nX is an instruction to skip (that is, do not read) n columns. Thus, 6X skips 6 columns, X skips 1 column, and so forth.

2. / is an instruction to go to the next card, // is an instruction to skip one card and go to the following card, and so forth.

3. Iw is an instruction to read an integer variable consisting of w columns. Thus, I6 reads an integer variable of 6 columns. Integer variables are called *fixed-point* variables. Decimals are not allowed for these variables.

4. nIw is an instruction to read n integer variables, each variable consisting of w columns. Now suppose we have the following data:

Col.	1	2	3	4	5	6
	8	2	1	3	6	4

Then

(a)   I6 would read 821364;
(b)   3I2 would read 82, 13, and 64;
(c)   2I3 would read 821 and 364;
(d)   2I2, 2I1 would read 82, 13, 6, and 4.

5. F$w.d$ is an instruction to read a decimal variable consisting of w columns with $d \leq w$ columns after the decimal point. Thus, for the above example

(a)   F6.3 would read 821.364;
(b)   F6.5 would read 8.21364;
(c)   F6.0 would read 821364.

Decimal variables are called *floating-point* variables. The decimal points may or may not be punched. If a decimal point is punched, then d in this instruction is overridden.

6. nF$w.d$ is an instruction to read n decimal variables, each variable consisting of w columns with $d \leq w$ columns after the decimal place. For the above example

(a)   2F3.1 would read 82.1 and 36.4;

(b)   3F2.0 would read 82., 13., and 64.;

(c)   F2.1, 2F2.2 would read 8.2, 0.13 and 0.64.

7.   E$w.d$ is an instruction to read a floating-point variable with $d < w$ digits after the decimal point and an exponent of the form E$\pm z$, where z is a one- or two-digit integer. For example, 102.36 may be written as 0.10236×10^3 and coded as 0.10236E$+03$, where E$+03$ represents 10^3. The format to read this is E11.5, since there are five places after the decimal and 11 columns in all, including the leading zero, decimal point, and exponent. This format allows d significant places, regardless of the magnitude of the number. To accommodate a negative number, this format would be E12.5.

8.   Tc is an instruction to begin reading in column c. Thus, T23 indicates that the next variable starts in column 23.

9. nAw is an instruction to read n groups of w alphanumeric characters, $1 \leqslant w \leqslant k$, where k is the maximum word size for the computer. For example, if we have the data

Col.	1	2	3	4	5	6	7	8	9	10
	M	A	L	E	F	E	M	A	L	E

then

(a)   A4,A6 would read MALE,FEMALE;

(b)   2A3,A4 would read MAL,EFE,MALE.

EXAMPLE 1.4.1 (*continued*)   The format

FØRMAT (I4, 2F4.0, I3, I1, I4, 4F4.0, F4.1, 2F4.2, 2F4.1, F4.0, 4F4.1, T80, I1, /)

would read only the first card and skip the second card for each patient. For the first patient, the data would be read as 517, 68., 165., 1, 1, 2, 114., 88., 95., 73., 1.7, 1.41, 0.66, 11.5, 22.5, 110, 56.2, 20.6, 11.3, 34.0, and 1. This format specifies some fixed-point variables (patient ID number, sex, survival, shock type, and card sequence number), while the rest are floating-point variables. Note that the number of columns specified for some of the variables is more than the maximum number needed; for example, age and height are allocated 4 columns each, while at most 3 are needed. Thus we could have used (...,1X,F3.0,1X,F3.0,...) instead of (...,2F4.0,...). This format can be abbreviated as (...,2(1X,F3.0),...) which is an example of parentheses within parentheses.

A format which would read all of these variables as floating-point variables is

FØRMAT (3F4.0, F3.0, F1.0, 5F4.0, F4.1, 2F4.2, 2F4.1, F4.0, 4F4.1, T80, F1.0, /)

This would read all numbers with a decimal point, for example, 517., 68.,...,1. for the first patient. To read the second card and skip the first card the format would be

FØRMAT (/, I4,..., T80, I1).

Now suppose that the investigator wishes to read only initial and final heart rates (HR). Since these two quantities appear in columns 29–32 on cards 1 and 2, respectively, the format would be

$$\text{FØRMAT (T29, F4.0,/, T29, F4.0)}$$

1.5 What to Look for in a Packaged Statistical Program

In this section we discuss some of the properties of a good packaged statistical program. Each of the characteristics discussed below is important, and no one program is universally good with respect to all of them. Furthermore, a program which is optimal for one problem may not be so for another. For this reason, the user must evaluate each program in the context of his problem and then make his choice according to these criteria. Briefly, a good packaged program has the following features:

1. *Accuracy*   By accuracy of the numerical answer to a given problem, we mean the number of correct significant places in the answer. Accuracy depends on two factors—the *precision* of the computer and the computational *algorithms* used in the program.

Precision is defined to be the number of digits in the mantissa that the computer is capable of recording (see Dorn and Greenberg, 1967). Most programs are written in *single precision* but some programs permit *double precision*, in which case twice as many digits are recorded in the mantissa. For statistical problems requiring the inversion of large matrices or the evaluation of determinants, double precision may be necessary in order to obtain acceptable results. (An excellent comparison of the accuracy of packaged least-squares programs is given in a paper by Longley, 1967.)

The *algorithm* refers to the procedure used to calculate a given quantity such as the mean of a set of observations or the eigenvalues of a square matrix. It is possible to use different algorithms to calculate the same quantity. For example, suppose we wish to calculate the sum of squares of deviations $S = \sum_{i=1}^{n} (x_i - \bar{x})^2$. One algorithm may proceed as follows:

    a. Compute $\bar{x} = (1/n)\sum_{i=1}^{n} x_i$;
    b. Compute each squared deviation $(x_i - \bar{x})^2$;
    c. Sum the n squared deviations.

Alternatively, we may express S in its algebraically equivalent form as $S_a = \sum_{i=1}^{n} x_i^2 - (\sum_{i=1}^{n} x_i)^2/n$, in which case the following algorithm would produce $S_a = S$:

   a. Compute $\sum_{i=1}^{n} x_i$;
   b. Compute $\sum_{i=1}^{n} x_i^2$;
   c. Substitute into the equation for S_a.

For manual computation on a desk calculator, this second algorithm is preferred, since $\sum_{i=1}^{n} x_i$ and $\sum_{i=1}^{n} x_i^2$ may be obtained simultaneously. Note that for large n this algorithm may yield a large sum of squares $\sum_{i=1}^{n} x_i^2$. For manual computation on a desk calculator, the investigator at some step $k < n$ may *round off* his sum $\sum_{i=1}^{k} x_i^2$ to stay within the capacity of the calculator. He thereby loses accuracy on the right, that is, the least significant places in the answer. On the other hand, if this algorithm is used in a packaged program and the sum of squares $\sum_{i=1}^{k} x_i^2$ becomes too large for the capacity of the computer, it causes an *overflow* to occur. The investigator now loses accuracy on the left, that is, the most significant places in the answer. Thus, for a packaged program the first algorithm may be preferable, since subtracting $\bar{x}$ from each observation keeps the magnitude of the square $(x_i - \bar{x})^2$ less than the square $x_i^2, i = 1, \ldots, n$.

The *order* of the calculations in the algorithm is also important. For example, to calculate $(\sum_{i=1}^{n} x_i)^2/n$ we may first square $\sum_{i=1}^{n} x_i$ and then divide by n, or we may first divide $\sum_{i=1}^{n} x_i$ by n and then multiply this result by $\sum_{i=1}^{n} x_i$. For a packaged program this latter procedure is preferred, since there is less chance for an overflow to occur.

For more complex operations such as the inversion of matrices, the evaluation of determinants, the extraction of eigenvalues, and so forth, the choice of the algorithm becomes critical. Algorithms used for manual operations are generally not appropriate for computer implementation. Instead, algorithms especially written for computers are used. Discussion of these algorithms are given in books on *numerical analysis* (for example, McCracken and Dorn, 1964; Ralston and Wilf, 1960). The manual accompanying the packaged program usually describes the algorithms used for the more complex quantities. This aids the investigator in evaluating the program for his purposes.

2. Efficiency   An efficient packaged program is one which minimizes the computation time and expense. Efficiency as well as accuracy depends upon the algorithms used. The time it takes to calculate the desired quantity is a function of the number of multiplications and additions in the algorithm. For the example above, it takes $n+1$ multiplications and $3n$ additions to calculate $S = \sum_{i=1}^{n} (x_i - \bar{x})^2$, while it takes $n+2$ multiplications but only $2n+1$ additions to calculate $S_a = \sum_{i=1}^{n} x_i^2 - (\sum_{i=1}^{n} x_i)^2/n$. Thus, the first algorithm is less efficient (but more accurate) than the second algorithm.

The manual accompanying a packaged program usually gives estimates of the calculation time as a function of the parameters of the problem and the computer used. For example, the estimate of running time on the IBM 7094 for BMD02R (stepwise regression) is $(p+q) \times n/100$, where $p+q$ is the total number of variables and n is the sample size.

3. *Flexibility*   By flexibility we mean the kinds of options that are available to the user. Examples of useful options are:

a.   Optional input and output media—a good packaged program allows card, tape, disc, and graphic input and/or output.

b.   Optional output format—a good packaged program allows the user to specify the output format. Commonly used FØRTRAN output formats are the F and E formats described in the previous section. For the F format a fixed number of decimal places is printed, while for the E format a fixed number of significant places is printed. The difference between them is seen in this example: Suppose that the quantity $\bar{x} = 0.00132$ is to be printed. In the F10.3 format it is printed as 0.001, while an E10.3 format would print $0.132E-02$. Obviously, the E format is more desirable in this case.

c.   Scaling the input—some programs permit changing the decimal point of the input data via the input format. For the above example, moving the decimal point two or three places to the right for the input data would result in at least 3 significant places for $\bar{x}$ using the F10.3 output format.

d.   Transformation of variables—a good packaged program permits transforming any of the variables by standard transformations such as

$$\log x, \quad e^x, \quad ax+b, \quad \frac{ax+b}{cx+d}, \quad \sqrt{x}, \quad \text{and so forth.}$$

Transformations are called *transgenerations* by some programs. Note that for the above example, transforming x to $x' = 1000x$ would result in at least 3 significant places for $\bar{x}'$ using the F10.3 output format.

e.   Selection of subproblems—another option is to permit more than one problem to be solved for a given run using the same set of data. This option enables the user to select subsets of the data and/or the variables for each subproblem. For example, it may be desired to analyze a data set consisting of p variables and n observations as a sequence of m subproblems, each with $p_i \leqslant p$ variables and $n_i \leqslant n$ observations, $i = 1, ..., m$. Ideally this should be accomplished in one run of the program.

f.   Intermediate output—a good packaged program has the option of intermediate output in the event that the investigator wants to: (1) check his results; or (2) perform additional analyses. For example, a descriptive program which calculates the means and standard deviations of each variable should also calculate and print the covariances and/or correlation matrix between all of the variables. As will be seen in Section 3.1, this information can be used to calculate linear regression equations.

4. *Easy identification of errors*   Three types of errors that occur are *round off errors*, *program errors*, and *arithmetic-check errors*. Roundoff error depends on the algorithm used, the precision of the computer, and the magnitude of the

observations. In general, the user is not able to detect these errors. He can, however, do a *perturbation study* to get a feeling for the stability of his solution. To do this he increases or decreases his observations by small quantities representing possible errors in measurement. He then determines the effects of these changes on his results. Ideally his final answers should not be greatly affected.

Program errors, on the other hand, are errors made by the user on the control cards or the data cards. Examples are incorrectly specifying the number of variables, observations, or the format of the data cards. A good program is one which prints comprehensible *error messages* directing the user to the source of his error. Usually in this case no calculations are made, so that the error messages are the only output to the program.

Arithmetic-check errors result in messages indicating overflows, division by zero, square roots of negative numbers, and so forth. Since these messages are not particular to the program used, they are usually given special codes. It may be necessary to obtain the assistance of a programming consultant to locate the source of error.

5. *Checks on the program* A good program is one which prints checks on the calculations of the program. These include:

a. Input checks—the parameters and at least some of the data should be printed to ascertain that the input was correctly read.

b. Intermediate checks—for example, if a matrix $\mathbf{A}$ is inverted, $\mathbf{A}\mathbf{A}^{-1}$ should be printed as a check on the accuracy of inversion.

c. Intermediate results—a good program gives the user the option of printing intermediate results which may be used for checking the final results.

* *1.6 Other Uses of the Computer as a Statistical Tool*

There are many other ways of utilizing the computer in a statistical analysis. One way is the random selection of a set of items from a larger set. This procedure entails randomly selecting a number z from the $U(0, 1)$ distribution. Programs which perform this operation are called *pseudo-random number generators* and are usually found in the *program library* of the computer center. The problem of generating pseudo-random numbers which duplicate the properties of random numbers has been investigated by many computer scientists (for example, Lewis *et al.*, 1969; Chen, 1971).

The generation of random numbers is needed for performing *simulation studies*. For example, it is often difficult to perform the necessary mathematical manipulations in order to determine the sampling distribution of some complex test statistic. The statistician may choose to simulate the situation by repetitively selecting random samples from a specified underlying theoretical distribution and then calculate the value of the test statistic for each sample in order to produce a simulated distribution. The distribution would then be studied and its characteristics would be considered representative of the unknown theoretical sampling distribution. This procedure is known as *Monte Carlo sampling* (Hammersley and Handscomb, 1964). An example of such is given in Azen and Derr (1968).

If a theoretical distribution with known cdf $z = F(x)$ has an inverse $x = F^{-1}(z)$ expressible in closed form, then it is easy to obtain a random sample of size n from this distribution. Since $F(x)$ is $U(0, 1)$, we first select random numbers $z_1, z_2, ..., z_n$ from the $U(0, 1)$ distribution. The values $x_1, ..., x_n$, where $x_i = F^{-1}(z_i)$, then represent a random sample from the distribution with cdf $F(x)$. Other methods exist for obtaining samples from distributions with general inverses (see Hastings, 1955). Box and Mueller (1958) show methods for obtaining a random sample from the $N(0, 1)$ distribution.

Another way to obtain random samples from a given distribution is to exploit the relationship between the desired distribution and the distribution for which random generators are available. For example, to obtain a random sample of size n from the $\chi^2(v)$ distribution, we select n independent samples of v randomly selected values from the $N(0, 1)$ distribution. Denoting the ith sample by $u_{i1}, ..., u_{iv}, i = 1, ..., n$, then $x_1, ..., x_n$, where $x_i = \sum_{k=1}^{v} u_{ik}^2$, is the desired sample.

To obtain a random sample of size n from the $t(v)$ distribution, we select $u_1, ..., u_n$ randomly from $N(0, 1)$ and $x_1, ..., x_n$ randomly from $\chi^2(v)$. Then $t_1, ..., t_n$, where

$$t_i = \frac{u_i}{\sqrt{x_i/v}}$$

is the desired sample.

To obtain a random sample of size n from the $F(v_1, v_2)$ distribution, we select $u_1, ..., u_n$ randomly from $\chi^2(v_1)$ and $v_1, ..., v_n$ randomly from $\chi^2(v_2)$. Then $w_1, ..., w_n$, where

$$w_i = \frac{u_i/v_1}{v_i/v_2}$$

is the desired sample.

Another valuable use of the computer is to calculate percentiles of theoretical distributions. If the density $f(x)$ is known, but the cdf $F(x)$ is not expressible in closed form, then we can use a numerical integration routine (usually available in the program library) to find the percentiles. If $f(x)$ is not expressible in closed form, percentiles may be obtained by a Monte Carlo sampling study. This is done by randomly selecting a large number of values of x, and then estimating the qth percentile by finding the value x_q below which $q\%$ of the sample values fall.

Finally, another important use of computers is the numerical evaluation of maximum likelihood estimates. One such method called *scoring*, is discussed in Rao (1965). A survey of the many numerical techniques is given in Ralston and Wilf (1960).*

2 PRELIMINARY DATA ANALYSIS

In this chapter we discuss the use of packaged computer programs as a tool for preliminary data analysis. We indicate how maximum information can be obtained from the output of the program and how the computer can be used to gain additional insight into the data. Thus, in Section 2.1 we discuss the *frequency count program* for analyzing discrete observations, and in Section 2.2 we present the *descriptive program* for analyzing one continuous variable. In Sections 2.3 and 2.4 we study the *descriptive program with strata* for analyzing more than one continuous random variable. Finally, in Section 2.5 we illustrate the *cross-tabulation program* for analyzing contingency tables. Appropriate tests of hypotheses are given throughout the chapter.

2.1 Frequency Count Programs—
The Analysis of Discrete Variables

In this section we discuss a commonly used packaged program called a *frequency count* or *tabulation* program. For any discrete variable X, this program scans a set of n observations and tabulates the frequency f_x of occurrence of each realization (or value) x of X. It then prints a *frequency table* consisting of the variable name and each value of the variable with its associated frequency, such as the accompanying tabulation, where f_{x_i} is the frequency of occurrence of $x_i, i = 1, ..., k$.

Variable name

Values	Frequency
x_1	f_{x_1}
x_2	f_{x_2}
$\vdots$	$\vdots$
x_k	f_{x_k}

Some programs will permit alphanumeric values for X. They tabulate and print the frequency of occurrence of numbers as well as letters and special symbols, for example, *, $, /, and so forth. For these programs, a variable measured on the nominal or ordinal scale may be coded as letters or as numbers.

There are three main uses for such a packaged program. The first use is to determine if there are any errors in the data deck. For example, suppose we have a dichotomous variable X coded as 1 or 2. Then if the frequency table for a sample of $n = 25$ observations is

Variable name

Values	Frequency
1	13
2	11
$	1

we can safely assume that $ was a mispunch. We then locate this observation and change it to its correct value. Such errors are called *blunders*.

The second use of this program is to locate *outliers*. Outliers are not blunders, but rather they are observations which so differ from the rest of the observations in magnitude that they are regarded as having come from another population. For example, suppose we have an ordinal variable X coded from 1 to 5 (where 1 is low and 5 is high). Then, if the frequency table for a sample of $n = 25$ observations is

Variable name

Values	Frequency
1	19
2	5
3	0
4	0
5	1

we see that all but one observations are at the low end of the scale. If the observation coded as $x = 5$ is not a mispunch, the investigator may choose to remove it from the sample, thus restricting the population to individuals with small values of X. In effect, he regards the observation $x = 5$ as having come from another population—namely, the population defined by large values of X. Of course, this decision is dependent on the design of the experiment and the objective of the analysis.

The third use of this program is to determine the *empirical distribution* of X. This distribution is given by a table listing each value x of X and the proportion $\hat{p}_x = f_x/n$ of occurrences of x. Thus, we have the accompanying tabulation, where $\hat{p}_{x_i} = f_{x_i}/n, i = 1, ..., k$.

Values	Variable name Proportion
x_1	$\hat{p}_{x_1}$
x_2	$\hat{p}_{x_2}$
$\vdots$	$\vdots$
x_k	$\hat{p}_{x_k}$

Once the empirical distribution is obtained, it is natural to make statistical inferences about the population from which the sample is taken. In the remainder of this section we consider two possible populations—one characterized by only two proportions, and the other having more than 2 proportions.

2.1.1 The Analysis of Dichotomous Observations

In this situation, a population W is divided into two disjoint and exhaustive categories A and B. Thus, an individual in the population belongs to A or to B, but not both A and B. Let p be the proportion of individuals in the population belonging to A, and let $q = 1 - p$ be the proportion of individuals in W belonging to B. Hence, statistical inference about this population reduces to the study of the parameter p. We wish to estimate p, and to test hypotheses about it.

From a sample of size n, a frequency count program generates the accompanying tabulation. Here, r and $n-r$ are the frequencies associated with A and B,

Values	Variable name Frequency
A	r
B	$n-r$

respectively. From this tabulation we obtain the maximum likelihood (m.l.) estimate $\hat{p}$ of p, that is,

$$\hat{p} = \frac{r}{n}. \tag{2.1.1}$$

It follows that the m.l. estimate $\hat{q}$ of q is

$$\hat{q} = 1 - \hat{p} = \frac{n-r}{n}. \tag{2.1.2}$$

To test the hypothesis that the population proportion p is equal to a given constant p_0 against a one- or two-tailed alternative, we may calculate the P value using the binomial distribution $b_n(i, p) = \binom{n}{i} p^i (1-p)^{n-i}$ (Table 1, Appendix II).

The accompanying table lists the P values in terms of the alternative hypothesis.

Null hypothesis	Alternative	P value
H_0: $p = p_0$	H_1: $p > p_0$	$P = \sum\limits_{i=r}^{n} b_n(i, p_0)$
	H_1: $p < p_0$	$P = \sum\limits_{i=0}^{r} b_n(i, p_0)$
	H_1: $p \neq p_0$	$P = 2\min\left(\sum\limits_{i=0}^{r} b_n(i, p_0), \sum\limits_{i=r}^{n} b_n(i, p_0) \right)$

In any case, if the calculated P value is less than the significance level α, we reject the null hypothesis H_0.

The traditional presentation of the binomial distribution is to define a random variable X as

$$X = \begin{cases} 1 & \text{if individual is in category } A, \\ 0 & \text{otherwise.} \end{cases} \tag{2.1.3}$$

Then the observations in the sample are denoted by $x_1, \ldots, x_n$, where each x_i is 0 or 1, $i = 1, \ldots, n$. Thus, the m.l. estimate of p is $\hat{p} = \bar{x}$ the sample mean, and $\hat{q} = 1 - \bar{x}$. This representation leads to the use of the central limit theorem to derive an approximation to the sampling distribution of $\bar{x}$. The result is that as n approaches ∞, the sampling distribution of $\bar{x}$ approaches $N(p, p(1-p)/n)$. This limiting distribution can be used as an approximation to the distribution of $\bar{x}$ for any n. The rule of thumb is that the approximation is "good" if $np(1-p) > 9$.

Using this normal approximation we may test the hypothesis $H_0: p = p_0$ by the test statistic

$$z_0 = \frac{\hat{p} - p_0}{(p_0(1-p_0)/n)^{1/2}}. \tag{2.1.4}$$

Under H_0, z_0 is approximately distributed as $N(0, 1)$ tabulated in Table 2, Appendix II. We summarize the alternatives and P values in the following table. For $H_1: p > p_0$, P is the area to the right of z_0 under the $N(0, 1)$ frequency curve (see Fig. 2.1.1a); for $H_1: p < p_0$, P is the area to the left of z_0 (Fig. 2.1.1b); and for $H_1: p \neq p_0$, P is twice the area to the right of $|z_0|$ (Fig. 2.1.1c). We reject H_0 if the P value is less than α.

Null hypothesis	Alternative	P value
H_0: $p = p_0$	H_1: $p > p_0$	$P = Pr(z > z_0)$
	H_1: $p < p_0$	$P = Pr(z < z_0)$
	H_1: $p \neq p_0$	$P = 2 Pr(z > |z_0|)$

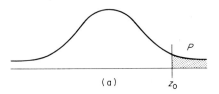

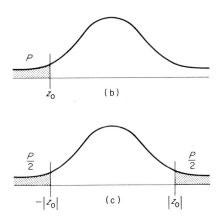

FIG. 2.1.1. *Critical regions for* H_0: $p = p_0$ *using the normal approximation to the binomial distribution.* (a) *Alternative* H_1: $p > p_0$; (b) *Alternative* H_1: $p < p_0$; (c) *Alternative* H_1: $p \neq p_0$.

The normal distribution may be used to calculate an approximate $100(1 - \alpha)\%$ *confidence interval for p.* Thus, we have

$$\hat{p} \pm z_{1 - (\alpha/2)} \sqrt{\frac{\hat{p}(1 - \hat{p})}{n}}, \tag{2.1.5}$$

where $z_{1 - (\alpha/2)}$ is the $100[1 - (\alpha/2)]$th percentile of the $N(0, 1)$ distribution.

EXAMPLE 2.1.1   Persons with a certain type of cold are known to recover without treatment within a week on the average; that is, half the population of persons with this cold recover within one week without treatment. A drug manufacturer claimed that Brand Schlimm is beneficial in curing this kind of cold. In particular, the manufacturer claimed that if the drug is used, more than half the population will be cured within one week. Dr. Gesund became interested in testing this claim. Letting p be the proportion of individuals with this cold who take Brand Schlimm and recover within one week, he then tested the hypothesis $H_0: p = 0.5$ versus the alternative $H_1: p > 0.5$ at significance level $\alpha = 0.05$. He administered the drug to ten randomly selected patients and observed that $r = 7$ persons recovered within one week. Thus, $\hat{p} = 0.7$ and $\hat{q} = 0.3$. To determine if $\hat{p}$ is significantly different from $p = 0.5$, he calculated the P value using the binomial distribution. Thus, using Table 1 in Appendix II,

$$P = \sum_{i=7}^{10} b_{10}(i, 0.5) = 0.1172 + 0.0439 + 0.0098 + 0.0010 = 0.1719.$$

To interpret this P value, he referred to this book, found to his delight that he was mentioned in Example 2.1.1, and learned that the hypothesis could not be rejected since $P > \alpha = 0.05$. Realizing that the results may have been inconclusive due to the small sample size, he then treated a random sample of 100 patients (not including the original 10 patients to avoid biasing his results). Of the 100 patients, 64 recovered within one week. Using the normal approximation [Eq. (2.1.4)], he calculated the test statistic

$$z_0 = \frac{0.64 - 0.50}{\sqrt{0.50(0.50)/100}} = 2.80.$$

From Table 2 in Appendix II, he found $P = 0.0026$. Since $P < \alpha$, he rejected H_0.

The reader should note that although a smaller percentage of patients was cured in the sample of 100 patients than in the sample of 10 patients (64% versus 70%), it was possible to reject the hypothesis for the larger sample and not for the smaller sample. This is due to the fact that the power of the test (that is, the probability of correctly rejecting the null hypothesis) increases with sample size.

Since H_0 was rejected, Dr. Gesund desired to obtain an estimate of the true proportion p. The point estimate of p is $\hat{p} = 0.64$. The approximate 95% confidence interval is

$$0.640 \pm 1.96 \sqrt{\frac{0.64(0.36)}{100}} = 0.640 \pm 0.094 = (0.546, 0.734).$$

Thus, with approximately 95% confidence, this interval covers the true proportion of treated individuals recovering within one week.

2.1.2 The Analysis of Observations Belonging to One of k Categories

In this situation, a population W is divided into $k \geqslant 2$ disjoint and exhaustive categories $A_1, A_2, \ldots, A_k$, so that an individual in this population belongs to one and only one category. For $i = 1, \ldots, k$, let p_i be the proportion of individuals in the population who belong to A_i with $p_1 + \cdots + p_k = 1$. Statistical inference about this population reduces to the study of the parameters $p_1, \ldots, p_k$.

From a sample of size n, a frequency count program generates the accompanying frequency tabulation where r_i is the frequency associated with the category

Variable name

Values	Frequency
A_1	r_1
A_2	r_2
$\vdots$	$\vdots$
A_k	r_k

A_i. Note that $r_1 + \cdots + r_k = n$. From this table we obtain m.l. estimates $\hat{p}_i$ of p_i, that is,

$$\hat{p}_i = \frac{r_i}{n}, \qquad i = 1, \ldots, k. \tag{2.1.6}$$

To test the null hypothesis that the population proportions p_i are equal to given fractions $p_i^{(0)}$ with

$$\sum_{i=1}^{k} p_i^{(0)} = 1,$$

we use a χ^2 test. We first calculate the *expected frequencies* $e_1, \ldots, e_k$ under the null hypothesis $H_0: p_1 = p_1^{(0)}, \ldots, p_k = p_k^{(0)}$. These are

$$e_i = np_i^{(0)}, \qquad i = 1, \ldots, k. \tag{2.1.7}$$

The χ^2 test statistic is

$$\chi_0^2 = \sum_{i=1}^{k} \frac{(r_i - e_i)^2}{e_i}, \tag{2.1.8}$$

which under H_0 has an approximate χ^2 distribution with $\nu = k - 1$ degrees of freedom. The appropriate alternative hypothesis H_1 is that some of the equalities $p_i = p_i^{(0)}$ specified under H_0 are incorrect. The P value is the area to the right of χ_0^2 under the $\chi^2(k-1)$ frequency curve (Table 3, Appendix II). We reject H_0 if $P < \alpha$.

COMMENTS 2.1.1   1.   When $k = 2$, this χ^2 test procedure is another way of testing the hypothesis $H_0: p = p_0$ against $H_1: p \neq p_0$ discussed in Section 2.1.1. In fact, it can be shown that χ_0^2 in Eq. (2.1.8) is the square of z_0 given in Eq. (2.1.4).
    2.   The accuracy of the χ^2 approximation depends on the sample size. The test becomes exact as each $e_i \to \infty$. It suffices to insure that $e_i \geqslant 5$, but the approximation is reasonable even if a few $e_i \geqslant 2$ and the remaining $e_i \geqslant 5$ (see Maxwell, 1961).

EXAMPLE 2.1.2   The foreman of a production line noticed (to his dismay) that there was a tendency for his workers to become sloppy around lunch time and quitting time. In his factory each worker is given 100 parts to assemble each hour. To test his suspicion, the foreman examined the output of 10 randomly selected workers for one day. He then counted the number of parts poorly assembled for each hour of the day. The results are:

Hour	1	2	3	4	5	6	7	8
Number of poorly assembled parts	19	23	27	42	42	26	18	43,

Total $n = 240$.

The population for this problem consists of all the poorly assembled parts under the supervision of this foreman. The $k = 8$ categories $A_1, ..., A_8$ correspond to the hours 1 through 8. A part poorly assembled in the ith working hour belongs to category A_i and occurs with probability p_i. The null hypothesis that the number of poorly assembled parts does not depend on the hour of day can therefore be stated as $H_0: p_1 = p_2 = \cdots = p_8 = \frac{1}{8}$. The expected frequency $e_i = 240(\frac{1}{8}) = 30$, $i = 1, ..., 8$, and the calculated χ^2 is

$$\chi_0^2 = \frac{(19 - 30)^2}{30} + \frac{(23 - 30)^2}{30} + \cdots + \frac{(43 - 30)^2}{30} = 26.5.$$

The P value, which is the area to the right of $\chi_0^2 = 26.5$ under the $\chi^2 (7)$ frequency curve, is less than 0.001. Hence, the null hypothesis is rejected, and the foreman then decides to administer snacks throughout the day as a possible solution.

2.2 Descriptive Programs—The Analysis of Continuous Variables

In this section we discuss a commonly used packaged program called a *descriptive program*. For any variable X, discrete or continuous, a typical descriptive program scans a set of n observations and calculates a frequency table, plots a histogram and calculates sample statistics such as the mean, median, variance, and so forth. From this information the investigator may make certain inferential statements about the population. For example, he can test hypotheses about population means or variances, he can estimate population percentiles, he can test whether the population distribution is normal, and so forth. Since the theory underlying most of the material in this section assumes that the variable X is continuous, we will emphasize the analysis of continuous observations. We first discuss the following.

2.2.1 Getting the Most from a Histogram

We assume that we have a continuous random variable X and a set of n observations denoted by $x_1, ..., x_n$. A typical descriptive program selects k disjoint and equal sized intervals $[c_1, c_2), [c_2, c_3), ..., [c_k, c_{k+1})$, such that each observation x_i belongs to exactly one of these intervals. These intervals are called *class intervals*, and the number of intervals is either determined in advance by the investigator through a control card or is computed by the program. For example, one computer algorithm for determining k is

$$k = \text{largest integer in } (10 \log_{10} n), \tag{2.2.1}$$

where k is restricted to $5 \leqslant k \leqslant 30$.

Once the intervals have been determined, the number f_i of observations in the class interval $[c_i, c_{i+1})$ is tallied, $i = 1, ..., k$. The results may be printed as a *frequency table*, such as the accompanying tabulation, and/or it may be printed

Variable name	
Class interval	Frequency
$[c_1, c_2)$	f_1
$[c_2, c_3)$	f_2
$\vdots$	$\vdots$
$[c_k, c_{k+1})$	f_k

as a graph. In this graph one axis specifies the class intervals while the other axis specifies the frequencies. Such a graph is called a (*frequency*) *histogram* and appears in many styles in computer output. For example, it may appear as a line of asterisks [(a) and (b)], or as a bar of 1's (c), or as an actual bar (d). For the examples in this section we will use this more conventional bar shape.

(a)	(b)	(c)	(d)
f_1		f_1	f_1
* f_2		11 f_2	f_2
* * f_3	c_3 ** f_3	11 11 f_3	f_3
* * *	c_2 *** f_2	11 11 11	
* * *	c_1 **** f_1	11 11 11	
c_1 c_2 c_3		c_1 c_2 c_3	c_1 c_2 c_3

As in the case of the frequency table for a discrete variable, two uses of the frequency table and the histogram for a continuous variable are to locate *blunders* and *outliers*.

EXAMPLE 2.2.1   A common technique for measuring the amount of blood pumped per hour in a human being or an animal (called *cardiac output*) is the indicator dilution method. This technique is performed by injecting a known amount of dye or radioactive isotope into an arm vein and determining the concentration of the indicator in serial samples of arterial blood. The time that it takes for the indicator to begin to appear in the arterial blood samples is called the *appearance time*. The actual method of computing cardiac output may be found in many clinical physiology books (for example, Weil and Shubin, 1967).

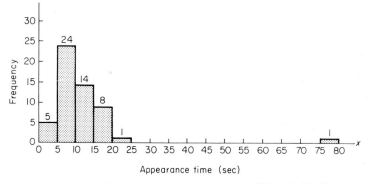

FIG. 2.2.1. *Frequency histogram of appearance times of 53 critically ill patients.*

Initial measurements of $X =$ appearance time (sec) were made on $n = 53$ critically ill patients. A descriptive program was used to graph a histogram of these measurements (see Fig. 2.2.1). Inspection of this graph shows that there are $f_1 = 5$ patients whose appearance time was greater than or equal to $c_1 = 0$ seconds, but less than $c_2 = 5$ seconds, that is, in the interval $[0, 5)$. Similarly, there were $f_2 = 24$ patients in the interval $[5, 10)$, $f_3 = 14$ in $[10, 15)$, etc. Note that there is one observation in the interval $[75, 80)$. Since the remainder of the observations fell between 0 and 25, it was suspected that this observation was a blunder. Examination of the data revealed that the observation $x = 7.8$ was mispunched on the data card as $x = 78$.

COMMENTS 2.2.1   1.   From this histogram two other kinds of histograms may be generated. The first kind, the *relative frequency histogram*, replaces each f_i by its *relative frequency* $\hat{p}_i = f_i/n$, $i = 1, ..., k$. The other kind of histogram, the *percent frequency histogram*, multiplies each $\hat{p}_i$ by 100. The advantage of using either of these two scales is that they enable us to compare histograms constructed on the same class intervals for different samples from the same population.

2.   Another graphical aid, the *frequency polygon*, is obtained from any histogram by connecting the midpoints of the bars by straight lines.

3.   The sample *mode* is approximated from the histogram by taking the midpoint of the class interval with the largest frequency. That is, if $[c_i, c_{i+1})$ has the largest frequency f_i, then the sample mode is approximately $(c_i + c_{i+1})/2$.

EXAMPLE 2.2.2   Another clinical measurement of cardiac function is $X =$ *cardiac index* (liters/min/m^2) which is defined as cardiac output (liters/min) divided by body surface area (m^2).

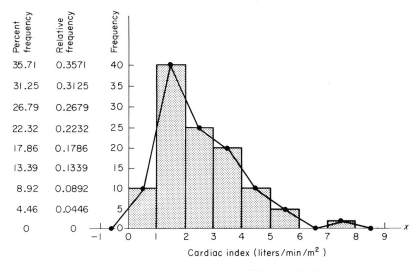

FIG. 2.2.2. *Histograms of cardiac index of 112 critically ill patients.*

Figure 2.2.2 shows the three types of histograms of X for $n = 112$ critically ill patients. The frequency table used to graph the histogram is shown in the accompanying tabulation. There is no evidence of blunders, since all of the data

X = cardiac index

Class interval	Frequency
[0, 1)	10
[1, 2)	40
[2, 3)	25
[3, 4)	20
[4, 5)	10
[5, 6)	5
[6, 7)	0
[7, 8)	2

is in the usual range of this measurement. The frequency polygon is also drawn on the graph. The sample mode is approximated by $x = 1.5$.

A third use of the histogram is to determine the *empirical distribution* as an estimate of the population distribution. This can be done directly from the histogram or it can be done from the *empirical cumulative distribution function* (*cdf*). In Section 2.2.2 we discuss goodness-of-fit tests based on either a histogram or an empirical cdf. In this section we describe the construction of this cdf and the estimation of population percentiles.

Define

$$F_1 = 0, \qquad F_2 = f_1, \qquad F_3 = f_1 + f_2, \qquad ..., \qquad F_{k+1} = \sum_{i=1}^{k} f_i.$$

Then F_i, the number of individuals with values less than c_i, is called the *cumulative frequency* at c_i, $i = 1, ..., k$. The empirical cdf is a line graph connecting the points $(c_1, F_1), (c_2, F_2), ..., (c_{k+1}, F_{k+1})$ (see Fig. 2.2.3). A plot of $(c_1, F_1/n)$, $(c_2, F_2/n), ..., (c_{k+1}, F_{k+1}/n)$ is called a *relative cdf*, and if each F_i/n is expressed as a percent, we have a *percent cdf*. One important use of the percent cdf is that we may obtain approximations of sample *percentiles* and sample *percentile ranks*. These quantities are defined as follows. The qth percentile is the number x below which $q\%$ of the sample fall. The inverse quantity, the percentile rank of a number x, is the percentage q of the sample with value less than x. A percentile of particular interest is the 50th percentile, called the sample *median*, which is the number m below which half the observations fall. Other percentiles of interest are the 25th and the 75th percentiles, called the 1st and 3rd *quartiles*, respectively,

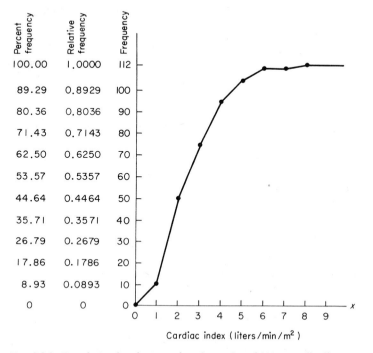

FIG. 2.2.3. *Cumulative distributions of cardiac index of 112 critically ill patients.*

and the 10th, 20th, ..., 90th percentiles called the 1st, 2nd, ..., 9th *deciles*, respectively. Example 2.2.3 illustrates how these quantities are read from the percent cdf.

EXAMPLE 2.2.2 (*continued*)   From the histogram in Fig. 2.2.2 we generated the empirical cdf. The points which were plotted in Fig. 2.2.3 are $(0,0)$, $(1,10)$, $(2,50)$, $(3,75)$, $(4,95)$, $(5,105)$, $(6,110)$, $(7,110)$, $(8,112)$. In this figure, scales also are given for the relative and percent cumulative frequencies. For example, there are $F_2 = 10$ persons (8.93%) with cardiac index less than 1, $F_3 = 50$ persons (44.64%) with cardiac index less than 2, and so forth.

EXAMPLE 2.2.3   To illustrate the calculation of percentiles and percentile ranks, the percent cumulative frequency histogram of a hypothetical sample is presented in Fig. 2.2.4. To obtain the $q = 70$th percentile, a horizontal line is extended from 70 on the vertical axis until it intersects the graph. A vertical line is drawn from the point of intersection to the horizontal axis. The point of intersection with the horizontal axis $x = 4.4$ is the approximated 70th percentile. To obtain the percentile rank of $x = 2$, the inverse operation is performed. The intersection with the vertical axis $q = 25$ is the percentile rank of 2. The median, the 50th percentile (or the 5th decile), is seen to be $m = 3.3$.

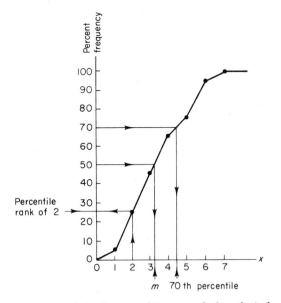

FIG. 2.2.4. *Percent cumulative frequency histogram of a hypothetical sample.*

A fourth use of a histogram or a frequency table is to obtain *estimates of the population moments*. Usually, a descriptive program calculates sample statistics from the raw data, including some measures of central tendency and variability. For example, a descriptive program normally calculates and prints the *maximum* x_{max}, the *minimum* x_{min}, the *range* $= x_{max} - x_{min}$, the *sample mean* $\bar{x} = \sum_{i=1}^{n} x_i/n$, the *sample variance* $s^2 = \sum_{i=1}^{n} (x_i - \bar{x})^2/(n-1)$, and the *sample standard deviation* $s = +\sqrt{s^2}$. Some programs will also calculate the *sample median* m from the raw data. This is given by the $(n+1)/2$ largest observation if n is odd, or by the average of the $n/2$ and the $(n/2)+1$ observations if n is even.

If any of these quantities are not given by a histogram program, then they may be approximated from the frequency table, histogram, or cdf. We have already seen how to obtain the approximate mode from the histogram and the approximate median and percentiles from the cdf. The *approximate sample mean* is calculated from the histogram or frequency table by

$$\bar{x} = \frac{1}{n} \sum_{i=1}^{k} f_i \bar{c}_i, \tag{2.2.2}$$

where $\bar{c}_i = (c_i + c_{i+1})/2$, that is, the midpoint of the ith interval, $i = 1, ..., k$, and the *approximate sample variance* is

$$s^2 = \frac{1}{n-1} \left(\sum_{i=1}^{k} f_i \bar{c}_i^2 - \frac{(\bar{x})^2}{n} \right). \tag{2.2.3}$$

In general, the *approximate jth moment about the origin* is given by

$$m_j' = \frac{1}{n} \sum_{i=1}^{k} f_i (\bar{c}_i)^j, \tag{2.2.4}$$

and the *approximate jth moment about the mean* is given by

$$m_j = \frac{1}{n} \sum_{i=1}^{k} f_i (\bar{c}_i - \bar{x})^j. \tag{2.2.5}$$

The following comments discuss some additional quantities obtained from descriptive programs.

COMMENTS 2.2.2   1. A function of $\bar{x}$ and s^2, called the *coefficient of variation*, is defined to be $s/\bar{x}$. It expresses the standard deviation as a proportion of the mean.

2. The transformation $z = (x - \bar{x})/s$ transforms the variable X to the dimensionless *standardized variable* Z. Histograms of more than one standardized variable may be compared provided that class limits of each histogram are the same. To graph the histogram of Z requires two computer runs; the first to obtain $\bar{x}$ and s, the second to transform X to Z and obtain its histogram.

3. By transforming variables we may obtain other useful statistics. These statistics are given as the sample means of the transformed observations. The accompanying table presents these transformations along with their sample means.

Transformation	Sample mean
$\frac{1}{x}$	Harmonic mean
$\log x$	Log of geometric mean
x^i	ith moment about the origin
$(x-\bar{x})^i$	ith moment about the mean
$\|x-\bar{x}\|$	Mean absolute deviation

4. Other descriptive measures of the population are the *coefficient of skewness*

$$\beta_1 = \frac{\mu_3}{(\sigma^2)^{3/2}}$$

and the *coefficient of kurtosis*

$$\beta_2 = \frac{\mu_4}{(\sigma^2)^2},$$

where μ_i is the ith population moment about the mean. When the population density is symmetric, then $\beta_1 = 0$. When the population density has a long tail to the right, then $\beta_1 > 0$, and when it has a long tail to the left, $\beta_1 < 0$. The value of β_2 for any normal distribution is $\beta_2 = 3$, but if the distribution is concentrated towards the mean, then $\beta_2 < 3$, otherwise $\beta_2 > 3$.

2.2.2 Goodness-of-Fit

As mentioned in the previous section, the histogram or the cdf of the observations may be used to estimate the underlying distribution of the variable X. In this section we discuss two test statistics which may be used to test the hypothesis that the observations are distributed according to some theoretical cdf $F_0(x)$. Since standard tests of hypotheses about μ and σ^2 assume that X is normally distributed, we are especially interested in this case.

1. *The χ^2 goodness-of-fit test*   Suppose that we have a random sample of size n. Further, suppose that we have selected k class intervals $[c_1, c_2), [c_2, c_3),$ $..., [c_k, c_{k+1}),$ with $c_1 = -\infty$ and $c_{k+1} = +\infty$. Let f_i be the observed frequency in the interval $[c_i, c_{i+1}),$ and let

$$F_i = nPr(c_i \leqslant x < c_{i+1}) = n(F_0(c_{i+1}) - F_0(c_i)) \tag{2.2.6}$$

be the expected frequency in this interval, $i = 1, ..., k$. Then, under $H_0: F(x) = F_0(x)$, the test statistic

$$\chi_0^2 = \sum_{i=1}^{k} \frac{(F_i - f_i)^2}{F_i} \qquad (2.2.7)$$

has an approximate χ^2 distribution with

$$v = k - 1 - m \qquad (2.2.8)$$

degrees of freedom when n is large. The quantity m is the number of independent parameters of the hypothesized distribution which are estimated from the data. The P value is the area to the right of χ_0^2 under the $\chi^2(v)$ frequency curve (Table 3, Appendix II). We reject H_0 and accept $H_1: F(x) \neq F_0(x)$ if $P < \alpha$.

If the hypothesized cdf is $N(\mu, \sigma^2)$, then

$$F_i = n \left(\Phi \left(\frac{c_{i+1} - \mu}{\sigma} \right) - \Phi \left(\frac{c_i - \mu}{\sigma} \right) \right), \qquad (2.2.9)$$

where $\Phi(x)$ is the standard normal cdf. Since μ and σ are usually not known, we estimate them from our sample by $\bar{x}$ and s and substitute them into Eq. (2.2.9). Hence, in Eq. (2.2.8), $m = 2$ and $v = k - 3$. The accuracy of the χ^2 approximation improves as F_i increases. Hence, the class intervals should be chosen so that each F_i is "not small," that is, it suffices to insure that each $F_i \geq 5$, but the approximation is reasonable even when a few $F_i \geq 2$ and the remaining $F_i \geq 5$.

2. *The Kolmogorov–Smirnov (K–S) test*   In this case we have n observations $x_1, ..., x_n$ which we order from smallest to largest. Denoting by $x_{(i)}$ the ith smallest observation in the sample, $i = 1, ..., n$, we construct the empirical cdf $\hat{F}(x)$ defined by

$$\hat{F}(x) = \begin{cases} 0, & -\infty < x < x_{(1)}, \\ \dfrac{i}{n}, & x_{(i)} \leq x < x_{(i+1)}, \quad i = 1, ..., n-1, \\ 1, & x_{(n)} \leq x < \infty. \end{cases} \qquad (2.2.10)$$

Note that this cdf has a "jump" of $1/n$ at each x_i, while the cdf discussed in Section 2.2.1 has jumps of varying magnitude at each class limit. The test statistic is

$$D = \max_x |\hat{F}(x) - F_0(x)|. \qquad (2.2.11)$$

The hypothesis $H_0: F(x) = F_0(x)$ is rejected if the P value associated with D is less than α. The P values for $n \leq 100$ and approximate formulas for computing the P value for $n > 100$ are given in Table 4 in Appendix II.

COMMENTS 2.2.3   1.   Some packaged programs compute the K–S statistic D and the corresponding P value, for example, subroutine KØLMØ of the IBM SSP. In this program the D computed is slightly different from that of Eq. (2.2.11). It is

$$D = \max_i |\hat{F}(x_{(i)}) - F_0(x_{(i)})|,$$

and the P value computed uses the approximate formula mentioned above.

2.   In order to choose between these two test statistics, it is necessary to know the power of each test. Since the distribution under the alternative hypothesis is usually not known, it is not possible to determine the power exactly. Massey (1951) and Kac *et al.* (1955) compared the power of the two tests and found that the K–S test has higher power than the χ^2 test for some alternatives. In particular, the K–S test is more powerful than the χ^2 test in the case of testing for normality with μ and σ^2 estimated by $\bar{x}$ and s^2.

3.   In the case when the parameters are estimated from the sample, the P values for the K–S test are not exact (see Lilliefors, 1967).

EXAMPLE 2.2.2 (*continued*)   The null hypothesis that the cardiac index data of Example 2.2.2 comes from a normal distribution was tested at the $\alpha = 0.05$ level using both goodness-of-fit tests. Since μ and σ^2 are unknown, we estimated them by the sample estimates $\bar{x} = 2.45$ and $s^2 = 1.74$. Letting X be cardiac index, the null hypothesis is stated as H_0: X is $N(2.45, 1.74)$. The alternative is H_1: X is not $N(2.45, 1.74)$. For the χ^2 test we used the class intervals shown in Table 2.2.1.

TABLE 2.2.1

Table of Observed and Expected Frequencies of Cardiac Index for 112 Critically Ill Patients

Class interval	Observed frequency f_i	Expected frequency F_i
$[-\infty, 0.5)$	1	7.85
$[0.5, 1.0)$	9	7.38
$[1.0, 1.5)$	23	11.20
$[1.5, 2.0)$	17	14.67
$[2.0, 2.5)$	13	16.58
$[2.5, 3.0)$	12	16.46
$[3.0, 3.5)$	10	14.00
$[3.5, 4.0)$	9	10.42
$[4.0, 4.5)$	9	6.61
$[4.5, 5.0)$	3	3.81
$[5.0, \infty)$	6	3.02

Except for the first and last interval, note that the interval length is 0.5. The first and last intervals were chosen so that the expected frequency in each is equal to or greater than the recommended 5.0. Calculating

$$\chi_0{}^2 = \frac{(7.85-1)^2}{7.85} + \frac{(7.38-9)^2}{7.38} + \cdots + \frac{(3.02-6)^2}{3.02},$$

we obtained $\chi_0{}^2 = 26.7$ with $v = 11 - 3 = 8$ degrees of freedom. Since $P < 0.001$, we rejected H_0.

For the K–S test a packaged program was used to calculate $D = 0.161$ from the raw data. Since the 95th percentile of the asymptotic distribution of D is $1.36/(112)^{1/2} = 0.129 < 0.161$ (Table 4, Appendix II), this implies $P < 0.05$. Again we rejected H_0.

2.2.3 Transformations to Induce Normality

The standard tests of hypotheses about means and variances assume that the underlying populations are normally distributed. If for a particular sample, we reject the hypothesis of normality, then we may proceed to make statistical inferences in one of several ways. For example, if our sample size n is sufficiently large, we may still choose to use the standard tests as approximations. Another way is to use a *nonparametric procedure* (see, for example, Noether, 1967), and a third way is to find a transformation of the variable which induces normality.

The specific transformation to induce normality is often difficult to determine, but the data may suggest an appropriate transformation. Certain variables have standard transformations associated with them, for example, a logarithmic transformation may be appropriate for measurements on plants and animals. Sometimes the shape of the frequency histogram will suggest a transformation. For example, a histogram which is highly skewed with a long tail to the right suggests a lognormal or χ^2 distribution, so that a logarithmic or square root transformation might be appropriate. If the empirical distribution is bimodal, the experimenter, suspecting a mixture of two distributions, may treat each independently (see Bliss, 1967, Ch. 7).

In addition, statisticians have worked out certain transformations which are appropriate when the standard deviation is functionally related to the mean. For example, if the standard deviation is proportional to the mean, the logarithmic transformation is used; if the variance is proportional to the mean, the square root transformation will induce approximate normality. One way of detecting these relationships between μ and σ is to divide the sample into subsamples, calculate the mean and standard deviation for each, and plot them. (For a more complete discussion, see Brownlee, 1965, pp. 144–146.)

A packaged descriptive program, with the option of transforming the variables, produces in a single run histograms of the observations as well as histograms of any desired transformations of the observations. The investigator examines each histogram to find one which seems to be normally distributed, and then tests the goodness-of-fit of the transformed observations using either the χ^2 or the K–S test.

EXAMPLE 2.2.2 (*continued*)   The skewness of the histogram in Fig. 2.2.2 suggests that the data comes from a population with a lognormal distribution. Hence, the transformation $y = \log_{10} x$ was used. The histogram of the transformed data is shown in Fig. 2.2.5. On the graph are also shown the expected frequencies calculated using the mean $\bar{y} = 0.335$ and the standard deviation $s_y = 0.261$ of the transformed data. The data seem to approximate a normal distribution (shown by the dotted line), and, indeed, the χ^2 statistic calculated on twelve class intervals has a nonsignificant value of $\chi_0^2 = 7.64$ with $v = 9$ degrees of freedom.

Having found an appropriate transformation, we may now test hypotheses about the population mean and variance and form confidence intervals for these parameters.

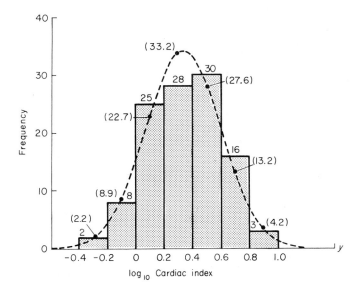

FIG. 2.2.5. *Frequency histogram of* $\log_{10}$ *cardiac index of 112 critically ill patients. Expected frequencies are shown in parentheses.*

2.2.4 Tests of Hypotheses and Confidence
Intervals for μ and σ^2

In this section we discuss the standard tests of hypotheses about the population mean μ and the population variance σ^2. As mentioned in Section 2.2.3 these tests assume that the underlying distribution is normal. Thus, for both tests we assume that $x_1, \ldots, x_n$ is a random sample from the $N(\mu, \sigma^2)$ distribution. To test the hypothesis that the mean μ is equal to some constant μ_0 (denoted by $H_0: \mu = \mu_0$), we use the z test when σ^2 is known (or when σ^2 is unknown but n is large, that is, $n \geqslant 30$) and the t *test* when σ^2 is unknown. (The z test is discussed in Section I.4 in Appendix I.) The t test statistic is

$$t_0 = \frac{(\bar{x} - \mu_0)}{s} \sqrt{n}, \qquad (2.2.12)$$

which under H_0 has a Student's t distribution with $\nu = n - 1$ degrees of freedom (Table 5, Appendix II). The P value depends on the alternative hypothesis and is summarized in the accompanying table. (See Fig. 2.2.6 for a graph of the critical regions.) Thus, for example, if the alternative hypothesis is one sided of the form $H_1: \mu > \mu_0$, the P value is the area under the $t(\nu)$ frequency curve to the right of t_0. In all cases if the P value is less than α, the null hypothesis is rejected.

Null hypothesis	Alternative	P value
$H_0: \quad \mu = \mu_0, \; \sigma^2$ unknown	$H_1: \quad \mu > \mu_0$	$P = Pr(t(\nu) > t_0)$
	$H_1: \quad \mu < \mu_0$	$P = Pr(t(\nu) < t_0)$
	$H_1: \quad \mu \neq \mu_0$	$P = 2Pr(t(\nu) > |t_0|)$

The t test is used when μ_0 is a known or assumed mean for a given population and we wish to test whether the mean μ of the population under consideration is different from μ_0, that is, greater than, less than, or not equal to μ_0. (An important use of the t test is in the *matched pairs* or "before-after treatment" design discussed in Section 2.3.1.)

An interval estimate for μ is given by the $100(1 - \alpha)\%$ *confidence interval*

$$\bar{x} - t_{1-(\alpha/2)}(n-1)\frac{s}{\sqrt{n}}, \qquad \bar{x} + t_{1-(\alpha/2)}(n-1)\frac{s}{\sqrt{n}}, \qquad (2.2.13)$$

where $t_{1-(\alpha/2)}(n-1)$ is the $100[1-(\alpha/2)]$th percentile of the Student's t distribution with $\nu = n - 1$ degrees of freedom (Table 5, Appendix II). This interval may be used to test $H_0: \mu = \mu_0$ against $H_1: \mu \neq \mu_0$. We reject H_0 at level α if μ_0 falls outside of this confidence interval.

The t test is known to be *robust*, that is, insensitive to moderate deviations from the assumption of normality, when the sample is random. In contrast we

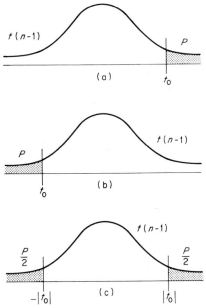

FIG. 2.2.6. *Critical regions for* H_0: $\mu = \mu_0$, σ^2 *unknown.* (a) *Alternative* H_1: $\mu > \mu_0$. (b) *Alternative* H_1: $\mu < \mu_0$. (c) *Alternative* H_1: $\mu \neq \mu_0$.

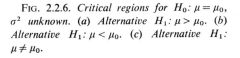

now discuss a test which is highly sensitive to this assumption. This is the χ^2 test for the hypothesis that the variance σ^2 is equal to some constant σ_0^2 (denoted by H_0: $\sigma^2 = \sigma_0^2$). The statistic is

$$\chi_0^2 = \frac{(n-1)s^2}{\sigma_0^2}, \qquad (2.2.14)$$

which under H_0 has a χ^2 distribution with $v = n-1$ degrees of freedom (Table 3, Appendix II).

The P value depends on the alternative hypothesis and is summarized in the accompanying table. (See Fig. 2.2.7 for a graph of the critical regions.) For example, for the alternative H_1: $\sigma^2 \neq \sigma_0^2$ the P value is twice the smaller of the areas to the right and to the left of χ_0^2 under the $\chi^2(v)$ frequency curve. In all cases, H_0 is rejected if the P value is less than α.

Null hypothesis	Alternative	P value
H_0: $\sigma^2 = \sigma_0^2$	H_1: $\sigma^2 > \sigma_0^2$	$P = Pr(\chi^2(v) > \chi_0^2)$
	H_1: $\sigma^2 < \sigma_0^2$	$P = Pr(\chi^2(v) < \chi_0^2)$
	H_1: $\sigma^2 \neq \sigma_0^2$	$P = 2\min[Pr(\chi^2(v) < \chi_0^2), Pr(\chi^2(v) > \chi_0^2)]$

This test is used when σ_0^2 is a known variance for a given population. Then for the population under consideration we wish to test whether its variance σ^2 is

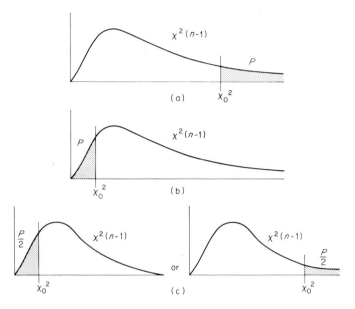

FIG. 2.2.7. *Critical regions for* $H_0: \sigma^2 = \sigma_0^2$. (a) *Alternative* $H_1: \sigma^2 > \sigma_0^2$. (b) *Alternative* $H_1: \sigma^2 < \sigma_0^2$. (c) *Alternative* $H_1: \sigma^2 \neq \sigma_0^2$.

different from σ_0^2. Since this test is so sensitive to normality, we recommend that it be used with caution. For alternative tests for σ^2 and references see Nemenyi (1969).

A $100(1-\alpha)\%$ *confidence interval for* σ^2 is

$$\frac{(n-1)s^2}{\chi^2_{1-(\alpha/2)}(n-1)} < \sigma^2 < \frac{(n-1)s^2}{\chi^2_{\alpha/2}(n-1)}. \qquad (2.2.15).$$

EXAMPLE 2.2.2 (*continued*)   In this example X is a random variable measuring cardiac index (liters/min/m²). In the population of healthy individuals it is known that the mean cardiac index μ_0 is 3.5 liters/min/m². Since critically ill patients are characterized by reduced blood flow, an interesting question is whether the mean cardiac index μ_x in the population of critically ill patients is less than 3.5. Thus, we test $H_0: \mu_x = 3.5$ against the one-sided alternative $H_1: \mu_x < 3.5$ at the $\alpha = 0.05$ level. The statistics from the sample $x_1, ..., x_{112}$ are $\bar{x} = 2.45$ and $s_x = 1.32$, and the value of t is

$$t_0 = \frac{(2.45 - 3.50)\sqrt{112}}{1.32} = -8.4.$$

Since the P value is less than 0.001, H_0 is rejected.

A 95% confidence interval for the mean cardiac index μ_x is

$$2.45 \pm \frac{1.98(1.32)}{\sqrt{112}} = (2.20, 2.70).$$

Hence, we have 95% confidence that this interval includes the true mean cardiac index μ_x of critically ill patients.

Now recall that it was shown above that X was not normally distributed but that $Y = \log_{10} X$ was. Therefore, it might be argued that the hypothesis should be tested using the transformed observations $y_1, \ldots, y_{112}$. It can be shown that the relationship between μ_y and μ_x is $\mu_y = \log_{10}\mu_x - 1.15\sigma_y^2$. Since a good estimate of σ_y^2 is $s_y^2 = 0.068$, the hypothesized value of μ_y is $H_0 : \mu_y = \log_{10} 3.5 - 1.15(0.068) = 0.466$. The sample statistics are $\bar{y} = 0.335$ and $s_y = 0.261$, and the value of t is

$$t_0 = \frac{(0.335 - 0.466)\sqrt{112}}{0.261} = -5.3.$$

Again H_0 is rejected at $P < 0.001$.

For testing hypotheses about variances we should use the transformed variable Y since it was shown to be more normally distributed than X. Previous experience with healthy patients has shown that $\sigma_y = 0.3$. To determine whether these patients came from a population with the same variance, we test $H_0 : \sigma_y^2 = 0.09$. Since there is no logical basis for a one-sided alternative, we chose $H_1 : \sigma_y^2 \neq 0.09$. The value of χ^2 then is

$$\chi_0^2 = \frac{111(0.261)^2}{0.09} = 84.0$$

with $\nu = 111$ degrees of freedom. Since $P \simeq 0.20$, we accept H_0.

Note that $\sigma_y^2 = 0.09$ lies in the 95% confidence interval

$$\left(\frac{111(0.261)^2}{130}, \frac{111(0.261)^2}{74}\right) = (0.059, 0.102),$$

which is equivalent to accepting H_0 at the $\alpha = 0.05$ level.

2.3 Descriptive Programs with Strata— The Analysis of Two Continuous Variables

In this section we discuss the use of a descriptive program for analyzing data on two continuous random variables. We first consider the case where two different random variables X_1 and X_2 are defined on the same population W.

Thus, we can study the *covariance* and *correlation* between X_1 and X_2. A random sample from this population consists of n pairs of observations where each pair is measured on the same individual in the sample. We describe how a descriptive program prints a histogram and sample statistics for each X_i and calculates and prints estimates of the population covariances and correlations. Furthermore, if X_1 and X_2 can be compared, that is, if they measure the same or a similar characteristic, we can compare the means of X_1 and X_2 using the *paired sample t test*.

After this we discuss the case where the same random variable X is defined on two different populations W_1 and W_2. These populations may be regarded as subpopulations, or *strata*, from a larger population W. In this case two independent random samples are selected from each population. We describe how we may use a *descriptive program with strata* to graph a histogram of both X_1 and X_2 and discuss how the sample statistics may be used to test the hypothesis that the means of X_1 and X_2 are equal. This introduces the *two-sample t test* and the *Welch t test*.

2.3.1 One Population—Two Random Variables

In this situation we have two random variables X_1 and X_2 defined on a population W. Let μ_i and σ_i^2 be the mean and variance of X_i in the population, and let $\sigma_{ij} = \sigma_{ji}$ be the covariance between X_i and X_j for $i, j = 1, 2$ (see Section I.6, Appendix I). Note that $\sigma_{ii} = \sigma_i^2$, $i = 1$ or 2. The *population correlation coefficient* ρ_{ij} between X_i and X_j is defined to be

$$\rho_{ij} = \frac{\sigma_{ij}}{\sigma_i \sigma_j} \qquad \text{for} \quad i, j = 1, 2. \tag{2.3.1}$$

Note that $\rho_{11} = \rho_{22} = 1$, and that $-1 \leqslant \rho_{12} = \rho_{21} \leqslant 1$. It will be shown in Section 3.1 that the correlation ρ_{12} is a measure of the linear association between X_1 and X_2—the closer $|\rho_{12}|$ is to 1, the larger the degree of linear association; the closer ρ_{12} is to 0, the smaller the degree of linear association.

From the population W we select a random sample of size n and observe both X_1 and X_2 on each individual in the sample. The resulting set of observations is denoted by $x_{11}, \ldots, x_{1n}$ and $x_{21}, \ldots, x_{2n}$, so that x_{ij} is the jth observation on the variable X_i, $i = 1, 2, j = 1, \ldots, n$. Using a packaged descriptive program with the input as in the accompanying tabulation, we obtain a histogram of

Individual	Data
1	x_{11}, x_{21}
2	x_{12}, x_{22}
⋮	⋮
n	x_{1n}, x_{2n}

$x_{11}, x_{12}, \ldots, x_{1n}$ and a histogram of $x_{21}, x_{22}, \ldots, x_{2n}$. In addition we obtain m.l. estimates for the population parameters. Thus, for $i, j = 1, 2$, we have

$$\bar{x}_{i.} = \frac{1}{n} \sum_{j=1}^{n} x_{ij},$$

$$s_i^2 = \frac{1}{n-1} \sum_{j=1}^{n} (x_{ij} - \bar{x}_{i.})^2,$$

$$s_{ij} = \frac{1}{n-1} \sum_{k=1}^{n} (x_{ik} - \bar{x}_{i.})(x_{jk} - \bar{x}_{j.}),$$

and

$$r_{ij} = \frac{s_{ij}}{s_i s_j}, \tag{2.3.2}$$

as estimates of μ_i, σ_i^2, σ_{ij}, and ρ_{ij}, respectively. (The dot notation for $\bar{x}_{i.}$ implies summation over the subscript which the dot replaces.)

The sample variances, covariances, and correlations are usually presented in *matrix* form and printed in the output as

covariance matrix

correlation matrix

$$\begin{bmatrix} s_{11} = s_1^2 & s_{12} \\ s_{21} & s_{22} = s_2^2 \end{bmatrix} \quad \text{and} \quad \begin{bmatrix} 1 & r_{12} \\ r_{21} & 1 \end{bmatrix}$$

(Since both matrices are symmetric, sometimes only the diagonal and upper off-diagonal elements of the matrix are printed.)

The analysis of covariance matrices is discussed in Section 5.6 and 5.7 and statistical inferences about correlations are presented in Section 3.1. For the purposes of this section we remind the reader that we may use each histogram to (a) locate blunders, (b) locate outliers, (c) determine the empirical distribution, (d) calculate sample statistics, (e) empirically induce normality, and (f) test hypotheses about μ_i and σ_i^2, $i = 1$ or 2, as described above. Furthermore, if X_1 and X_2 are comparable, that is, they measure the same or similar characteristics, we may test hypotheses about the difference $\mu_1 - \mu_2$. Before we discuss this test, we consider the following example.

EXAMPLE 2.3.1   In this example the population W consists of critically ill patients in circulatory shock. A sample of $n = 108$ patients was taken and $X_1 = $ venous pH and $X_2 = $ arterial pH was measured on each individual in the sample. A descriptive program plotted a histogram for both X_1 and X_2 and calculated the sample statistics $\bar{x}_1 = 7.373$ and $\bar{x}_2 = 7.413$, $s_1^2 = 0.1253$ and $s_2^2 = 0.1184$. The covariance matrix is

$$\begin{bmatrix} 0.1253 & 0.1101 \\ 0.1101 & 0.1184 \end{bmatrix}$$

and the correlation matrix is

$$\begin{bmatrix} 1.0000 & 0.9039 \\ 0.9039 & 1.0000 \end{bmatrix}$$

Note that $s_{12} = s_{21}$ and $r_{12} = r_{21}$. Also, note that $s_{11} = s_1^2$ and $s_{22} = s_2^2$. Finally, check that $r_{12} = s_{12}/s_1 s_2$. The large value of $r_{12} = 0.9039$ indicates a strong linear association between X_1 and X_2 (as expected). Further comments on this are given in Chapter 3.

It is medical knowledge that on the average venous pH is less than arterial pH in healthy individuals. Thus, it is reasonable to test this hypothesis for the above population of critically ill patients.

In general, when X_1 and X_2 are comparable measurements on the same individual, the null hypothesis $H_0: \mu_1 - \mu_2 = \delta$, where δ is a constant, may be tested using the *paired sample t test* (sometimes called the *correlated t test*). The test statistic is

$$t_0 = \frac{(\bar{d} - \delta)\sqrt{n}}{s_d}, \tag{2.3.3}$$

where

$$\bar{d} = \bar{x}_{1.} - \bar{x}_{2.} \quad \text{and} \quad s_d^2 = s_1^2 + s_2^2 - 2s_1 s_2 r_{12} = s_1^2 + s_2^2 - 2s_{12}.$$

Under H_0, t_0 has a Student's t distribution with $v = n-1$ degrees of freedom. The P value depends on the alternative hypothesis summarized by the accompanying table. If the packaged descriptive program does not print the covariance or correlation matrix, this t statistic may be calculated by making the transformation $D = X_1 - X_2$. The resulting sample statistics for the observations $d_i = x_{1i} - x_{2i}$, $i = 1, ..., n$, are the desired $\bar{d}$ and s_d^2. In this sense, the paired sample t test is the same as the one sample t test calculated on the differences d_i, $i = 1, ..., n$.

Null hypothesis	Alternative hypothesis	P value
$H_0: \quad \mu_1 - \mu_2 = \delta$	$H_1: \quad \mu_1 - \mu_2 > \delta$	$P = Pr(t(v) > t_0)$
	$H_1: \quad \mu_1 - \mu_2 < \delta$	$P = Pr(t(v) < t_0)$
	$H_1: \quad \mu_1 - \mu_2 \neq \delta$	$P = 2Pr(t(v) > |t_0|)$

EXAMPLE 2.3.1 (*continued*)   In this example we test the hypothesis that μ_1 = mean venous pH is less than μ_2 = mean arterial pH. Thus, we have H_0: $\mu_1 - \mu_2 = 0$ and $H_1: \mu_1 - \mu_2 < 0$. From Eq. (2.3.3),

$$t_0 = \frac{(7.373 - 7.413)\sqrt{108}}{\sqrt{0.1253 + 0.1184 - 2(0.1101)}} = -2.71,$$

with $v = 108 - 1 = 107$ degrees of freedom. Since the P value is less than 0.005 we reject H_0, thus corroborating medical fact. This example is discussed further in Section 3.1.

COMMENT 2.3.1　　　　Pairs of observations may arise in three ways. One way is to make two measurements on the same individual; for example, $X_1 = $ length of the right arm and $X_2 = $ length of the left arm. Another way is to measure a characteristic on the same individual before and after a treatment, for example, cardiac output before and after a drug. A third way is to measure the same variable on *matched pairs*, that is, pairs of individuals chosen for their similarity with respect to criteria related to the variable measured. For example, Dr. Gesund in his search for a cure for the common cold, may perform his experimentation on rats from the same litter. He gives an improved version of Brand Schlimm to one rat of the pair, leaves the second of the pair untreated, and measures recovery time for each. By thus controlling for external factors, he enhances the sensitivity of the experiment. [The treatment(s) should be assigned at random to the members of the pair.]

2.3.2　Two Populations—One Random Variable

In this situation we have data on a continuous random variable X defined on two populations W_1 and W_2. The two populations may be regarded as *subpopulations* or *strata* from a larger population W. In this case the population W is stratified into W_1 and W_2 by another variable Y defined on W. For example, $X = $ IQ score is defined on the population W of all college students in the United States. Then $Y = $ sex stratifies W into the two subpopulations $W_1 = $ all male college students and $W_2 = $ all female college students. Thus, X defined on W_1 is the IQ scores of males, and X defined on W_2 is the IQ score of females.

Let μ_i and σ_i^2 be the mean and variance of X in the population W_i, $i = 1$ or 2. (Note that in this situation covariances and correlations are meaningless.) From the population W_i a random sample of size n_i is taken, and X is observed for each individual in the sample, $i = 1$ or 2. Denote the resulting observations by $x_{11}, x_{12}, \ldots, x_{1n_1}$ for the sample from W_1 and $x_{21}, x_{22}, \ldots, x_{2n_2}$ for the sample from W_2. To obtain a histogram of each sample and estimates of the population statistics, we may run a descriptive program twice, once for each sample. An alternative way of obtaining this information is to use a *descriptive program with strata*, such as BMD07D. Such a program plots a histogram of X for both samples on the same page of output. Thus, the output takes the following form:

	Group 1	Group 2
c_3	$**f_3$	
c_2	$*****f_2$	$****f_2'$
c_1	$***f_1$	$**f_1'$

To obtain these histograms, the input data may be in the form

$$\text{1st sample} \left\{ \begin{array}{c} x_{11} \\ x_{12} \\ \vdots \\ x_{1n_1} \end{array} \right. \qquad \text{2nd sample} \left\{ \begin{array}{c} x_{21} \\ x_{22} \\ \vdots \\ x_{2n_2} \end{array} \right.$$

and n_1 and n_2 are specified on the control cards. Alternately, the user may define a variable Y, called the *base variable*, which specifies the subpopulation from which the sample is taken. In this case, the observations may be entered in any order as pairs (x_{ij}, y_i), for example,

$$\begin{array}{cc} x_{22}, & y_2 \\ x_{13}, & y_1 \\ x_{17}, & y_1 \\ \vdots & \vdots \end{array}$$

The program then classifies the observations into the correct sample according to a *cutpoint* for Y specified on a control card. For example, if $Y = 1$ for x_{1j} and $Y = 2$ for x_{2j}, then any value of Y between 1 and 2 is an appropriate cutpoint.

For each sample a histogram and the corresponding estimates of μ_i and σ_i^2 are calculated. These are given by

$$\bar{x}_{i.} = \frac{1}{n_i} \sum_{j=1}^{n_i} x_{ij}, \qquad \text{and} \qquad s_i^2 = \frac{1}{n_i - 1} \sum_{j=1}^{n_i} (x_{ij} - \bar{x}_{i.})^2 \qquad \text{for} \quad i = 1, 2.$$

$$(2.3.4)$$

We now consider the following example.

EXAMPLE 2.3.2   A useful measurement in many clinical situations is that of lactic acid concentration (called *lactate*, mM) in the arterial blood. Studies have shown that the logarithm of this variable is approximately normally distributed. In a study, $X = \log_{10}$ lactate was measured on the population W of critically ill patients in circulatory shock. This population was stratified into the subpopulation $W_1 =$ those patients who died and $W_2 =$ those patients who survived. Thus, a variable Y was defined as

$$Y = \begin{cases} 1 & \text{if the patient died,} \\ 2 & \text{if the patient survived.} \end{cases}$$

A sample of $n_1 = 41$ patients was selected from W_1 and a sample of $n_2 = 70$ patients was selected from W_2. The observations in the sample are the final values of X just before death or discharge from the critical care unit. A program printed the histograms for both samples (Fig. 2.3.1). The sample statistics are also given in the figure.

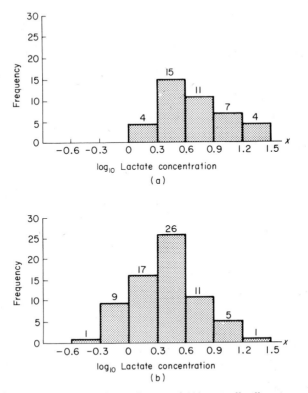

FIG. 2.3.1. *Frequency histograms of* $\log_{10}$ *lactate of 111 critically ill patients, stratified by survival.* (a) *Died:* $Y = 1$, $\bar{x}_{1.} = 0.695$, $s_1 = 0.326$, $n_1 = 41$. (b) *Survived:* $Y = 2$, $\bar{x}_{2.} = 0.399$, $s_2 = 0.383$, $n_2 = 70$.

In the above example it is of interest to test the hypothesis that the final mean $\log_{10}$ lactate for the nonsurvivors is greater than that for the survivors. In general, if $x_{11}, \ldots, x_{1n_1}$ and $x_{21}, \ldots, x_{2n_2}$ are random samples from $N(\mu_1, \sigma_1^2)$ and $N(\mu_2, \sigma_2^2)$, respectively, and $\sigma_1^2 = \sigma_2^2 = \sigma^2$, then the null hypothesis $H_0: \mu_1 - \mu_2 = \delta$, where δ is a constant, may be tested using the *two-sample t test*. The test statistic is

$$t_0 = \frac{(\bar{x}_{1.} - \bar{x}_{2.}) - \delta}{s_p \left(\dfrac{1}{n_1} + \dfrac{1}{n_2} \right)^{1/2}}, \tag{2.3.5}$$

where s_p^2, the *pooled sample variance*, is given by

$$s_p^2 = \frac{(n_1 - 1)s_1^2 + (n_2 - 1)s_2^2}{n_1 + n_2 - 2} \tag{2.3.6}$$

and is an unbiased estimator of the common variance σ^2. Under H_0, t_0 has a Student's t distribution with $v = n_1 + n_2 - 2$ degrees of freedom. The P value, which depends on the alternative hypothesis, is summarized in the accompanying table. In all cases we reject H_0 if $P < \alpha$.

Null hypothesis	Alternative hypothesis	P value
H_0: $\mu_1 - \mu_2 = \delta$	H_1: $\mu_1 - \mu_2 > \delta$	$P = Pr(t(v) > t_0)$
	H_1: $\mu_1 - \mu_2 < \delta$	$P = Pr(t(v) < t_0)$
	H_1: $\mu_1 - \mu_2 \neq \delta$	$P = 2Pr(t(v) > |t_0|)$

A $100(1-\alpha)\%$ *confidence interval for the difference between means* $\mu_1 - \mu_2$ is

$$(\bar{x}_1 - \bar{x}_2) \pm t_{1-(\alpha/2)}(n_1 + n_2 - 2) s_p \sqrt{\frac{1}{n_1} + \frac{1}{n_2}}, \qquad (2.3.7)$$

where $t_{1-(\alpha/2)}(n_1 + n_2 - 2)$ is the $100[1 - (\alpha/2)]$th percentile of the Student's t distribution with $n_1 + n_2 - 2$ degrees of freedom.

EXAMPLE 2.3.2 (*continued*)   For this example we tested the hypothesis that the final mean log lactate μ_1 for patients who died is larger than the final mean μ_2 for patients who recovered. The hypotheses are $H_0: \mu_1 - \mu_2 = 0$ and H_1: $\mu_1 - \mu_2 > 0$, and α was selected at 0.05. The sample statistics are $n_1 = 41$, $\bar{x}_1 = 0.695$, $s_1 = 0.326$, $n_2 = 70$, $\bar{x}_2 = 0.399$, and $s_2 = 0.383$. The pooled variance is

$$s_p{}^2 = \frac{(41-1)(0.326)^2 + (70-1)(0.383)^2}{41 + 70 - 2} = 0.132.$$

The t statistic is

$$t_0 = \frac{0.695 - 0.399}{\sqrt{0.132(\frac{1}{41} + \frac{1}{70})}} = 4.14,$$

with $v = 109$ degrees of freedom. Since the P value is less than 0.001, H_0 is rejected confirming medical knowledge (Weil and Afifi, 1970).

A 95% confidence interval for $\mu_1 - \mu_2$ is

$$(0.695 - 0.399) \pm 2.00 \sqrt{0.132(\tfrac{1}{41} + \tfrac{1}{70})} = (0.153, 0.439).$$

Thus, our confidence is 95% that this interval includes the true mean difference $\mu_1 - \mu_2$.

Note that the two-sample t test assumes equality of the population variances $\sigma_1{}^2$ and $\sigma_2{}^2$, that is, $\sigma_1{}^2 = \sigma_2{}^2 = \sigma^2$. This assumption may be tested by the *variance ratio* test. Thus, if $x_{11}, \ldots, x_{1n_1}$ and $x_{21}, \ldots, x_{2n_2}$ are random samples from $N(\mu_1, \sigma_1^2)$ and $N(\mu_2, \sigma_2^2)$, the null hypothesis $H_0: \sigma_1^2 = \sigma_2^2$ is tested by the test statistic

$$F_0 = \frac{s_1^2}{s_2^2}, \tag{2.3.8}$$

where s_i^2 is the ith sample variance, $i = 1$ or 2. Under H_0, F_0 has an F distribution with $v_1 = n_1 - 1$ and $v_2 = n_2 - 1$ degrees of freedom (Table 6, Appendix II). Since the lower percentiles for F are not tabulated in this table, the customary procedure in making this test is to renumber the populations, if necessary, so that $s_1^2 \geqslant s_2^2$. Thus, $F_0 \geqslant 1$ so that the P value depends only on the right tail of the F distribution. The P values are summarized for the two alternatives H_1: $\sigma_1^2 > \sigma_2^2$ and $H_1: \sigma_1^2 \neq \sigma_2^2$ in the accompanying table. It should be mentioned that this test is also very sensitive to deviations from the assumption of normality. If nonnormal distributions are suspected, this test should not be used.

Null hypothesis	Alternative	P value
$H_0: \sigma_1^2 = \sigma_2^2$	$H_1: \sigma_1^2 > \sigma_2^2$	$P = Pr(F(v_1, v_2) > F_0)$
	$H_1: \sigma_1^2 \neq \sigma_2^2$	$P = 2Pr(F(v_1, v_2) > F_0)$

EXAMPLE 2.3.2 (*continued*)   We now test the hypothesis that the variances are equal for this example. Since $s_2^2 > s_1^2$ we renumbered the populations so that W_1 is the population of patients who survived and W_2 is the population of patients who died. Thus, we test $H_0: \sigma_1^2 = \sigma_2^2$ against $H_1: \sigma_1^2 \neq \sigma_2^2$ at the $\alpha = 0.05$ level. The sample statistics are $n_1 = 70$, $s_1 = 0.383$, $n_2 = 41$, and $s_2 = 0.326$. Thus,

$$F_0 = \frac{(0.383)^2}{(0.326)^2} = 1.380.$$

Since $F_{.95}(69, 40) \cong 1.6$, the P value is greater than $2(0.05) = 0.10$ and H_0 is accepted.

When the equality of variances is not justified, a test developed by Welch (1937) is more appropriate than the two-sample t test. To test the null hypothesis $H_0: \mu_1 - \mu_2 = \delta$ against $H_1: \mu_1 - \mu_2 \neq \delta$, $\mu_1 - \mu_2 < \delta$, or $\mu_1 - \mu_2 > \delta$, the test statistic is

$$t_0 = \frac{\bar{x}_{1.} - \bar{x}_{2.} - \delta}{\sqrt{\dfrac{s_1^2}{n_1} + \dfrac{s_2^2}{n_2}}}. \tag{2.3.9}$$

Under H_0, t_0 has an approximate Student's t distribution for large samples. The approximate number of degrees of freedom is

$$v = \frac{\left(\dfrac{s_1^2}{n_1} + \dfrac{s_2^2}{n_2}\right)^2}{\dfrac{s_1^4}{n_1^2(n_1 - 1)} + \dfrac{s_2^4}{n_2^2(n_2 - 1)}}. \tag{2.3.10}$$

Since v is not necessarily an integer, the P value may be obtained by linearly interpolating in Table 5, Appendix II. This test is superior to the two-sample t test when the population variances are appreciably different since it achieves a significance level nearer the nominal α.

COMMENT 2.3.2   Three t tests for comparing two means have been introduced. The common assumptions for all three tests are that: (a) the underlying distributions are normal; and (b) within a population the individuals are selected at random. Differences in assumptions concern (a) equality of the variances σ_1^2 and σ_2^2, and (b) independence of the two samples. When the sample sizes n_1 and n_2 are equal, the accompanying table is a guide for choosing the appropriate t test.

Equal variances	Independent samples	
	Yes	No
Yes	Two-sample t test	Paired t test
No	Welch t test	Paired t test

2.4 Descriptive Programs with Strata—
The Analysis of $p \geqslant 2$ Continuous Variables

In this section we generalize the ideas of the previous section to $p \geqslant 2$ continuous random variables. As in that section, we first consider the case where p random variables $X_1, X_2, ..., X_p$ are defined on the same population W. Thus, we can study the $p(p+1)/2$ distinct covariances or correlations between X_i and X_j, $i, j = 1, ..., p$. A random sample from this population consists of n sets of p observations, where each set is measured on the same individual in the sample.

We then study the case where the same random variable X is defined on p populations $W_1, W_2, ..., W_p$. As before, these populations may be regarded as subpopulations (or as strata) from the larger population W. In this case, p independent random samples are selected from the p populations. We describe how testing the hypothesis of equality of the means of $X_1, ..., X_p$ leads to the one-way analysis of variance F *ratio*. We then discuss how comparisons between means may be made using *multiple comparison* techniques.

We first discuss the following case.

2.4.1 One Population—p Random Variables

In this situation we have data on p random variables $X_1, X_2, ..., X_p$ defined on a population W. Let μ_i and σ_i^2 be the mean and variance of X_i in the population, and let $\sigma_{ij} = \sigma_{ji}$ be the covariance between X_i and X_j for $i,j = 1, ..., p$. (Note that $\sigma_{ii} = \sigma_i^2, i = 1, ..., p$.) The *population correlation coefficient* ρ_{ij} between X_i and X_j is defined to be

$$\rho_{ij} = \frac{\sigma_{ij}}{\sigma_i \sigma_j} \quad \text{for} \quad i,j = 1, ..., p. \tag{2.4.1}$$

Note that $\rho_{11} = \rho_{22} = \cdots = \rho_{pp} = 1$, and that $-1 \leqslant \rho_{ij} = \rho_{ji} \leqslant 1$ for $i \neq j$. It is shown in Section 3.1 that ρ_{ij} is a measure of the linear association between X_i and X_j. In Section 3.3 it is shown how ρ_{ij}^2 plays an important role in picking the best predictor in a multiple regression situation.

From W a random sample of size n is selected and $x_{1j}, x_{2j}, ..., x_{pj}$ is observed on the jth individual, $j = 1, ..., n$. Using a packaged descriptive program with the input in the form of the accompanying tabulation, we obtain a histogram of

Individual	Data
1	$x_{11}, x_{21}, ..., x_{p1}$
2	$x_{12}, x_{22}, ..., x_{p2}$
$\vdots$	$\vdots$
n	$x_{1n}, x_{2n}, ..., x_{pn}$

$x_{11}, x_{12}, ..., x_{1n}$, a histogram of $x_{21}, x_{22}, ..., x_{2n}$, ..., and a histogram of $x_{p1}, x_{p2}, ..., x_{pn}$—$p$ histograms in all. Furthermore, we obtain estimates for the population parameters. Thus, for $i,j = 1, ..., p$ we obtain

$$\bar{x}_{i.} = \frac{1}{n} \sum_{j=1}^{n} x_{ij},$$

$$s_i^2 = \frac{1}{n-1} \sum_{j=1}^{n} (x_{ij} - \bar{x}_{i.})^2,$$

$$s_{ij} = \frac{1}{n-1} \sum_{k=1}^{n} (x_{ik} - \bar{x}_{i.})(x_{jk} - \bar{x}_{j.}),$$

and

$$r_{ij} = \frac{s_{ij}}{s_i s_j},$$ (2.4.2)

as estimates of μ, σ_i^2, σ_{ij}, and ρ_{ij}, respectively. The variances, covariances, and correlations may be presented in matrix form in the output as

Covariance matrix

$$\begin{bmatrix} s_{11} = s_1^2 & s_{12} & \cdots & s_{1p} \\ s_{21} & s_{22} = s_2^2 & \cdots & s_{2p} \\ \vdots & \vdots & \vdots & \vdots \\ s_{p1} & s_{p2} & \cdots & s_{pp} = s_p^2 \end{bmatrix}$$

and

Correlation matrix

$$\begin{bmatrix} 1 & r_{12} & \cdots & r_{1p} \\ r_{21} & 1 & \cdots & r_{2p} \\ \vdots & \vdots & \vdots & \vdots \\ r_{p1} & r_{p2} & \cdots & 1 \end{bmatrix}$$

In Sections 5.6 and 5.7 we study the factoring of the covariance or correlation matrix into components, and in Section 3.1 we present material on testing hypotheses about correlations. For $i \neq j$, if X_i and X_j are comparable, then we may test hypotheses about the mean difference $\mu_i - \mu_j$ using the paired sample t test of Section 2.3.

EXAMPLE 2.4.1   In a study, simultaneous measurements of $p = 5$ arterial pressures (mm Hg) were made on $n = 141$ patients using two different techniques. One technique used an *intra-arterial catheter* to measure $X_1 =$ systolic, $X_2 =$ diastolic, and $X_3 =$ mean arterial pressures. This is the more accurate, but more difficult, method of measuring pressures. The other technique was the common *cuff* *method* which measured $X_4 =$ systolic and $X_5 =$ diastolic pressures. The data collected was analyzed by a packaged descriptive program and is summarized in the accompanying table. The covariance and correlation matrices are shown on p. 69.

Method	Variable	Sample mean	Sample standard deviation
Intra-arterial	$X_1 =$ Systolic pressure	$\bar{x}_{1.} = 112.2$	$s_1 = 28.6$
	$X_2 =$ Diastolic pressure	$\bar{x}_{2.} = 59.4$	$s_2 = 17.1$
	$X_3 =$ Mean pressure	$\bar{x}_{3.} = 76.8$	$s_3 = 21.0$
Cuff	$X_4 =$ Systolic pressure	$\bar{x}_{4.} = 107.0$	$s_4 = 28.9$
	$X_5 =$ Diastolic pressure	$\bar{x}_{5.} = 66.8$	$s_5 = 19.3$

Covariance matrix

$$
\begin{array}{c c c c c c}
 & X_1 & X_2 & X_3 & X_4 & X_5 \\
X_1 & 817.9 & 410.3 & 556.8 & 719.9 & 415.6 \\
X_2 & & 292.4 & 347.2 & 384.5 & 273.3 \\
X_3 & & & 441.0 & 512.8 & 345.3 \\
X_4 & & & & 835.2 & 466.9 \\
X_5 & & & & & 372.5
\end{array}
$$

Correlation matrix

$$
\begin{array}{c c c c c c}
 & X_1 & X_2 & X_3 & X_4 & X_5 \\
X_1 & 1.000 & 0.839 & 0.927 & 0.871 & 0.753 \\
X_2 & & 1.000 & 0.967 & 0.778 & 0.828 \\
X_3 & & & 1.000 & 0.845 & 0.852 \\
X_4 & & & & 1.000 & 0.837 \\
X_5 & & & & & 1.000
\end{array}
$$

Note the high correlations between the five measurements.

From the table of sample means, note that on the average, the cuff method underestimates systolic pressure and overestimates diastolic pressure when compared to the corresponding exact measurements using the intra-arterial method. To test the significance of these differences we used the paired sample t test of Section 2.3.1. Thus, we first tested $H_0: \mu_4 - \mu_1 = 0$ against $H_1: \mu_4 - \mu_1 < 0$ at the $\alpha = 0.05$ level. The test statistic is

$$
t_0 = \frac{(107.0 - 112.2)\sqrt{141}}{\sqrt{835.2 + 817.9 - 2(719.9)}} = -4.2,
$$

which is significant at $P < 0.001$. We then tested $H_0: \mu_5 - \mu_2 = 0$ against $H_1: \mu_5 - \mu_2 > 0$ at the $\alpha = 0.05$ level. The test statistic is

$$
t_0 = \frac{(66.8 - 59.4)\sqrt{141}}{\sqrt{372.5 + 292.4 - 2(273.3)}} = 8.07,
$$

which is highly significant. Thus, the cuff measurements should not be regarded as accurate estimates of the intra-arterial measurements. This example will be expanded in Section 3.2.

2.4.2 p Populations—One Random Variable

In this section we have data on a continuous random variable X defined on p populations $W_1, W_2, ..., W_p$. As before, these populations may be thought of as p subpopulations or p strata from a larger population W. Another variable Y stratifies W into $W_1, W_2, ..., W_p$. Let μ_i and σ_i^2 be the mean and variance of X in population W_i, $i = 1, ..., p$. From the population W_i a random sample of size n_i is taken and X is observed for each individual in the sample, $i = 1, ..., p$. The resulting observations are denoted by $x_{11}, x_{12}, ..., x_{1n_1}$ for the sample from W_1; $x_{21}, x_{22}, ..., x_{2n_2}$ for the sample from W_2; ...; and $x_{p1}, x_{p2}, ..., x_{pn_p}$ for the sample from W_p.

To obtain p histograms of these samples from a descriptive program with strata, the observations may be arranged as follows. All the observations for the first sample are followed by all the observations for the second sample, ..., and followed by all the observations for the pth sample. The sample sizes are specified on a control card. Alternately, a base variable Y may be defined to specify the subpopulation from which the sample is taken. The program then classifies the observations into the correct sample according to a cutpoint for Y specified on a control card.

For each sample a histogram and the corresponding estimates of μ_i and σ_i^2 are calculated. These are given by

$$\bar{x}_{i.} = \frac{1}{n_i} \sum_{j=1}^{n_i} x_{ij}, \quad \text{and} \quad s_i^2 = \frac{1}{n_i - 1} \sum_{j=1}^{n_i} (x_{ij} - \bar{x}_{i.})^2 \quad \text{for} \quad i = 1, 2, ..., p.$$

(2.4.3)

We now consider the following example.

EXAMPLE 2.4.2   In an experiment[†] each of 21 drugs was administered to a random sample of rats to compare the effect of the drugs on $X = $ amount of hydrochloric acid (HCl) excreted in the rat's stomach. A 22nd sample served as a control group. Thus, there are $p = 22$ populations, where W_i is the population of all rats that take the ith drug, and W_{22} is the control population. The observations are denoted by $x_{ij}, j = 1, ..., n_i, i = 1, ..., 22$. By defining the variable $Y = i$ if x_{ij} is in the ith sample, $j = 1, ..., n_i$, $i = 1, ..., 22$, a descriptive program printed histograms of the samples and calculated the sample statistics. The sample means $\bar{x}_{i.}$, in increasing order, and the corresponding sample sizes n_i are listed in the accompanying table.

In this example it is of interest to test the hypothesis that the mean HCl excretion is the same for all 22 populations. In general, if $x_{11}, ..., x_{1n_1}$ is a random

† Personal communication from Dr. Alberto Rosenberg, UCLA, Los Angeles, California.

Sample number i	Sample size n_i	Sample mean $\bar{x}_{i.}$	Sample number i	Sample size n_i	Sample mean $\bar{x}_{i.}$
7	22	73.73	8	14	333.29
15	25	146.32	11	27	341.30
6	13	147.92	5	32	374.06
14	18	165.61	4	8	412.13
3	8	191.13	1 (control)	71	417.32
13	17	213.47	21	16	459.81
18	17	224.41	10	19	460.37
9	14	263.86	22	19	477.53
19	14	303.14	17	18	484.61
12	15	313.20	20	18	507.56
2	6	329.83	16	19	566.37

sample from $N(\mu_1, \sigma_1^2)$, $x_{21}, \ldots, x_{2n_2}$ is from $N(\mu_2, \sigma_2^2)$, ..., and $x_{p1}, \ldots, x_{pn_p}$ is from $N(\mu_p, \sigma_p^2)$ and $\sigma_1^2 = \sigma_2^2 = \cdots = \sigma_p^2 = \sigma^2$, then the null hypothesis $H_0: \mu_1 = \mu_2 = \cdots = \mu_p$ may be tested using an F ratio. The alternative hypothesis is H_1: not all of the μ_i are equal. The test statistic is

$$F_0 = \frac{\sum_{i=1}^{p} n_i (\bar{x}_{i.} - \bar{x}_{..})^2 / (p-1)}{\sum_{i=1}^{p} \sum_{j=1}^{n_i} (x_{ij} - \bar{x}_{i.})^2 / (n-p)}, \tag{2.4.4}$$

where $n = \sum_{i=1}^{p} n_i$ is the total sample size, $\bar{x}_{i.} = (1/n_i) \sum_{j=1}^{n_i} x_{ij}$ is the sample mean for the ith subpopulation, and $\bar{x}_{..} = (1/n) \sum_{i=1}^{p} \sum_{j=1}^{n_i} x_{ij}$ is the sample grand mean.

Under the null hypothesis, F_0 has an F distribution with $\nu_B = p-1$ and $\nu_W = n-p$ degrees of freedom. The P value is the area to the right of F_0 under the $F(\nu_B, \nu_W)$ frequency curve (Table 6, Appendix II). The hypothesis H_0 is rejected if P is less than the specified significance level α.

This F ratio is the result of a statistical method known as the *one-way analysis of variance*. (The topic of the *analysis of variance* is discussed in detail in Chapter 4.) It is customary to present the components of the numerator and the denominator of the F ratio in a table called an *analysis of variance table* similar to that of Table 2.4.1. The first column lists three *sources of variation*—the *between groups*, the *within groups*, and the *total*. The second column lists the *sum of squares* for these three sources of variation. Note that the quantities SS_B and SS_W are components of Eq. 2.4.4. This is also true for the between and within *degrees of freedom* ν_B and ν_W. Each *mean square* is computed by dividing its sum of squares by its degrees of freedom. (The mean square for the total is customarily not presented in the table.) Finally, the F ratio is the same as Eq. (2.4.4). In addition to the F ratio, two important quantities are MS_W, since it is an estimate of the common variance σ^2, and ν_W since it is the number of degrees of freedom used in confidence intervals.

TABLE 2.4.1. *One-Way Analysis of Variance Table*

Source of variation	Sum of squares	Degrees of freedom	Mean square	F ratio
Between sub-populations (or Groups)	$SS_B = \sum_{i=1}^{p} n_i(\bar{x}_{i.} - \bar{x}_{..})^2$	$\nu_B = p - 1$	$MS_B = \dfrac{SS_B}{\nu_B}$	$F = \dfrac{MS_B}{MS_W}$
Within sub-populations (or Groups)	$SS_W = \sum_{i=1}^{p} \sum_{j=1}^{n_i} (x_{ij} - \bar{x}_{i.})^2$	$\nu_W = n - p$	$MS_W = \dfrac{SS_W}{\nu_W}$	
Total	$SS_T = \sum_{i=1}^{p} \sum_{j=1}^{n_i} (x_{ij} - \bar{x}_{..})^2$	$\nu_T = n - 1$		

Thus, a $100(1-\alpha)\%$ *confidence interval for* μ_i *is*

$$\bar{x}_{i.} \pm t_{1-(\alpha/2)}(\nu_W)\sqrt{\frac{MS_W}{n_i}}, \qquad i = 1, \ldots, p, \qquad (2.4.5)$$

and a $100(1-\alpha)\%$ *confidence interval for* $\mu_i - \mu_j$ *is*

$$(\bar{x}_{i.} - \bar{x}_{j.}) \pm t_{1-(\alpha/2)}(\nu_W)\sqrt{MS_W\left(\frac{1}{n_i} + \frac{1}{n_j}\right)}, \qquad (2.4.6)$$

where $t_{1-(\alpha/2)}(\nu_W)$ is the $100[1-(\alpha/2)]$th percentile of the Student's t distribution with ν_W degrees of freedom.

EXAMPLE 2.4.2 (*continued*)   For this experiment we tested the hypothesis that the mean HCl excretion is the same in all 22 populations. (See accompanying analysis of variance tabulation.)

Source of variation	Sum of squares	Degrees of freedom	Mean square	F ratio
Between groups	7,536,412	21	358,877	6.49
Within groups	22,561,794	408	55,299	
Total	30,098,206	429		

To test $H_0: \mu_1 = \mu_2 = \cdots = \mu_{22}$, we compared $F_0 = 6.49$ with percentiles of $F(21, 408)$. Since the P value is less than 0.001, H_0 was rejected. Thus, the treatments do differ significantly in their effect on HCl excretion in a rat's stomach.

The estimate of the variance σ^2 is $MS_W = 55,299$. A 95% confidence interval for the mean μ_1 of the control group is

$$417.32 \pm t_{0.975}(408) \sqrt{\frac{55,299}{71}} = 417.32 \pm 1.96\,(27.9)$$

$$= (362.6, 472.0).$$

A 95% confidence interval for the difference between the control group mean μ_1 and the seventh treatment mean μ_7 is

$$(417.32 - 73.73) \pm t_{0.975}(408) \sqrt{55,299(\tfrac{1}{71} + \tfrac{1}{22})} = 343.59 \pm 1.96\,(57.31)$$

$$= (231.27, 455.91).$$

Note that this F test assumes that the p population variances $\sigma_1{}^2, \sigma_2{}^2, \ldots, \sigma_p{}^2$ are equal. The hypothesis $H_0: \sigma_1{}^2 = \cdots = \sigma_p{}^2$ may be tested using the *Bartlett test for equality of p variances*. Since this test is so highly sensitive to the assumption of normality, it is not discussed in this book. (For a treatment of this test, see Brownlee, 1965.)

Note also that the within mean square MS_W can also be expressed as

$$MS_W = \frac{(n_1 - 1)s_1{}^2 + (n_2 - 1)s_2{}^2 + \cdots + (n_p - 1)s_p{}^2}{n_1 + n_2 + \cdots + n_p - p}. \qquad (2.4.7)$$

Thus, it is a "pooled" variance. In particular, compare this equation for $p = 2$ with the pooled variance $s_p{}^2$ given in Eq. (2.3.6).

If we do not reject H_0, then all p means may be equal to some overall mean μ, that is, $\mu_1 = \mu_2 = \cdots = \mu_p = \mu$. The best estimator of this overall mean is $\hat{\mu} = \bar{x}_{..}$, the sample grand mean. On the other hand, if we do reject H_0, then we conclude that some of the means are not equal.

Since the F test does not indicate which means are unequal, the investigator must make additional tests of hypotheses among the means. He may test hypotheses about differences between the means, for example, $H_0: \mu_i - \mu_j = 0$, or he may test hypotheses about linear combinations of the means, for example, $H_0: 2\mu_1 + 3\mu_2 - 4\mu_5 = 0$. For a single test of the form $H_0: \mu_i - \mu_j = 0$, the investigator calculates a $100(1 - \alpha)\%$ confidence interval as in Eq. (2.4.6) and rejects H_0 at significance level α if the interval does not contain zero. For a single test about a linear combination of the means, the following test is appropriate.

Denote the linear combination by $c_1\mu_1 + c_2\mu_2 + \cdots + c_p\mu_p$, where the c_i's are constants. Then to test $H_0: \sum_{i=1}^{p} c_i \mu_i = 0$ against the alternative $H_1: \sum_{i=1}^{p} c_i \mu_i \neq 0$ at the significance level α, form the $100(1 - \alpha)\%$ *confidence interval for* $\sum_{i=1}^{p} c_i \mu_i$ given by

$$\sum_{i=1}^{p} c_i \bar{x}_{i.} \pm t_{1-(\alpha/2)}(v_W) \sqrt{MS_W \sum_{i=1}^{p} \frac{c_i{}^2}{n_i}}. \qquad (2.4.8)$$

If this interval contains zero, we accept H_0, otherwise we reject H_0 at the α level.

If the investigator wishes to make more than one test of these hypotheses, then the overall significance level (that is, the significance level of all of his tests combined) may be nowhere near the nominal α. Thus, he is unable to assert that all of his tests were simultaneously made at the α level. To circumvent this problem, he may use a *multiple comparisons procedure* for all of his tests so that he is able to assert that all of the tests are at the α level, that is, the overall level is the nominal α.

We consider here three multiple comparisons procedures. The first is the *Scheffé method* (Scheffé, 1953) which prescribes that to test $H_0: \sum_{i=1}^{p} c_i \mu_i = 0$ against $H_1: \sum_{i=1}^{p} c_i \mu_i \neq 0$, we form the confidence interval

$$\sum_{i=1}^{p} c_i \bar{x}_{i.} \pm S, \tag{2.4.9}$$

where

$$S^2 = p\,MS_W\,F_{1-\alpha}(p, n-p) \sum_{i=1}^{p} \frac{c_i^2}{n_i}, \tag{2.4.10}$$

and $F_{1-\alpha}(p, n-p)$ is the $100(1-\alpha)$th percentile of the $F(p, n-p)$ distribution. If this interval does not include zero, then H_0 is rejected at level α. We repeat the process for every other linear combination of interest. The overall level of all the tests is still α.

Most comparisons encountered in practice are *contrasts* in the means. A contrast is a linear combination $\lambda_1 \mu_1 + \cdots + \lambda_p \mu_p$ with the restriction that $\lambda_1 + \cdots + \lambda_p = 0$. Thus, contrasts are multiples of differences between weighted averages of the means, such as

$$\mu_1 - \mu_2, \quad \frac{\mu_1 + \mu_2}{2} - \frac{\mu_3 + \mu_4 + \mu_5}{3},$$

and so forth.

The Scheffé method applied to all contrasts takes the following form: To test $H_0: \sum_{i=1}^{p} \lambda_i \mu_i = 0$ against $H_1: \sum_{i=1}^{p} \lambda_i \mu_i \neq 0$, we form the confidence interval

$$\sum_{i=1}^{p} \lambda_i \bar{x}_{i.} \pm S, \tag{2.4.11}$$

where S^2 is now given by

$$S^2 = (p-1)\,MS_W\,F_{1-\alpha}(p-1, n-p) \sum_{i=1}^{p} \frac{\lambda_i^2}{n_i}, \tag{2.4.12}$$

and $F_{1-\alpha}(p-1, n-p)$ is the $100(1-\alpha)$th percentile of the $F(p-1, n-p)$ distribution. If this interval does not include zero, then H_0 is rejected at level α. We repeat this process for every linear combination of interest. The overall significance level of all tests is α.

The second multiple comparisons procedure is the *Tukey method* (see Scheffé, 1959 or Tukey, 1949b). This is applicable only for contrasts and only if the sample sizes are equal, that is, $n_1 = n_2 = \cdots = n_p = m$. To test $H_0 : \sum_{i=1}^{p} \lambda_i \mu_i = 0$ against $H_1 : \sum_{i=1}^{p} \lambda_i \mu_i \neq 0$, we form the confidence interval

$$\sum_{i=1}^{p} \lambda_i \bar{x}_{i.} \pm T, \tag{2.4.13}$$

where

$$T = \frac{1}{2} \sqrt{\frac{MS_W}{m}} \, q_{1-\alpha} \sum_{i=1}^{p} |\lambda_i|, \tag{2.4.14}$$

and $q_{1-\alpha}$ is the $100(1-\alpha)$th percentile of the *studentized range distribution* with p and $v = n - p$ degrees of freedom (Table 7, Appendix II).[†] If this interval does not include 0, H_0 is rejected at level α. Repeating this process for every contrast of interest gives an overall level of α.

A third technique is the *multiple t method*. Let k be the number of preselected contrasts of interest. Then to test $H_0 : \sum_{i=1}^{p} \lambda_i \mu_i = 0$ against $H_1 : \sum_{i=1}^{p} \lambda_i \mu_i \neq 0$, we form the approximate confidence interval

$$\sum_{i=1}^{p} \lambda_i \bar{x}_{i.} \pm t_{1-(\alpha/2k)}(v_W) \sqrt{MS_W \sum_{i=1}^{p} \frac{\lambda_i^2}{n_i}}, \tag{2.4.15}$$

where $t_{1-(\alpha/2k)}$ is the $100[1-(\alpha/2k)]$th percentile of the Student's t distribution with v_W degrees of freedom. If the interval does not include zero, we reject H_0.

COMMENTS 2.4.1   1.   Since the Scheffé, Tukey, and multiple t methods are based on different distributions (F, q, and t, respectively) they do not in general give the same confidence intervals. Scheffé's method permits unequal sample sizes and all linear combinations, while Tukey's method applies to equal sample sizes and only for contrasts. The multiple t method applies only to a set of contrasts selected *before* examining the data; for the other two, any contrast is permissible.

2.   In obtaining a confidence interval for a contrast, the user should choose the method which gives the shortest confidence interval. On the average, the Tukey method gives shorter confidence intervals than the Scheffé method for simple contrasts, that is, contrasts involving at most three means. On the other hand, Scheffé's method gives shorter confidence intervals on the average for contrasts involving 4 or more means.

---

† The *Studentized range* with p and v degrees of freedom is defined as follows. Let $Y_1, Y_2, \ldots, Y_p$ be independent random variables distributed as $N(\mu_y, \sigma_y^2)$ and let W be the range of these variables, that is, $W = \text{Max}_i(Y_i) - \text{Min}_i(Y_i)$. If s_y^2 with v degrees of freedom is an independent unbiased estimator of σ_y^2, then the distribution of W/s_y is the studentized range distribution with p and v degrees of freedom.

3. Concerning the multiple t method, if the number of preselected contrasts is "small," it may give the shortest confidence intervals. However, a common method of determining the contrasts is by inspection of the data.

4. Note that when $p = 2$, the F ratio in the analysis-of-variance table is equal to the square of the two-sample t statistic, that is, $F(1, v_W) = t^2(v_W)$.

5. The analysis of variance F test will be significant at a given level α if and only if $H_0: \sum_{i=1}^{p} \lambda_i \mu_i = 0$ is rejected for some contrast according to the Scheffé procedure. The particular contrast which is significant may not be easy to find or interpret. Hence, the situation may occur where the F test is significant at a level α but no interesting contrasts are found which are significant at this level. To identify interesting contrasts, we may use a larger α for the multiple comparison tests than that used for the F test. Thus, when $\alpha = 0.05$ is used for the F test, it may be advisable to find 90% multiple comparisons confidence intervals for the contrasts.

6. Some of the results of multiple comparisons may seem contradictory. For example, when $p = 3$, it is possible to obtain the conclusions that μ_1 is not significantly different from μ_2, μ_2 is not significantly different from μ_3, but μ_1 *is* significantly different from μ_3. If "not significantly different from" is interpreted as "equal to" and "significantly different from" is interpreted as "not equal to," these conclusions would indeed be contradictory. This interpretation however, is incorrect since significance statements are made with a nonzero probability of being false. The correct way of interpreting this example is that in the data there was sufficient evidence to show that μ_1 and μ_3 were significantly different, but not μ_1 and μ_2 or μ_2 and μ_3.

EXAMPLE 2.4.2 (*continued*)   To determine the significant differences between the means, the Scheffé multiple comparisons method was used to compare all of the $\binom{22}{2} = 231$ pairs of means at the $\alpha = 0.05$ level. Thus, for $i \neq j$ the hypothesis $H_0: \mu_i - \mu_j = 0$ was tested against the alternative $H_1: \mu_i - u_j \neq 0$ using Eqs. (2.4.11)–(2.4.12) (since these hypotheses are about contrasts). For example, to test $H_0: \mu_7 - \mu_5 = 0$, we calculated the 95% confidence interval as

$$(73.73 - 374.06) \pm \sqrt{21\,(55299)\,F_{0.95}(21, 408)(\tfrac{1}{22} + \tfrac{1}{32})}$$

$$= -300.33 \pm 365.33 = (-665.66, 65.00).$$

Since this interval includes 0, we accept H_0.

As another example, we tested $H_0: \mu_7 - \mu_4 = 0$. The 95% confidence interval is

$$(73.73 - 412.13) \pm \sqrt{21\,(55299)(1.5)(\tfrac{1}{22} + \tfrac{1}{8})} = (-883.29, -206.49).$$

Since this interval does not include 0, we reject H_0.

A useful technique for summarizing all of the results is as follows. List the treatments in order of increasing sample means. Then compare the smallest sample mean with each subsequent mean using the Scheffé procedure. Underscore all treatments with means not significantly different from the smallest treatment mean. Repeat the process for the treatment with the second smallest sample mean, that is, compare this sample mean with all subsequent sample means, underscoring all treatments with means not significantly different from it. Repeat it for the third smallest mean, etc. The results of this procedure for this example are shown in Table 2.4.2. Note that, for example, the mean of treatment 7 is not significantly different from the means of treatments 15, 6, 14, ..., 11 and 5, but it is significantly different from the means of treatments 4, 1, ..., 20 and 16.

TABLE 2.4.2. *Multiple Comparisons of Twenty-Two Means of HCl Stomach Excretion*

Treatment number	7	15	6	14	3	13	18	9	19	12	2	3	11	5	4	1	21	10	22	17	20	16

2.5 Cross-Tabulation Programs—
The Analysis of Contingency Tables

In this section we discuss packaged *cross-tabulation* programs which compute and display a two-dimensional table called a *two-way contingency table*. In this situation any individual or experimental unit in a population W can be classified by two different *factors* (or *criteria*) A and B. We assume that factor A has $r \geqslant 2$ *classes* (or *levels*) and factor B has $c \geqslant 2$ classes (or levels). Representing the r classes of A as the rows and the c classes of B as the columns, we obtain a two-

			B				Row
A	1	2	3	$\cdots$	c		totals
1	f_{11}	f_{12}	f_{13}	$\cdots$	f_{1c}		$f_{1.}$
2	f_{21}	f_{22}	f_{23}	$\cdots$	f_{2c}		$f_{2.}$
3	f_{31}	f_{32}	f_{33}	$\cdots$	f_{3c}		$f_{3.}$
$\vdots$	$\vdots$	$\vdots$	$\vdots$	$\vdots$	$\vdots$		$\vdots$
r	f_{r1}	f_{r2}	f_{r3}	$\cdots$	f_{rc}		$f_{r.}$
Column totals	$f_{.1}$	$f_{.2}$	$f_{.3}$	$\cdots$	$f_{.c}$		n

way contingency table as shown in the accompanying table. Such a two-way table is called an $r \times c$ *contingency table* and the intersection of a row and a column is called a *cell*. For a sample of size n from W, the number of individuals f_{ij} with the ith level of A and the jth level of B is entered in the ij cell, $i = 1, ..., r, j = 1, ..., c$. Each f_{ij} is called the *observed frequency* of the ijth cell.

A packaged cross-tabulation program scans a sample of n observations and tallies the observed frequencies in each cell. It then computes the total $f_{i.}$ for the ith row, and the total $f_{.j}$ for the jth column, $i = 1, ..., r, j = 1, ..., c$. These quantitites are called the *row* and *column totals*, respectively. Note that $\sum_{i=1}^{r} f_{i.} = \sum_{j=1}^{c} f_{.j} = n$, the sample size. The resulting table is printed in the output.

After the table has been generated it is possible to test hypotheses about the factors A and B. All of these hypotheses can be stated in terms of the *independence* of the two factors. In this context, independence means that the proportion of each row total belonging to a given column is the same for all rows, and that the proportion of each column total belonging to a given row is the same for all columns.

In some situations the levels of one factor (A, say) are disjoint subpopulations $W_1, W_2, ..., W_r$ of the population W. In this case the hypothesis of independence can also be stated as a hypothesis about the *homogeneity* of the factor B with respect to the levels of A. In order to make these distinctions clearer we consider the following examples.

2.5.1 Hypothesis of Homogeneity

As stated previously, the levels of A in this situation stratify the population W into r disjoint subpopulations $W_1, W_2, ..., W_r$. Then, any individual from W_i is classified into one and only one of the c classes of factor B. We let p_{ij} be the proportion of individuals in the subpopulation W_i classified into the jth class of factor B. Then the hypothesis of homogeneity can be expressed as

$$H_0: p_{1j} = p_{2j} = \cdots = p_{rj} \qquad \text{for all} \quad j = 1, ..., c,$$

that is, for any class j the proportion of individuals in each subpopulation is the same for all subpopulations. The alternative hypothesis H_1 is that some of these proportions are not equal.

We note that the levels of A representing the subpopulations are measured on the nominal scale, but that the classes of B are measured on either a nominal or an ordinal scale. Furthermore, a continuous variable measured on an interval or ratio scale may be transformed into an ordinal scale. We now give examples of these situations.

EXAMPLE 2.5.1 ($r = c = 2$)   Let W be a population of adults stratified by sex, and let B classify an individual according to whether or not he has cancer. In this case B is measured on the nominal scale. The 2×2 contingency table is

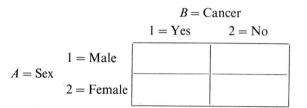

$B = $ Cancer

1 = Yes   2 = No

$A = $ Sex

1 = Male

2 = Female

If p_{11} is the proportion of males and p_{21} is the proportion of females with cancer, then the hypothesis is $H_0: p_{11} = p_{21}$. (Note that this implies $p_{12} = p_{22}$ also.) In terms of independence, this hypothesis of homogeneity could also be stated as H_0: occurrence of cancer is independent of sex.

EXAMPLE 2.5.2 ($r = 2$, $c = 3$)   Let the population W of critically ill patients be stratified by sex, and let B classify an individual according to his clinical status after a particular treatment. In this case B is measured on the ordinal scale. Then the 2×3 contingency table is

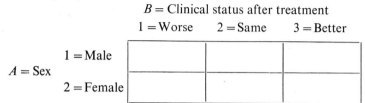

$B = $ Clinical status after treatment

1 = Worse   2 = Same   3 = Better

$A = $ Sex

1 = Male

2 = Female

The hypothesis of homogeneity is $H_0: p_{11} = p_{21}, p_{12} = p_{22}, p_{13} = p_{23}$, and the hypothesis of independence is H_0: clinical status after treatment is independent of sex.

EXAMPLE 2.5.3 ($r = 5$, $c = 3$)   Let the population W of critically ill patients in circulatory shock be stratified into 5 subpopulations according to shock type, and let B classify an individual according to his clinical status after a particular treatment. Then the 5×3 contingency table is

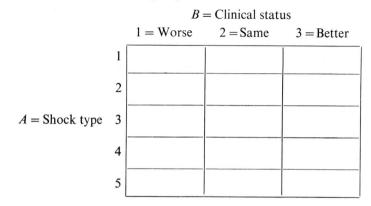

$B = $ Clinical status

1 = Worse   2 = Same   3 = Better

$A = $ Shock type

1

2

3

4

5

The hypothesis is $H_0: p_{11} = \cdots = p_{51}, \ldots, p_{13} = \cdots = p_{53}$, or alternatively, H_0: clinical status after treatment is independent of shock type. Note that a population may be stratified into more than two subpopulations.

EXAMPLE 2.5.4 ($r = 2$, $c = 3$)   Let the population W of critically ill patients be stratified by sex, and let X be a random variable measuring the age of an individual in this population. Let B define the age groups: 1 for $X < 30$, 2 for $30 \leqslant X \leqslant 45$, and 3 for $X > 45$. Thus X has been transformed into an ordinal scale. The 2×3 contingency table is

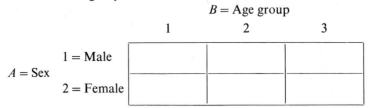

The hypothesis is that $H_0: p_{11} = p_{21}, p_{12} = p_{22}$, and $p_{13} = p_{23}$, or that H_0: age group and sex are independent for these patients.

2.5.2 Hypotheses of Independence

In this case we have one population W and each individual from this population is doubly classified according to two factors A and B. Here the null hypothesis can only be stated in terms of the independence of A and B. The alternative hypothesis is that A and B are not independent.

In this situation both A and B can be measured on the nominal or ordinal scales. The most general case is where we have two continuous variables X and Y and each variable is transformed into an ordinal scale. We now give examples of these situations.

EXAMPLE 2.5.5 ($r = 2$, $c = 3$)   Let the individuals in the population W of patients be classified as either cyanotic or not (factor A) and by their reaction to a particular treatment (factor B). In this case both A and B are nominal. The 2×3 contingency table is

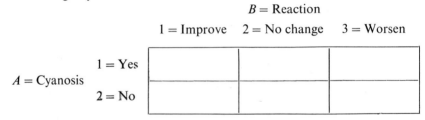

The test of hypothesis is H_0: the reaction is independent of cyanosis.

EXAMPLE 2.5.6 ($r = 3, c = 4$)   In this example the population W consists of patients with a certain disease who were given a new treatment. For each patient the random variable X measured his age in years (factor A) and the random variable Y measured the duration of his fever in days (factor B). The ranges of these variables were divided into 3 and 4 classes, respectively. The 3×4 contingency table was

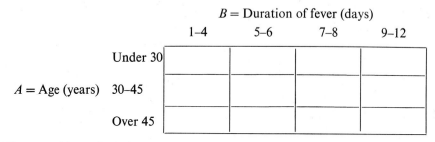

The test of hypothesis is H_0: the duration of the fever is independent of the age of the patient.

In Example 2.5.6 it is seen that each of the two continuous variables X and Y is transformed into classes. Some cross-tabulation programs will automatically determine the classes for each of these variables. This is done in much the same way that a histogram program determines the classes for one variable.

In the case where we have two random variables X and Y, the contingency table can be used to estimate the joint distribution of the two variables. Thus, the two-way frequency table generalizes the concept of a histogram. Furthermore, the test of independence for the factors A and B is a test for independence of the random variables X and Y. When the underlying distribution is bivariate normal, then it is more desirable to calculate the correlation coefficient between X and Y and test independence using the sample correlation coefficient (see Section 3.1).

2.5.3 The Chi-Square Test for Contingency Tables

To test either the hypothesis of homogeneity or the hypothesis of independence, we use the same test procedure. This procedure is to first calculate the *expected* frequency F_{ij} for the ijth cell, $i = 1, ..., r, j = 1, ..., c$. This quantity is given by

$$F_{ij} = \frac{f_{i.}}{n}(f_{.j}), \qquad i = 1, ..., r, \quad j = 1, ..., c. \qquad (2.5.1)$$

We then calculate the χ^2 statistic

$$\chi_0^2 = \sum_{i=1}^{r} \sum_{j=1}^{c} \frac{(f_{ij} - F_{ij})^2}{F_{ij}}. \qquad (2.5.2)$$

Under the null hypothesis, χ_0^2 has an approximate χ^2 distribution with $v = (r-1)(c-1)$ degrees of freedom. The P value is the area to the right of χ_0^2 under the $\chi^2(v)$ frequency curve (Table 3, Appendix II). We reject H_0 if P is less than a preselected significance level α.

EXAMPLE 2.5.7   Let the population W of critically ill patients be stratified into two subpopulations according to whether or not they are in shock. A sample of 112 critically ill patients was taken and each patient was classified according to survival and existence or absence of shock. This is summarized in the accompanying table where the entries not in parantheses are the observed

| | Survival | | Row |
Shock	Yes	No	totals
Yes	40 (49.5)	37 (27.5)	77
No	32 (22.5)	3 (12.5)	35
Column totals	72	40	112

$$\chi_0^2 = 16.34, \qquad v = 1$$

frequencies, and the entries in parentheses are the expected frequencies. Thus, for example, there are 40 patients who were in shock and survived but 37 patients in shock who did not survive. The hypotheses may be stated as H_0: survival is independent of being in shock, or H_0: the proportion of shock patients who survive is equal to the proportion of nonshock patients who survive. The calculated χ^2 statistic is $\chi_0^2 = 16.34$ with $v = (2-1)(2-1) = 1$ degree of freedom. Since the P value is less than 0.001, we conclude that the probability of death is significantly greater for shock patients than for nonshock patients.

If the investigator were interested in comparing survival for shock types, he would stratify the population of shock patients into five shock types obtaining a 5×2 table.

| | Survival | | Row |
Shock type	Yes	No	totals
Hypovolemic shock	7 (7.79)	8 (7.21)	15
Cardiogenic shock	11 (11.43)	11 (10.57)	22
Neurogenic shock	10 (8.31)	6 (7.69)	16
Septic shock	9 (8.31)	7 (7.69)	16
Endochrine shock	3 (4.16)	5 (3.84)	8
Column totals	40	37	77

$$\chi_0^2 = 1.71, \qquad v = 4$$

The χ^2 statistic for testing H_0: proportion of survivors is the same for each shock type is $\chi_0^2 = 1.71$ with $v = (5-1)(2-1) = 4$ degrees of freedom. Since the test is nonsignificant, there is no evidence of dependence of survival on shock type.

COMMENTS 2.5.1   1.   As in the analysis of variance, when the null hypothesis of independence is rejected by the χ^2 test, the researcher does not have indications of which alternatives hold. Examination of the observed and theoretical frequencies, however, might reveal some of these alternatives. Cochran (1954) and Maxwell (1961) discuss some techniques appropriate for this purpose.

2.   If the investigator performs more than one χ^2 test on the same set of data, then the joint significance level is no longer the nominal α. A more sophisticated technique is to partition the total χ^2 into components; see Maxwell (1961, Chapter 3).

3.   As discussed earlier in Section 2.1, the χ^2 test is an approximate test yielding good results when the expected frequencies F_{ij} are large. In the case of a 2×2 table with small F_{ij}, the difficulty can be overcome by applying the *Yates correction for continuity*. This is done by adding $\frac{1}{2}$ to the negative differences $(f_{ij} - F_{ij})$ and subtracting $\frac{1}{2}$ from the positive differences. However, recent studies (Grizzel, 1967) have shown that this correction makes the test too conservative, that is, H_0 is rejected less often than it should be.

Problems

Notes: 1.   In this and the following chapters, problems are grouped by section number. Within a section, problems are numbered sequentially.

2.   The problems in this chapter utilize either: Data set A: Example 1.4.1, Tables 1.4.1 and 1.4.2; or Data set B: Example 1.4.2, Tables 1.4.3 and 1.4.4. Problems using either of these sets are so identified.

3.   Problems using data sets A or B may be modified by performing the analysis on only a subset of randomly selected individuals rather than on the whole set. Choosing a small subset makes it easier to do most of the problems by hand, if desired.

Section 2.1

2.1.1. (Data Set A)   Test the hypothesis that males constitute 50% of the population. Compute a 95% confidence interval for the proportion of males.

2.1.2. (Data Set A)   Compute a 90% confidence interval for the *mortality rate* (proportion of deaths) among critically ill patients.

2.1.3. (Data Set A)   Test the hypothesis that the population is uniformly distributed among the shock types.

2.1.4. (Data Set B)   Test at the 0.05 level the hypothesis that the individuals in the population are uniformly distributed among the five socio-economic status categories.

Section 2.2

2.2.1. (Data Set A)   Using the initial data, graph a histogram of each variable. Identify which variables are continuous, dichotomous, or discrete.

2.2.2. (Data Set A)   For patients who survived, graph the empirical cdf of the initial and final MAP and compute the median, P_{25}, P_{75}, and *interquartile mean* $= (P_{25} + P_{75})/2$ for each variable. Note any changes from initial to final.

2.2.3. (Data Set A)   Repeat Problem 2.2.2 for patients who died.

2.2.4. (Data Set A)   Using the initial data for all patients, test whether HR, DP, AT, and MCT are (a) normal, (b) log normal.

2.2.5. (Data Set A)   The average value of hematocrit (Hct) for healthy adults is 40. Does the initial data for these patients show that they have "healthy" Hct? How about the final data on survivors?

2.2.6. (Data Set B)   Using a packaged descriptive program, graph a histogram for each of the variables. Which of the variables are continuous? discrete? dichotomous?

2.2.7. (Data Set B)   Can you conclude that more than 4% of the population had heart disease in 1950? Make a 95% confidence interval for this proportion. Interpret your results.

2.2.8. (Data Set B)   Graph the empirical cdf for systolic pressure in 1950 and in 1962. Use these to estimate the median, $P_{25} = $ 25th percentile, $P_{75} = $ 75th percentile, and the *interquartile mean* $= (P_{25} + P_{75})/2$ for each variable. Note any changes from 1950 to 1962. Now, the average systolic pressure in healthy young adults is 120. Find the percentile rank of 120 for 1950 and 1962. Comment.

2.2.9. (Data Set B)   For systolic pressure (1950) calculate estimates of the coefficients of skewness and kurtosis β_1 and β_2. [Hint: see Comments 2.2.2.3 and 2.2.2.4.] Do these estimates suggest a near normal distribution?

Section 2.3

2.3.1. (Data Set A)   For patients who survived, test whether there is a significant average change in MAP, DP, and CI from initial to final.

2.3.2. (Data Set A)   For patients who died, test whether the average changes in AT and MCT are significant.

2.3.3. (Data Set A)   Test whether the variance of the initial SP is the same for patients who survived and those who died.

2.3.4. (Data Set B)   Let $X_1 = $ weight in 1950, $X_2 = $ weight in 1962, $X_3 = X_2 - X_1$. Test the hypothesis that each of these three variables is normally distributed. [Hint: a reasonable procedure for choosing the intervals is to determine ten intervals, each having an expected frequency of 10% of the data.]

2.3.5. (Data Set B)   Repeat Problem 2.3.4 using systolic pressure.

2.3.6. (Data Set B)   Repeat Problem 2.3.4 using diastolic pressure.

2.3.7. (Data Set B)   Repeat Problem 2.3.4 using serum cholesterol.

2.3.8. (Data Set B)   Which of the variables—weight, systolic pressure, diastolic pressure, and serum cholesterol—showed a significant average increase from 1950 to 1962? Obtain 95% confidence intervals for the mean differences. What assumptions are you making? [Hint: all of the necessary statistics may be obtained by a single run of a descriptive program.]

Section 2.4

2.4.1. (Data Set A)   Use a descriptive program with strata to obtain histograms of the initial and final MAP, MCT, UO, and Hgb stratifying on survival. Which variables show a significant difference between the two groups?

2.4.2. (Data Set A)   Graph histograms of age stratified on shock type. Is there a significant difference between these types in terms of age? Using the Scheffé multiple comparisons method identify the pairs of mean ages which are different from each other (use $\alpha = 0.05$).

2.4.3. (Data Set B)   Graph a histogram of each continuous variable stratifying on survival. Do the mean values of these variables differ significantly between surviving patients and those who died?

2.4.4. (Data Set B)   Repeat Problem 2.4.3 by stratifying on socio-economic status. Compute multiple confidence intervals for systolic pressure (1950) for (a) $\mu_1 - \mu_5$, and (b) $(\mu_1 + \mu_2 + \mu_3)/3 - (\mu_4 + \mu_5)/2$, using three different methods. Comment.

2.4.5. (Data Set B)   Repeat Problem 2.4.3 stratifying on examining doctor (1950). Are there differences among examining doctors?

Section 2.5

2.5.1. (Data Set B)   Test the independence of survival and socio-economic status.

2.5.2. (Data Set B)   Test the independence of socio-economic status and clinical status (1950).

2.5.3. (Data Set B)   Test the independence of survival and clinical status (1950).

2.5.4. (Data Set B)   Using the estimated 20th, 40th, 60th, and 80th percentiles, divide the range of systolic pressure (1950) into 5 intervals. Then use χ^2 to test whether this variable is independent of (a) survival, and (b) socio-economic status.

2.5.5. (Data Set B)   Use χ^2 to test whether systolic pressure 1950 and 1962 are independent.

2.5.6. (Data Set B)   Test whether weight and serum cholesterol (1950) are independent. [Hint: use intervals similar to those in problem (2.5.4.).]

3 REGRESSION AND CORRELATION ANALYSIS

In this chapter we discuss two techniques, called *regression and correlation analysis*, which are concerned with the interrelationship among two or more continuous variables. In regression analysis we study the relationship between one variable, called the *dependent variable*, and several other variables, called the *independent variables*. This relationship is represented by a *mathematical model*, that is, an equation which associates the dependent variable with the independent variables along with a set of relevant assumptions. The independent variables are related to the dependent variable by a function, called the *regression function*, which involves a set of unknown *parameters*. When the function is linear in the parameters (but not necessarily linear in the independent variables), we say that we have a *linear regression model*. Otherwise, the model is called *nonlinear*. In either case, the dependent variable is *regressed* on the independent variables.

The statistical problems involved in regression analysis are: (a) to obtain the best point and interval estimators of these unknown regression parameters; (b) to test hypotheses about these parameters; (c) to determine the adequacy of the assumed model; and (d) to verify the set of relevant assumptions. The choice of the appropriate model is not a statistical one, but rather, it should be derived from the underlying physical situation. Some analytical tools, helpful in determining the relationship between the variables, will be discussed in the chapter.

The reasons for regression analysis are twofold. The first reason is to obtain a *description* of the relationship between the variables as an indication of possible causality. The second reason is to obtain a *predictor* of the dependent variable, that is, the regression equation can be used to predict the value of the dependent variable from a set of values of the independent variables. This use is particularly valuable when the dependent variable is costly or difficult to measure.

The strength of the linear relationship between two variables is measured by the *simple correlation coefficient*, while the strength of the linear relationship between one variable and a set of variables is measured by the *multiple correlation coefficient*. Another measure of association, the *partial correlation coefficient*, measures the linear association between two variables after removing the linear effect of a set of other variables. The technique of *correlation analysis* is concerned with statistical inferences about these three measures of linear association. As will be seen in this chapter, the techniques of regression and correlation analysis are intimately connected.

In this chapter, Section 3.1 discusses the linear regression of a dependent variable on one independent variable. This is called *simple linear regression analysis* and is related to *simple correlation analysis*. Section 3.2 discusses *multiple linear regression analysis* along with multiple and partial correlation analysis. In this case there is more than one independent variable. Section 3.3 is concerned with a procedure, called *stepwise regression*, for selecting the best independent variables for predicting the dependent variable. Finally, Section 3.4 discusses *nonlinear regression analysis*.

The theory underlying the linear regression model is obtained from the theory of the *general linear model*. Since this theory also includes the foundations of the analysis of variance, it is presented in Chapter 4.

3.1 Simple Linear Regression and Simple Correlation Analysis

In this section we discuss the situation where two variables are linked together by a linear relationship. Let Y be the dependent variable and let X be the independent variable.

We assume that we have a sample of pairs of observations $(x_1, y_1), (x_2, y_2), \ldots, (x_n, y_n)$ from a population W. This sample may arise in one of two ways. The first way is to arbitrarily fix values of X at $X = x_1, X = x_2, \ldots, X = x_n$, say, so that for $X = x_i$ we have a subpopulation W_i of W consisting of all individuals with $X = x_i, i = 1, \ldots, n$. In W_i we randomly select an individual and observe $Y = y_i$, $i = 1, \ldots, n$. Thus, in this situation Y is the only random variable.

For the second method of sampling, we randomly select n individuals from W and observe both X and Y for each individual. In this situation both X and Y are random variables. The advantage of this method of sampling is that we can make statistical inferences about the population correlation coefficient between X and Y; for the first method, we cannot do this.

Regardless of the method of sampling, there are two preliminary steps in determining the existence and degree of linear association between X and Y. The first step is to plot the points $(x_1, y_1), (x_2, y_2), \ldots, (x_n, y_n)$ in the xy plane.

Such a graph is called a *scattergram*. From the scattergram we can empirically decide whether a linear relationship between X and Y should be assumed. The second step is to calculate the sample correlation coefficient

$$r = \frac{\sum_{i=1}(x_i - \bar{x})(y_i - \bar{y})}{[\sum_{i=1}^{n}(x_i - \bar{x})^2 \sum_{i=1}^{n}(y_i - \bar{y})^2]^{1/2}}. \tag{3.1.1}$$

Then, if the correlation coefficient is large in absolute value (as will be discussed in Section 3.1.4), it indicates a reasonably strong linear relationship between the variables.

Some packaged *correlation programs* plot a scattergram as well as calculate the correlation coefficient between X and Y. These programs are particularly useful in finding a linear relationship if they permit transformations of the variables. Thus, in one run of such a program the investigator may obtain correlations and scattergrams of any combinations of transformations of X and Y, for example, $(X, \log Y)$, $(\log X, Y)$, $(\log X, \log Y)$, $(\sqrt{X}, \log Y)$, and so forth. The transformation(s) which produce the largest correlation coefficient in absolute value is the one for which the strongest linear relationship exists. Thus, if the correlation coefficient between X and $\log Y$ is the largest in absolute value, then the scattergram of X and $\log Y$ exhibits the strongest empirical linear relationship. We now give three examples which will be analyzed in this chapter.

EXAMPLE 3.1.1   An instrument which measures lactic acid concentration in the blood is to be calibrated. The investigator uses $n = 20$ samples of known concentration and then computes the concentration determined by the instrument. Let $X =$ known lactic acid concentration (mM) and let $Y =$ lactic acid concentration (mM) determined by the instrument. The data obtained are given in the accompanying tabulation. Note that these data arise from the first sampling situation, that is, X was fixed in advance at one of five values, $X = 1$, $X = 3$, $X = 5$, $X = 10$, or $X = 15$. The sample correlation coefficient $r = 0.987$ indicates a very strong linear relationship between X and Y. This is also evident from the scattergram given in Fig. 3.1.1.

X	Y	X	Y	X	Y
1	1.1	5	7.3	15	18.7
1	0.7	5	8.2	15	19.7
1	1.8	5	6.2	15	17.4
1	0.4	10	12.0	15	17.1
3	3.0	10	13.1		
3	1.4	10	12.6		
3	4.9	10	13.2		
3	4.4				
3	4.5				

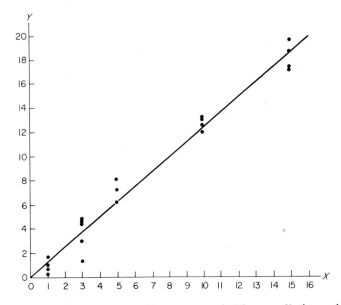

FIG. 3.1.1. $Y = $ *lactic acid concentration from instrument* (*mM*) *versus* $X = $ *known lactic acid concentration* (*mM*).

EXAMPLE 3.1.2   In this example we extend the analysis of Example 2.3.1. In this situation $X = $ venous pH and $Y = $ arterial pH were measured simultaneously on $n = 108$ critically ill patients. Note that these observations arise from the second sampling situation, that is, both X and Y are random variables. The sample correlation coefficient of $r = 0.9039$ again indicates a strong linear relationship between X and Y. The data is plotted in Fig. 3.1.2. Note that overlapping data points are designated by their multiplicity. (This is done by many packaged plotting programs.)

EXAMPLE 3.1.3   In determining cardiac output an additional derived measurement, called *mean circulation time*, is usually made. This measures the average time it takes an injected dye to appear at the arterial sampling site. Figure 3.1.3 is a scattergram of $Y = $ cardiac index (liters/min/m^2) versus $X = $ mean circulation time (sec) for 107 critically ill patients. This scattergram exhibits an exponential relationship between Y and X, but by taking the logarithm of Y, a linear relationship with X may be obtained. If a transformation of the variables produces a linear relationship, then we say the model is *intrinsically linear*. Note, however, that we may apply the techniques of *nonlinear regression* to the original data (see Section 3.4).

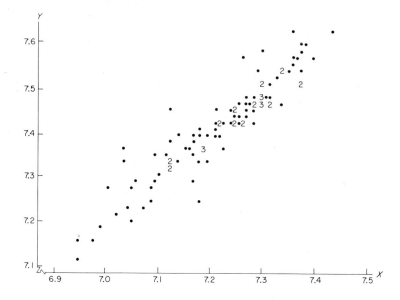

FIG. 3.1.2. *Y=arterial pH versus X=venous pH for 108 critically ill patients. The numbers 2 and 3 indicate two or three points at that location.*

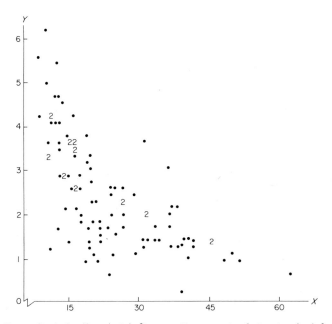

FIG. 3.1.3. *Y=cardiac index (liters/min/m²) versus X=mean circulation time (sec) for 107 critically ill patients.*

COMMENT 3.1.1   A correlation program may be used to determine the best predictor of Y from a set of p variables $X_1, X_2, ..., X_p$. The variable X_i having the largest correlation with Y (in absolute value) has the strongest linear relationship with Y. This procedure is the first step in the so-called *stepwise regression* procedure discussed in Section 3.3.

We now discuss the simple linear regression model. The theory underlying the next two sections is based on the assumption that X is fixed at given values. The results, however, are applicable for the case where X is random as will be seen in Section 3.1.3.

3.1.1 The Simple Linear Regression Model and the Least-Squares Estimators

If a linear relationship between Y and X is assumed, then the theoretical model given by

$$y_i = \beta_0 + \beta_1 x_i + e_i, \qquad i = 1, ..., n, \qquad (3.1.2)$$

is called the *simple linear regression model* of Y on X. The quantities β_0 and β_1 are unknown parameters, and $e_1, e_2, ..., e_n$ are uncorrelated error random variables with mean 0 and unknown variance σ^2, that is,

$$E(e_i) = 0, \quad \text{and} \quad V(e_i) = \sigma^2, \qquad i = 1, ..., n. \qquad (3.1.3)$$

Figure 3.1.4 presents this model graphically. For any value of $X = x_i$, there is a distribution of Y (not necessarily normal) whose mean is $\beta_0 + \beta_1 x_i$ and whose variance is σ^2, $i = 1, ..., n$.

We now find estimators of the unknown quantities β_0 and β_1 based on our sample of size n. The best estimators b_0 and b_1 of β_0 and β_1 are obtained by minimizing the *sum of squares of deviations*

$$S = \sum_{i=1}^{n} (y_i - \beta_0 - \beta_1 x_i)^2, \qquad (3.1.4)$$

with respect to β_0 and β_1. These estimators, called the *least-squares estimators*, are given by

$$b_0 = \bar{y} - b_1 \bar{x} \qquad (3.1.5)$$

and

$$b_1 = \frac{\sum_{i=1}^{n} (x_i - \bar{x}) y_i}{\sum_{i=1}^{n} (x_i - \bar{x})^2} \equiv \frac{\sum_{i=1}^{n} (x_i - \bar{x})(y_i - \bar{y})}{\sum_{i=1}^{n} (x_i - \bar{x})^2}. \qquad (3.1.6)$$

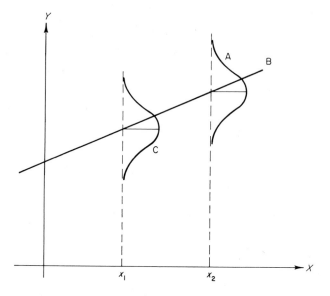

FIG. 3.1.4. *Simple linear regression model. (a) Distribution of Y at $X = x_2$, mean $= \beta_0 + \beta_1 x_2$, variance $= \sigma^2$; (b) $y = \beta_0 + \beta_1 x$; (c) distribution of Y at $X = x_1$, mean $= \beta_0 + \beta_1 x_1$, variance $= \sigma^2$.*

Note that S is a measure of the error incurred in fitting the sample with a straight line; the estimates b_0 and b_1 minimize this error. As a consequence of the theory of Section 4.1, these estimators are unbiased and have minimum variance among all unbiased estimators of β_0 and β_1 which are linear in the observations $y_1, y_2, \ldots, y_n$.

The *estimated regression equation* (or the *least-squares line*) is then given by

$$\hat{y} = b_0 + b_1 x, \tag{3.1.7}$$

so that the estimate of Y at $X = x_i$ is $\hat{y}_i = b_0 + b_1 x_i$. The difference between the observed and the estimated values of Y at $X = x_i$ is the *deviation* (or *residual*) $d_i = y_i - \hat{y}_i$. The least-squares line gives the minimum sum of squared deviations $\hat{S} = \sum_{i=1}^{n} d_i^2$. Figure 3.1.5 depicts the relationship between the theoretical regression line, the least-squares line, and the sample points.

There are many packaged simple linear regression programs which calculate the least-squares estimates b_0 and b_1 from a set of observations. In the output of these programs, the estimate b_1 is usually called the *regression coefficient*, and the estimate b_0 is often called the *intercept*.

EXAMPLE 3.1.1 (*continued*)   For this example $Y =$ lactic acid concentration determined by the instrument was regressed on $X =$ known lactic acid concentration. From a packaged regression program the estimates of β_0 and β_1 were calculated from the $n = 20$ observations. These are given by $b_0 = 0.159$ and

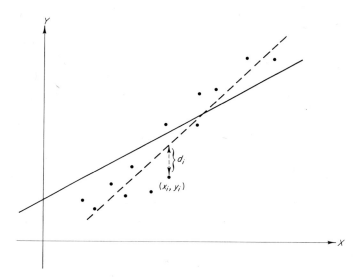

FIG. 3.1.5. *Theoretical and estimated simple linear regression lines showing the ith deviation* $d_i = y_i - \hat{y}_i$. *The least-squares line minimizes S. Dashed curve: estimated least-squares line* $\hat{y} = b_0 + b_1 x$; *solid curve: unknown theoretical line* $y = \beta_0 + \beta_1 x$.

$b_1 = 1.227$. Thus, the least-squares line is $\hat{y} = 0.159 + 1.227x$. When $X = 1$, $\hat{y} = 1.386$, and when $X = 10$, $\hat{y} = 12.43$. The line is plotted in Fig. 3.1.1.

In practice we would like to predict the true concentration X from the observed concentration Y. Thus, we would use the estimated regression equation *in reverse*. That is, the estimate of X based on Y is $\hat{x} = (y - 0.159)/1.227$.

COMMENTS 3.1.2   1.   The estimates may also be obtained from a *multiple linear regression program* in which the number of variables p is specified as $p = 2$ on a control card. These programs are discussed in Section 3.2.

2.   There are two interpretations for the predicted value $\hat{y}$. The first arises when the investigator is interested in estimating the value of Y for an individual whose value of X is x. In this case $\hat{y}$ is the best estimate of a *single value* of Y corresponding to a given $X = x$. The second interpretation is when the investigator is making inferences about the mean of Y in the subpopulation specified by $X = x$. Then the same $\hat{y}$ is the best estimate of the *mean* of Y given $X = x$. The distinction between the two interpretations is important when confidence intervals are formed. This is discussed in Section 3.1.2.

3.   We may "center" the regression model as

$$y_i = \beta_0' + \beta_1 w_i + e_i, \quad i = 1, \ldots, n,$$

where

$$w_i = x_i - \bar{x} \quad \text{and} \quad \beta_0' = \beta_0 + \beta_1 \bar{x}.$$

In this case the least-squares estimator of β_1 is still b_1 but the least-squares estimator of β_0' is now $b_0' = \bar{y}$. This representation has the practical and theoretical advantage that the two estimators $\bar{y}$ and b_1 are uncorrelated.

4. If it is known that the intercept $\beta_0 = 0$, then a model of the form $y_i = \beta x_i + e_i$, $i = 1, ..., n$, may be assumed. Many packaged programs permit analyzing this model by using the option of *forcing the regression through the origin*.

3.1.2 Confidence Intervals and Testing of Hypotheses

To make statistical inferences about β_0, β_1, or $\hat{y}$, we first need an estimator of the error variance σ^2, and then we need to describe the distribution of the error random variables e_i, $i = 1, ..., n$. From the theory of the general linear model, the usual unbiased estimator for σ^2 is given by the *variance of the estimate*

$$s^2 = \frac{\sum_{i=1}^{n} (y_i - b_0 - b_1 x_i)^2}{n-2}. \tag{3.1.8}$$

The positive square root of this quantity is called the *standard error of the estimate*. Both of these quantities are typically printed in the output of a packaged regression program. The variance of the estimate can also be found in the *analysis of variance table* which often appears in the output.

TABLE 3.1.1

Analysis of Variance Table for Simple Linear Regression

Source of variation	Sum of squares	Degrees of freedom	Mean square	F ratio
Due regression	$SS_D = \sum_{i=1}^{n} (\hat{y}_i - \bar{y})^2$	$\nu_D = 1$	$MS_D = SS_D$	$F_0 = \dfrac{MS_D}{MS_R}$
Deviations about regression (residual)	$SS_R = \sum_{i=1}^{n} (y_i - \hat{y}_i)^2$	$\nu_R = n - 2$	$MS_R \equiv s^2$ $= \dfrac{SS_R}{\nu_R}$	
Total	$SS_T = \sum_{i=1}^{n} (y_i - \bar{y})^2$	$\nu_T = n - 1$		

Table 3.1.1 shows the form of this table. The quantity s^2 is identical to MS_R, the *deviation about regression* (or *residual*) *mean square*. The *residual sum of*

squares SS_R and *residual degrees of freedom* ν_R are the numerator and denominator of Eq. (3.1.8), respectively. The *due regression sum of squares* SS_D gets its name from the fact that it can also be written as a function of the estimated regression coefficient b_1, namely,

$$SS_D = b_1{}^2 \sum_{i=1}^{n} (x_i - \bar{x})^2. \tag{3.1.9}$$

Thus, the larger the regression coefficient is, the larger is the sum of squares "due to the regression."

The last column, the *F ratio*, can be used for hypothesis testing if the errors $e_1, e_2, ..., e_n$ are assumed to be normally distributed. Thus, the simple linear regression model is now

$$y_i = \beta_0 + \beta_1 x_i + e_i, \qquad i = 1, ..., n, \tag{3.1.10}$$

where $e_1, e_2, ..., e_n$ are independent $N(0, \sigma^2)$ error random variables. To test the hypothesis that there is no simple linear regression of Y on X, that is, H_0: $\beta_1 = 0$ against H_1: $\beta_1 \neq 0$, we use the F ratio from the analysis of variance table given by

$$F_0 = \frac{MS_D}{MS_R} \equiv \frac{MS_D}{s^2}. \tag{3.1.11}$$

Under H_0, F_0 has an F distribution with $\nu_D = 1$ and $\nu_R = n - 2$ degrees of freedom. The P value is the area to the right of F_0 under the $F(\nu_D, \nu_R)$ frequency curve. We reject H_0 if P is less than the significant level α. If H_0 is accepted, then the best estimate of Y given any $X = x$ is then assumed to be the mean $\bar{y}$.

If the errors are assumed to be normal, we may test additional hypotheses and form confidence intervals. To test H_0: $\beta_1 = \beta_1^{(0)}$ where $\beta_1^{(0)}$ is a constant, we use the test statistic

$$t_0 = \frac{b_1 - \beta_1^{(0)}}{[\hat{V}(b_1)]^{1/2}}, \tag{3.1.12}$$

where

$$\hat{V}(b_1) = \frac{s^2}{\sum_{i=1}^{n} (x_i - \bar{x})^2}. \tag{3.1.13}$$

In the output of a regression program the quantity $[\hat{V}(b_1)]^{1/2}$ is often called the *standard error of the regression coefficient*. Under H_0, t_0 has a Student's t distribution with $\nu_R = n - 2$ degrees of freedom. The P value depends on the form of the alternative hypothesis as shown in the accompanying table. For some

Null hypothesis	Alternative hypothesis	P value
H_0: $\beta_1 = \beta_1^{(0)}$	H_1: $\beta_1 > \beta_1^{(0)}$	$P = Pr(t(\nu_R) > t_0)$
	H_1: $\beta_1 < \beta_1^{(0)}$	$P = Pr(t(\nu_R) < t_0)$
	H_1: $\beta_1 \neq \beta_1^{(0)}$	$P = 2Pr(t(\nu_R) > |t_0|)$

programs the value of t_0 is printed in the output. A $100(1-\alpha)\%$ *confidence interval for* β_1 is

$$b_1 \pm \sqrt{\hat{V}(b_1)}\, t_{1-(\alpha/2)}(n-2). \qquad (3.1.14)$$

To test $H_0: \beta_0 = \beta_0^{(0)}$, where $\beta_0^{(0)}$ is a constant, we use the test statistic

$$t_0 = \frac{b_0 - \beta_0^{(0)}}{[\hat{V}(b_0)]^{1/2}}, \qquad (3.1.15)$$

where

$$\hat{V}(b_0) = \frac{s^2 \sum_{i=1}^{n} x_i^2}{n \sum_{i=1}^{n}(x_i - \bar{x})^2}. \qquad (3.1.16)$$

In the output of a regression program the quantity $[\hat{V}(b_0)]^{1/2}$ may be called the *standard error of the intercept*. Under H_0, t_0 has a Student's t distribution with $v_R = n-2$ degrees of freedom. The P value depends on the alternative in the same way as the above test. A $100(1-\alpha)\%$ *confidence interval for* β_0 is

$$b_0 \pm \sqrt{\hat{V}(b_0)}\, t_{1-(\alpha/2)}(n-2). \qquad (3.1.17)$$

We now give two confidence intervals based on the estimate $\hat{y}$ (see Comment 3.1.2.2). If $\hat{y} = b_0 + b_1 x$ is interpreted as an estimate of a *single* value of Y at $X = x$, then a $100(1-\alpha)\%$ *confidence interval for* Y is

$$\hat{y} \pm s \left[1 + \frac{1}{n} + \frac{(x-\bar{x})^2}{\sum_{i=1}^{n}(x_i - \bar{x})^2} \right]^{1/2} t_{1-(\alpha/2)}(n-2). \qquad (3.1.18)$$

If, on the other hand, $\hat{y}$ is interpreted as the estimate of the *mean* of Y given $X = x$, then a $100(1-\alpha)\%$ *confidence interval for this mean* is

$$\hat{y} \pm s \left[\frac{1}{n} + \frac{(x-\bar{x})^2}{\sum_{i=1}^{n}(x_i - \bar{x})^2} \right]^{1/2} t_{1-(\alpha/2)}(n-2). \qquad (3.1.19)$$

The choice of the confidence interval depends on the investigator's use of $\hat{y}$ as an estimate. Note that as x moves away from $\bar{x}$, the confidence intervals become larger. Thus, our estimate becomes less precise. Also note that if n and $\sum_{i=1}^{n}(x_i - \bar{x})^2$ are large, then Eq. (3.1.18) is approximated by the "quick" confidence interval $\hat{y} \pm s t_{1-(\alpha/2)}(n-2)$. Thus, s is indeed the "standard error of the estimate $\hat{y}$"—hence, the name.

COMMENT 3.1.3   If a regression program is not available in the program library, the investigator may calculate all quantities of the regression model using a packaged descriptive program. From a typical descriptive program he obtains

$$\bar{x}, \quad \bar{y}, \quad s_x^2 = \frac{1}{n-1}\sum_{i=1}^{n}(x_i-\bar{x})^2, \quad s_y^2 = \frac{1}{n-1}\sum_{i=1}^{n}(y_i-\bar{y})^2,$$

$$s_{xy} = \frac{1}{n-1}\sum_{i=1}^{n}(x_i-\bar{x})(y_i-\bar{y}).$$

Thus,

$$b_1 = \frac{s_{xy}}{s_x^2}, \qquad b_0 = \bar{y} - b_1\bar{x},$$

$$SS_D = (n-1)b_1^2 s_x^2, \qquad SS_T = (n-1)s_y^2, \qquad SS_R = SS_T - SS_D,$$

$$s^2 \equiv MS_R = \frac{SS_R}{(n-2)},$$

$$\hat{V}(b_1) = \frac{s^2}{(n-1)s_x^2}, \qquad \hat{V}(b_0) = \frac{s^2[(n-1)s_x^2 + n\bar{x}^2]}{n(n-1)s_x^2}.$$

Note that all terms to calculate the analysis of variance table, test statistics, and confidence intervals have thus been extracted.

EXAMPLE 3.1.1 (*continued*)   From a packaged regression program the analysis of variance table is given by the accompanying table. The estimate of σ^2 is $s^2 \equiv MS_R = 1.164$, and the standard error of the estimate is $s = 1.079$. Since $F = 681.5$, the test for $H_0: \beta_1 = 0$ is rejected at $P < 10^{-3}$.

Source of variation	Sum of squares	Degrees of freedom	Mean square	F ratio
Due regression	793.099	1	793.099	681.5
Residual	20.945	18	1.164	
Total	814.044	19		

To test $H_0: \beta_1 = 1.0$ against $H_1: \beta_1 \neq 1.0$, we need the standard error of the regression coefficient $[\hat{V}(b_1)]^{1/2} = 0.047$. Thus,

$$t_0 = \frac{1.227 - 1.000}{0.047} = 4.83.$$

This is significant at $P < 0.001$.

To test that the regression line goes through the origin, that is, $H_0: \beta_0 = 0$ against $H_1: \beta_0 \neq 0$, we form a 95% confidence interval for β_0. This is $0.159\pm$

$2.10(0.396) = (-0.671, 0.989)$, where $[\hat{V}(b_0)]^{1/2} = 0.396$ and $t_{0.975}(18) = 2.10$. Since this interval includes 0, we accept H_0.

A 95% confidence interval for the mean of Y at $X = 7.7$ is obtained as follows: The estimate of the mean of Y is $\hat{y} = 0.159 + 1.227(7.7) = 9.66$. With $\bar{x} = 6.7$ we calculate

$$1.079 \sqrt{\frac{1}{20} + \frac{(7.7 - 6.7)^2}{526}} = 0.246$$

from Eq. (3.1.19). The confidence interval is $9.66 \pm 2.10(0.246) = (9.14, 10.18)$. Thus, with 95% confidence this interval includes the true mean of Y at $X = 7.7$.

3.1.3 Testing the Adequacy of the Linear Model

In this section we discuss testing for the adequacy of the simple linear regression model. By *adequacy* of the simple linear model, we mean that no other model significantly improves the prediction of Y. For example, the investigator may wish to test whether or not the *polynomial regression model* $y_i = \beta_0 + \beta_1 x + \beta_2 x^2 + \cdots + \beta_m x^m + e_i$ significantly improves the prediction of Y for some $m \geq 2$. The null hypothesis in this case is $H_0: \beta_2 = \cdots = \beta_m = 0$ (see Section 3.2).

If all of the n values $x_1, x_2, \ldots, x_n$ of X are unequal (that is, no two values of Y have the same value of X), then only limited tests for the adequacy of the linear model (such as the one given) may be made. On the other hand, if more than one value of Y is observed for some values of X, then we may test the hypothesis that no alternative model significantly improves the prediction of Y over the linear model. The test statistic is another F ratio which is obtained from the analysis of variance table as follows.

Suppose that we have k distinct values of X, that is, $x_1, x_2, \ldots, x_k$. Furthermore, suppose that at each x_i we have n_i observations $y_{i1}, y_{i2}, \ldots, y_{in_i}$ of Y, $i = 1, \ldots, k$. Let $n_i > 1$ for some i, and let $\sum_{i=1}^{k} n_i = n$. Then the simple linear regression model can now be written as

$$y_{ij} = \beta_0 + \beta_1 x_i + e_{ij}, \quad j = 1, \ldots, n_i, \quad i = 1, \ldots, k, \quad (3.1.20)$$

where e_{ij} are independent $N(0, \sigma^2)$ error random variables.

From a packaged regression program we may obtain the least-squares estimates b_0 and b_1 of β_0 and β_1 by treating the sample as n pairs of observations (x_1, y_{11}), $(x_1, y_{12}), \ldots, (x_1, y_{1n_1}), \ldots, (x_k, y_{k1}), (x_k, y_{k2}), \ldots, (x_k, y_{kn_k})$. In terms of this notation, the estimates are

$$b_0 = \bar{y}_{..} - b_1 \bar{x}, \quad \text{and} \quad b_1 = \frac{\sum_{i=1}^{k} n_i (x_i - \bar{x}) \bar{y}_{i.}}{\sum_{i=1}^{k} n_i (x_i - \bar{x})^2}, \quad (3.1.21)$$

where

$$\bar{y}_{i.} = \frac{1}{n_i} \sum_{j=1}^{n_i} y_{ij}, \qquad \bar{y}_{..} = \frac{1}{n} \sum_{i=1}^{k} \sum_{j=1}^{n_i} y_{ij}, \qquad \text{and} \qquad \bar{x} = \frac{1}{k} \sum_{i=1}^{k} x_i. \quad (3.1.22)$$

The least-squares line is $\hat{y} = b_0 + b_1 x$, so that $\hat{y}_i = b_0 + b_1 x_i$ is the estimate of Y at $X = x_i$.

The sums of squares in the analysis of variance table are

$$\text{SS}_\text{D} = \sum_{i=1}^{k} \sum_{j=1}^{n_i} (\hat{y}_i - \bar{y}_{..})^2, \qquad \text{and} \qquad \text{SS}_\text{R} = \sum_{i=1}^{k} \sum_{j=1}^{n_i} (y_{ij} - \hat{y}_i)^2, \quad (3.1.23)$$

with $v_\text{D} = 1$ and $v_\text{R} = n - 2$ degrees of freedom, respectively.

To test the hypothesis of adequacy of the linear model, the residual sum of squares SS_R and the degrees of freedom v_R are partitioned into two sources of variation—the *about regression* and the *within groups*. The corresponding sums of squares SS_A and SS_W and degrees of freedom v_A and v_W are shown in Table 3.1.2.

TABLE 3.1.2. *Expanded Analysis of Variance for Simple Linear Regression*

Source of variation	Sum of squares	Degrees of freedom	Mean square	F ratio
Due regression	$\text{SS}_\text{D} = \sum_{i=1}^{k} \sum_{j=1}^{n_i} (\hat{y}_i - \bar{y}_{..})^2$	$v_\text{D} = 1$	$\text{MS}_\text{D} = \text{SS}_\text{D}$	
About regression	$\text{SS}_\text{A} = \sum_{i=1}^{k} \sum_{j=1}^{n_i} (\bar{y}_{i.} - \hat{y}_i)^2$	$v_\text{A} = k - 2$	$\text{MS}_\text{A} = \dfrac{\text{SS}_\text{A}}{v_\text{A}}$	$F_0 = \dfrac{\text{MS}_A}{\text{MS}_W}$
Within groups	$\text{SS}_\text{W} = \sum_{i=1}^{k} \sum_{j=1}^{n_i} (y_{ij} - \bar{y}_{i.})^2$	$v_\text{W} = n - k$	$\text{MS}_\text{W} = \dfrac{\text{SS}_\text{W}}{v_\text{W}}$	
Total	$\text{SS}_\text{T} = \sum_{i=1}^{k} \sum_{j=1}^{n_i} (y_{ij} - \bar{y}_{..})^2$	$v_\text{T} = n - 1$		

Note the similarity between the within groups sum of squares in this table and in the one-way analysis of variance table discussed in Section 2.4. The test statistic for testing H_0: the simple linear model is adequate against H_1: the simple linear model is not adequate, is

$$F_0 = \frac{\text{MS}_\text{A}}{\text{MS}_\text{W}}, \quad (3.1.24)$$

where MS_A and MS_W are the about regression and within groups mean squares, respectively. Under H_0, F_0 has an F distribution with $v_\text{A} = k - 2$ and $v_\text{W} = n - k$

degrees of freedom. The P value is the area to the right of F_0 under the $F(v_A, v_W)$ frequency curve.

If H_0 is accepted, the residual sum of squares SS_R and degrees of freedom v_R are recomputed, that is, $SS_R = SS_A + SS_W$ and $v_R = v_A + v_W$. Then the hypothesis $H_0: \beta_1 = 0$ may be tested by the F ratio of Eq. (3.1.11).

COMMENT 3.1.4   We may compute this expanded analysis of variance table from the combined output of a regression program and a descriptive program with strata as follows: From the regression program we obtain SS_D, v_D, MS_D, SS_R, SS_T, v_R, and v_T (see Table 3.1.1). Using a descriptive program with strata, we stratify Y by the k values of X and obtain the within groups sum of squares SS_W and degrees of freedom v_W from the one-way analysis of variance table. By subtraction, we obtain $SS_A = SS_R - SS_W$ and $v_A = v_R - v_W$. These quantities then generate Table 3.1.2.

EXAMPLE 3.1.1 (continued)   Since for this example multiple measurements were made at each of $k = 5$ values of X, we can test the adequacy of the simple linear model at $\alpha = 0.05$. The expanded analysis of variance table is given in the accompanying table. Note that $SS_R = 20.945 = SS_A + SS_W$ and $v_R = 18 = v_A + v_W$. Since $F_0 = 1.27 < F_{0.95}(3, 15)$ the null hypothesis is accepted.

Source of variation	Sum of squares	Degrees of freedom	Mean square	F ratio
Due regression	793.099	1	793.099	
About regression	4.251	3	1.417	1.27
Within groups	16.694	15	1.113	
Total	814.044	19		

3.1.4 The Correlation Coefficient

In this section we discuss the population and sample correlation coefficients. These quantities were introduced in Chapter 2 as measures of linear association between two variables. It was pointed out earlier that statistical inferences about the population correlation coefficient could only be made if both X and Y are random variables. In particular, when the joint distribution of X and Y is bivariate normal, then the population correlation coefficient and linear regression model are related to this distribution in a very interesting way. We now discuss this theory.

We assume that the random variables X and Y have a bivariate normal distribution. Let μ_x and μ_y be the population means and σ_x^2 and σ_y^2 be the population variances of X and Y. The population covariance of X and Y is denoted by σ_{xy}. Then the *simple* (or *product-moment*) *correlation coefficient* between X and Y is

$$\rho = \frac{\sigma_{xy}}{\sigma_x \sigma_y}. \tag{3.1.25}$$

This coefficient is a measure of linear association between X and Y. The range of ρ is $-1 \leqslant \rho \leqslant 1$, where positive ρ implies that Y tends to increase with X, while negative ρ implies that Y tends to decrease with X. The extreme cases of $\rho = \pm 1$ indicate perfect linear association between X and Y, that is, given $X = x$, then the value of Y is determined exactly.

For a given value of $X = x$, there is a subpopulation of values of Y corresponding to $X = x$. Its distribution, the *conditional distribution of Y given X = x*, is univariate normal with mean

$$\mu_{y.x} = \mu_y + \frac{\sigma_{xy}}{\sigma_x^2}(x - \mu_x), \tag{3.1.26}$$

called the *conditional expectation of Y given X = x* (or the *regression of Y on X*). The variance of this distribution, called the *conditional variance of Y given X = x*, is

$$\sigma^2 = \sigma_y^2(1 - \rho^2). \tag{3.1.27}$$

This last equation lends itself to an important interpretation for ρ. Note that σ_y^2 is the unconditional variance of Y, that is, it is the variance of Y when the value of X is not known. On the other hand, σ^2 is the conditional variance of Y, that is, it is the variance of Y when we know that the corresponding $X = x$. Thus, from Eq. (3.1.27) the reduction in the variance of Y due to knowledge of X is

$$\sigma_y^2 - \sigma^2 = \rho^2 \sigma_y^2. \tag{3.1.28}$$

From this equation we obtain

$$\rho^2 = \frac{\sigma_y^2 - \sigma^2}{\sigma_y^2}, \tag{3.1.29}$$

which implies that the squared correlation coefficient is the proportion of the variance of Y due to or "explained" by knowledge of X.

Now define the random variable $e = Y - \mu_{y.x}$, which measures the deviation of Y from its mean at $X = x$. The conditional distribution of e given $X = x$ is normal with mean 0 and variance σ^2. Thus, we may write

$$Y = \mu_{y.x} + e = \mu_y - \frac{\sigma_{xy}}{\sigma_x^2}\mu_x + \frac{\sigma_{xy}}{\sigma_x^2}x + e$$

$$= \beta_0 + \beta_1 x + e, \tag{3.1.30}$$

where

$$\beta_0 = \mu_y - \beta_1 \mu_x, \qquad \beta_1 = \frac{\sigma_{xy}}{\sigma_x^2}, \qquad (3.1.31)$$

and e is $N(0, \sigma^2)$. Note that this equation is in the same form as the simple linear regression of Y on X given by Eq. (3.1.10). Thus, the theory developed in Sections 3.1.1–3.1.3 applies to this model.

Note from Eq. (3.1.29) that the *squared correlation coefficient is the proportion of the variance of Y "explained" by the linear regression of Y on X*. When $\rho = 0$, we know from Eq. (3.1.29) that $\sigma^2 = \sigma_y^2$. This implies that none of the variance of Y is explained by the regression of Y on X. When $\rho = \pm 1$, then $\sigma^2 = 0$. This implies that all the variance of Y is explained by the regression of Y on X, that is, the relationship between Y and X is perfectly linear.

We now discuss estimation of the population parameters. We assume that we have a random sample $(x_1, y_1), (x_2, y_2), \ldots, (x_n, y_n)$ which was obtained in the second manner described in the beginning of this section, that is, both X and Y are random variables. The estimators of μ_x and μ_y are $\bar{x}$ and $\bar{y}$, of σ_x^2 and σ_y^2 are s_x^2 and s_y^2, and of σ_{xy} and ρ are s_{xy} and r. Furthermore, the estimators of β_0, β_1, and σ^2 are b_0, b_1, and s^2. Thus, these quantities are given in a packaged regression program as the sample means, variances, covariance, correlation, intercept, regression coefficient and variance of the estimate, and the kitchen sink, respectively.

EXAMPLE 3.1.2 (*continued*)   In this example $X =$ venous pH and $Y =$ arterial pH were measured on $n = 108$ critically ill patients. The sample estimates obtained from a descriptive program were given in Example 2.3.1 and are summarized here for convenience. Thus

$$\bar{x} = 7.373, \quad \bar{y} = 7.413, \quad s_x^2 = 0.1253, \quad s_y^2 = 0.1184,$$

$$s_{xy} = 0.1101, \quad r = 0.9039.$$

Applying Comment 3.1.3, we calculate from this data

$$b_1 = 0.879, \quad b_0 = 0.934, \quad SS_D = 10.359, \quad SS_T = 12.669, \quad SS_R = 2.310,$$

$$[\hat{V}(b_1)]^{1/2} = 0.04032, \quad [\hat{V}(b_0)]^{1/2} = 0.2976.$$

Thus, we may form the analysis of variance table as in the accompanying table. Since $F = 470.9$, the hypothesis $H_0: \beta_1 = 0$ is rejected at $P < 10^{-3}$, implying that arterial pH does change linearly with venous pH. The test of $H_0: \beta_1 = 1$, Eq. (3.1.12) yields a statistic $t_0 = (0.879 - 1.000)/0.04032 = -3.00$. For the alternative $H_1: \beta_1 \neq 1$, the P value is less than 0.001, so that H_0 is rejected.

Source of variation	Sum of squares	Degrees of freedom	Mean square	F ratio
Due regression	10.359	1	10.359	470.9
Residual	2.310	106	0.022	
Total	12.669	107		

A 95% confidence interval for β_0 is $0.943 \pm 0.2976(1.98) = (0.354, 1.514)$. Since this interval does not include 0, the hypothesis $H_0: \beta_0 = 0$ is rejected at $\alpha = 0.05$.

Finally, a 95% confidence interval for the mean of Y at $x = 7.395$ is $7.395 \pm 0.0014(1.98) = (7.392, 7.398)$.

The remainder of this section discusses statistical inferences concerning the population correlation coefficient. Note that

$$\rho = \frac{\beta_1 \sigma_x}{\sigma_y} \tag{3.1.32}$$

expresses the relationship between ρ and β_1. Hence, $\rho = 0$ if and only if $\beta_1 = 0$. Thus, we may test $H_0: \rho = 0$ by using (a) the F ratio of Eq. (3.1.11), (b) the t test of Eq. (3.1.12) with $\beta_1^{(0)} = 0$, or (c) the Fisher transformation to be discussed. Note also that $\rho = 0$ implies independence of the variables X and Y.

In general, to test $H_0: \rho = \rho_0$, where ρ_0 is a constant $\neq \pm 1$, we make the *Fisher transformation* given by

$$v = \frac{1}{2} \log_e \frac{1+r}{1-r} = 1.1513 \log_{10} \frac{1+r}{1-r}. \tag{3.1.33}$$

Values of v for a range of r are tabulated in Table 8 in Appendix II. Under the null hypothesis, v is approximately normal with mean

$$\mu_v = \frac{1}{2} \log_e \frac{1+\rho_0}{1-\rho_0} \tag{3.1.34}$$

and variance

$$\sigma_v^2 = \frac{1}{n-3}. \tag{3.1.35}$$

The test statistic is then computed as

$$z = \frac{v - \mu_v}{\sigma_v}, \tag{3.1.36}$$

which under H_0, is approximately $N(0, 1)$ when n is large. The P value depends on the alternative hypothesis, and H_0 is rejected if $P < \alpha$.

A $100(1-\alpha)\%$ confidence interval for μ_v is (v_1, v_2), where

$$v_1 = v - \sigma_v z_{1-(\alpha/2)}, \quad \text{and} \quad v_2 = v + \sigma_v z_{1-(\alpha/2)}. \quad (3.1.37)$$

Using the inverse Fisher transformation

$$r = \frac{e^{2v}-1}{e^{2v}+1}, \quad (3.1.38)$$

or looking up Table 8 (Appendix II) in reverse, the confidence limits for ρ are obtained. The confidence interval may also be used to test $H_0: \rho = \rho_0$ against $H_1: \rho \neq \rho_0$, i.e., H_0 is rejected at level α if the interval excludes ρ_0.

An equivalent way to obtain 95% or 99% confidence intervals for ρ is to use the chart given in Table 9 in Appendix II. This chart is computed according to the exact distribution of r (David, 1938). To use this chart, we draw a vertical line from the computed r (on the horizontal axis) until it intersects the two bands corresponding to n. The projections of these two points of intersection onto the vertical axis are the confidence limits.

COMMENTS 3.1.5   1.   The following relationships hold between the estimates s_x, s_y, s, and r:

$$s^2 = \frac{n-1}{n-2} s_y^2 (1-r^2) \quad \text{and} \quad r = \frac{b_1 s_x}{s_y}.$$

2.   The correlation coefficient is invariant to changes in location or scale of X and/or Y. Hence, the correlation coefficient for the centered model is the same as for the original model. Also, since $\hat{y}_i$ is a linear function of x_i, the correlation computed from the observed y_i and the predicted $\hat{y}_i$, $i = 1, \ldots, n$, is equivalent to the absolute value of the correlation coefficient r.

3.   The test statistic t_0 given by Eq. (3.1.12) for testing $H_0: \rho = 0$ can be written equivalently as

$$t_0 = \frac{r\sqrt{n-2}}{\sqrt{1-r^2}}.$$

EXAMPLE 3.1.2 (continued)   The sample correlation coefficient between venous and arterial pH is $r = 0.9039$. A test of $H_0: \rho = 0$ (that is, independence of X and Y) is rejected on the basis of the analysis of variance F ratio $F = 470.9$, or the t test (Comment 3.1.5.3).

$$t_0 = \frac{0.9039\sqrt{106}}{\sqrt{1-0.8170}} = 21.7.$$

The 95% confidence interval for ρ is obtained by first finding (Table 8, Appendix II)

$$v = \frac{1}{2} \log_e \frac{1.904}{0.096} = 1.493.$$

This gives, from Eq. (3.1.37),

$$v_1 = 1.493 - \frac{1.96}{\sqrt{105}} = 1.302 \quad \text{and} \quad v_2 = 1.493 + \frac{1.96}{\sqrt{105}} = 1.684.$$

Looking up Table 8 in reverse gives the interval $(0.86, 0.93)$ for ρ. Thus, this interval includes the true correlation ρ with 95% confidence.

Alternatively, we enter Table 9 (Appendix II) with $r = 0.90$ and obtain $(0.85, 0.93)$ showing agreement between the two methods. Either interval may be used to test $H_0: \rho = \rho_0$; for example, $H_0: \rho = 0.5$ would be rejected since the interval does not include 0.5.

3.1.5 Examination of Residuals

Three assumptions have been made in the preceding discussion of simple linear regression. They are assumptions about the form of the model, the distribution, and randomness of the error variable e. One method for examining the adequacy of the linear model was discussed in Section 3.1.3. All three assumptions may be checked by examining plots of the *residuals* $d_i = y_i - \hat{y}_i$, $i = 1, \ldots, n$. Such plots are included in the output of many packaged programs.

To test the adequacy of the model, we may plot d_i versus x_i or y_i for $i = 1, \ldots, n$. If the residuals fall in a horizontal band centered around the abscissa, the model may be judged as adequate (see Fig. 3.1.6a). If the band widens as x or y increases (see Fig. 3.1.6b), this indicates *heteroscedasticity* (that is, lack of constant variance σ^2). Specifically, σ may be a function of $\beta_0 + \beta_1 x$, necessitating a transformation of the Y variable. A plot showing a linear trend (Fig. 3.1.6c) suggests the addition of another independent variable to the model (see Section 3.2 on multiple regression). A plot such as that shown in Fig. 3.1.6d indicates that a linear or quadratic term should be added to the model.

To test normality of e_i, $i = 1, \ldots, n$, a histogram of d_i is appropriate. Normality can then be tested using a goodness-of-fit test.

If data is collected in some order (for example, successive points in time or in neighboring locations), then a plot of the residuals d_i in the same order as collected is appropriate for checking randomness. If no trend is apparent, we cannot reject the hypothesis of randomness. On the other hand, seasonal and linear trends may be exhibited as in Fig. 3.1.7a and b.

Further discussion and references are given by Anscombe (1961), Anscombe and Tukey (1963), Box and Watson (1962), and Draper and Smith (1968).

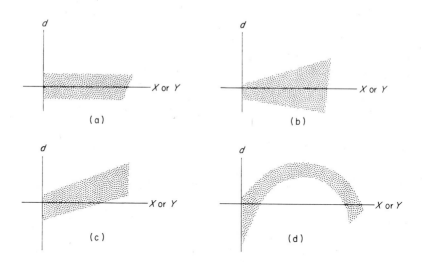

FIG. 3.1.6. *Examples of residual plots:* (a) *adequate model;* (b) *heteroscedasticity;* (c) *linear independent variable;* (d) *linear or quadratic independent variable.*

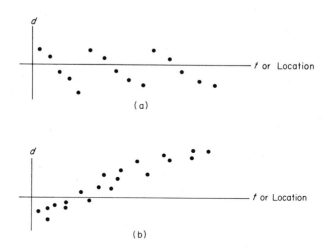

FIG. 3.1.7. *Examples of nonrandomness:* (a) *seasonal trend;* (b) *linear trend.*

3.2 *Multiple Linear Regression, Multiple and Partial Correlations*

We now consider the problem of predicting one variable Y from p variables $X_1, X_2, ..., X_p, p > 1$. Traditionally, the variable Y is called the *dependent variable*, while variables $X_1, ..., X_p$ are called the *independent variables*. This use of the word "independent" should not be confused with the concept of "statistical independence" as defined in Section I.1.6. In fact, in some cases, the independent variables $X_1, ..., X_p$ are random variables which are not statistically independent as will be seen.

As mentioned in the introduction to this chapter, we may approximate values of Y by a regression function $f(\ )$ containing unknown parameters. The model equation expressing the relationship between the dependent and independent variables is then given by

$$y = f(x_1, ..., x_p; \beta_1, ..., \beta_m) + e, \tag{3.2.1}$$

where $\beta_1, \beta_2, ..., \beta_m$ are unknown parameters and e is an error variable representing the error incurred in approximating y by the regression function. In particular, if $m = p+1$ and $f(x_1, ..., x_p; \beta_0, \beta_1, ..., \beta_p) = \beta_0 + \beta_1 x_1 + \cdots + \beta_p x_p$, then we have the *multiple linear regression model*

$$y = \beta_0 + \beta_1 x_1 + \cdots + \beta_p x_p + e. \tag{3.2.2}$$

In this equation, some of the independent variables may be functions of other variables or of each other. For example, $y = \beta_0 + \beta_1 \sin Z_1 + \beta_2 \cos Z_1 + e$ is in the form of the multiple linear regression model with $X_1 = \sin Z_1$ and $X_2 = \cos Z_1$. In particular, if $X_i = X^i$, $i = 1, ..., p$, then we have the *polynomial regression model*

$$y = \beta_0 + \beta_1 x + \beta_2 x^2 + \cdots + \beta_p x^p + e. \tag{3.2.3}$$

Finally, it must be remembered that "linear" refers to linearity in the parameters and not in the independent variables. Thus, $y = \beta_0 + \sin(\beta_1 X_1) + \beta_2 X_2$ is not a linear function in the parameters.

This section discusses the multiple linear regression model given in the general form by Eq. (3.2.2). Although many real-life situations are more appropriately described by a nonlinear model (Section 3.4), this model has been shown to be useful, at least as a first approximation to nonlinear models. Section 3.2.1 discusses the estimation of the parameters, and Section 3.2.2 presents the various tests of hypotheses and confidence intervals concerning these parameters. Sections 3.2.3–3.2.5 present the concepts, theory, and estimation of two measures of association or dependence between Y and the independent variables—called the multiple and partial correlation coefficients. Finally, since many of the equations in this chapter are too complex to express in a simple form, the more mathematically sophisticated reader will find matrix representations in the starred sections.

3.2.1 *Estimation of the Parameters*

We estimate the parameters of the model from a sample of size n taken from a population W. As in Section 3.1, this sample may arise in one of two ways. In the first way, we fix values for $X_1, ..., X_p$, and in the subpopulation defined by these restrictions, we then observe one or more values of Y. We then fix new values for $X_1, ..., X_p$ and observe one or more values of Y in this subpopulation, and so forth, until n observations are obtained. In this sampling situation Y is the only random variable. In the second way, we randomly select n individuals from the population W and observe all $p+1$ variables simultaneously. In this case all $p+1$ variables are random variables. Although the estimation procedures are the same for either sampling situation, the underlying theory of least-squares estimation (Section 4.1) assumes that the sample was collected in the first way. On the other hand, the underlying theory for multiple and partial correlations assumes that the sample was collected in the second way from a multivariate normal population.

For this and the next two sections, we assume that $x_{1i}, ..., x_{pi}$, $i = 1, ..., n$, are fixed values of the independent variables $X_1, ..., X_p$ (denoted by $X_1 = x_{1i}, ..., X_p = x_{pi}$) and we let y_i be the observed value of the random variable Y. Thus, our sample consists of the set of n observations $(y_1, x_{11}, ..., x_{p1}), ..., (y_n, x_{1n}, ..., x_{pn})$. Assuming a multiple linear regression model, we have

$$y_i = \beta_0 + \beta_1 x_{1i} + \cdots + \beta_p x_{pi} + e_i, \qquad i = 1, ..., n, \qquad (3.2.4)$$

where $\beta_0, \beta_1, ..., \beta_p$ are the unknown parameters, and $e_1, ..., e_n$ are independent $N(0, \sigma^2)$ error variables. To obtain the least-squares estimates $b_0, b_1, ..., b_p$ of the parameters, the user may use a multiple linear regression packaged program. These estimates, which minimize the *sum of squares of deviations*

$$S = \sum_{i=1}^{n} (y_i - \beta_0 - \beta_1 x_{1i} - \cdots - \beta_p x_{pi})^2, \qquad (3.2.5)$$

are usually called the (*partial*) *regression coefficients* in the computer output. Sometimes, the estimate b_0 is called the *constant, intercept,* or *y-intercept*. The estimated multiple regression equation (or *least-squares* plane) can then be written by the user as

$$\hat{y} = b_0 + b_1 x_1 + \cdots + b_p x_p. \qquad (3.2.6)$$

(Matrix expressions for the least-squares estimators are given in Comment 3.2.1.1.)

Note that the sum of squares of deviation S is a measure of the error incurred in "fitting" the sample with the multiple linear regression model; the least-squares estimates minimize this error. Furthermore, b_i is an unbiased estimator of β_i and is a linear function of the observations $y_1, ..., y_n$, $i = 0, 1, ..., p$. Finally, as a consequence of the Gauss–Markov Theorem (Section 4.1), the predicted value $\hat{y}$ has minimum variance among all linear predictors of Y for given values $x_1, ..., x_p$ of $X_1, ..., X_p$.

Four other useful quantities usually appear in the output of a multiple linear regression packaged program. The first, called the *residual* (or *error*) *sum of squares* SS_R, is the value of S with the least-squares estimates substituted for the parameters, that is,

$$SS_R = \sum_{i=1}^{n} (y_i - b_0 - b_1 x_{1i} - \cdots - b_p x_{pi})^2. \qquad (3.2.7)$$

This quantity, when divided by its degrees of freedom $\nu_R = n - p - 1$ (the *residual* or *error degrees of freedom*), gives an unbiased estimator of the error variance σ^2 called the *residual* (or *error*) *mean square* MS_R. Thus,

$$MS_R = \frac{SS_R}{\nu_R} \qquad (3.2.8)$$

and will sometimes be denoted by s^2 in this chapter. These three quantities usually appear in an analysis of variance table similar to that shown in Table 3.2.1. The fourth quantity (not appearing in the table) is the square root of MS_R and is called the *standard error of the estimate*. Thus, $s = \sqrt{MS_R}$ estimates the error standard deviation σ.

TABLE 3.2.1. *Analysis of Variance Table for Multiple Linear Regression Model*

Source of variation	Sum of squares	Degrees of freedom	Mean square	F ratio
Due regression	$SS_D = \sum_{i=1}^{p} b_i \sum_{j=1}^{n} (x_{ij} - \bar{x}_i) y_j$	$\nu_D = p$	$MS_D = \dfrac{SS_D}{\nu_D}$	$F = \dfrac{MS_D}{MS_R}$
Residual (error)	$SS_R = SS_T - SS_D$	$\nu_R = n - p - 1$	$MS_R \equiv s^2$ $= \dfrac{SS_R}{\nu_R}$	
Total	$SS_T = \sum_{i=1}^{n} (y_i - \bar{y})^2$	$\nu_T = n - 1$		

More about the table: Note that each mean square is equal to the sum of squares divided by the corresponding degrees of freedom. The total sum of squares and degrees of freedom are equal to the sum of the "due regression" and "residual" sums of squares and degrees of freedom, respectively. The F ratio is the ratio of the two mean squares and will be discussed in Section 3.2.2. The total sum of squares SS_T divided by its degrees of freedom ν_T is equal to the estimate of the variance of Y. Finally, the ratio $SS_D/SS_T = R^2$ (sometimes called the *coefficient of determination*) is the proportion of variance of Y "explained" by the regression of Y on $X_1, \ldots, X_p$. (As will be seen later, this quantity is the square of the multiple correlation coefficient.) Thus, R^2 is a measure of "goodness-of-fit," that is, the larger R^2 is, the better the model approximates Y.

COMMENTS 3.2.1   *1.   We now discuss the model and least-squares estimators in matrix notation. This presentation is a special case of the material
presented in Section 4.1.

Let $\boldsymbol{\beta} = (\beta_0, ..., \beta_p)'$ be the $(p+1) \times 1$ parameter vector, $\mathbf{y} = (y_1, ..., y_n)'$ be the
$n \times 1$ observation vector, $\mathbf{e} = (e_1, ..., e_n)'$ be the $n \times 1$ error vector, and

$$
\mathbf{X}' = \begin{bmatrix}
1 & x_{11} & \cdots & x_{p1} \\
1 & x_{12} & \cdots & x_{p2} \\
\vdots & \vdots & \vdots & \vdots \\
1 & x_{1n} & \cdots & x_{pn}
\end{bmatrix}
$$

be the $n \times (p+1)$ *design matrix*. Equation (3.2.4) can now be written as

$$
\mathbf{y} = \mathbf{X}'\boldsymbol{\beta} + \mathbf{e},
$$

where $\mathbf{e}$ is multivariate $N(0, \sigma^2 \mathbf{I})$. The matrix form of Eq. (3.2.5) is $S =$
$(\mathbf{y} - \mathbf{X}'\boldsymbol{\beta})'(\mathbf{y} - \mathbf{X}'\boldsymbol{\beta})$, and the vector of least-squares estimators $\mathbf{b} = (b_0, b_1, ..., b_p)'$
is obtained by solving the *normal equations* $(\mathbf{XX}')\boldsymbol{\beta} = \mathbf{Xy}$. The solution is
$\mathbf{b} = (\mathbf{XX}')^{-1}(\mathbf{Xy})$, and its covariance matrix is $\text{Cov}(\mathbf{b}) = \sigma^2(\mathbf{XX}')^{-1}$. Finally, the
unbiased estimator of the variance is

$$
MS_R \equiv s^2 = (\mathbf{y} - \mathbf{X}'\mathbf{b})'(\mathbf{y} - \mathbf{X}'\mathbf{b})/(n - p - 1).*
$$

2.   There is a "centered" form of the multiple linear regression model which
is a generalization of the centered, simple linear regression model presented in
Section 3.1. The centered model is given by

$$
y_i = \beta_0' + \beta_1(x_{1i} - \bar{x}_1) + \cdots + \beta_p(x_{pi} - \bar{x}_p) + e_i, \qquad i = 1, ..., n,
$$

where

$$
\bar{x}_j = \frac{1}{n} \sum_{k=1}^{n} x_{jk}, \qquad j = 1, ..., p, \qquad \text{and} \qquad \beta_0' = \beta_0 + \beta_1 \bar{x}_1 + \cdots + \beta_p \bar{x}_p.
$$

The least-squares estimators of $\beta_1, ..., \beta_p$ are the same as above, namely $b_1, ..., b_p$,
while the least-squares estimator of β_0' is $b_0' = \bar{y}$. The advantage of this model
is that the least-squares estimators $b_1, ..., b_p$ are uncorrelated with b_0'. For
Gesund, this means that it is easier to find confidence intervals based on the
predicted value $\hat{y} = \bar{y} + b_1(x_1 - \bar{x}_1) + \cdots + b_p(x_p - \bar{x}_p)$, as will be seen.

*3.   The centered model is written in matrix notation as follows. Let $\mathbf{A}$ be
the $p \times p$ *matrix of sums of squares and crossproducts of deviations* with ij element
$a_{ij} = \sum_{k=1}^{n}(x_{ik} - \bar{x}_i)(x_{jk} - \bar{x}_j)$, $i, j = 1, ..., p$. Let $\mathbf{g}$ be the $p \times 1$ vector with ith
element $g_i = \sum_{k=1}^{n}(y_k - \bar{y})(x_{ik} - \bar{x}_i)$, $i = 1, ..., p$. Then the vector of least-squares
estimators

$$
\tilde{\mathbf{b}} = (b_1, ..., b_p)' = \mathbf{A}^{-1}\mathbf{g}.
$$

In addition

$$\text{Cov}(\tilde{\mathbf{b}}) = \sigma^2 \mathbf{A}^{-1}, \quad \text{and} \quad \text{cov}(\bar{y}, b_i) = 0, \quad \text{for} \quad i = 1, \dots, p.*$$

4. When the sample mean values $\bar{x}_1, \dots, \bar{x}_p$ are substituted for $x_1, \dots, x_p$ in the estimated regression equation, the predicted value $\hat{y} = \bar{y}$.

EXAMPLE 3.2.1   In an experiment it was desired to estimate the octane count in gasolines containing various concentrations of compounds A and B. A multiple linear regression model between $Y =$ octane count and $X_1 = \%$ of compound A and $X_2 = \%$ of compound B was assumed, that is, $y = \beta_0 + \beta_1 x_1 + \beta_2 x_2 + e$. Each of the two independent variables was fixed at one of four values, and one observation of Y was made for each combination $X_1 = x_1$ and $X_2 = x_2$. The data to be analyzed are given in the accompanying tabulation. From a

X_1	X_2	Y	X_1	X_2	Y
2	2	96.3	4	2	96.2
	3	95.7		3	100.1
	4	99.9		4	103.2
	5	99.4		5	104.3
3	2	95.1	5	2	97.8
	3	97.8		3	102.2
	4	99.3		4	104.7
	5	104.9		5	108.8

packaged multiple regression program, the estimates $b_0 = 84.553$, $b_1 = 1.833$, and $b_2 = 2.683$ were obtained. Thus, the estimated multiple regression equation is $\hat{y} = 84.553 + 1.833x_1 + 2.683x_2$. The analysis of variance table was computed as in the accompanying table. Therefore, the unbiased estimate of the error variance σ^2 is $\text{MS}_R = 1.937$, and the standard error of the estimate is $s = \sqrt{1.937} = 1.392$. Finally, the ratio $R^2 = \text{SS}_D/\text{SS}_T = 211.084/236.266 = 0.893$ is the proportion of variance explained by the regression of Y on X_1 and X_2.

Source of variation	Sum of squares	Degrees of freedom	Mean square	F ratio
Due regression	211.084	2	105.542	54.485
Residual	25.182	13	1.937	
Total	236.266	15		

3.2.2 Confidence Intervals and Tests of Hypotheses

In addition to the least-squares estimates of the parameters $\beta_1, ..., \beta_p$, a packaged program also computes quantities which are used to form confidence intervals and to test hypotheses concerning these parameters. These quantities are called the *standard errors of the coefficients*. For each β_i the standard error of the coefficient, se(b_i), is the standard deviation of the estimate b_i of β_i, $i = 1, ..., p$. Since this quantity is a function of MS_R, it has $v_R = n - p - 1$ degrees of freedom. Thus, a $100(1 - \alpha)\%$ *confidence interval for* β_i is

$$b_i \pm \text{se}(b_i) t_{1 - (\alpha/2)}(v_R) \qquad \text{for} \quad i = 1, ..., p. \tag{3.2.9}$$

Some programs calculate the standard error of the intercept, se(b_0), in which case Eq. (3.2.9) may be used with $i = 0$. Otherwise, see starred Comment 3.2.2.1.

Tests of hypotheses about the coefficients $\beta_1, ..., \beta_p$ fall into three categories: we may test that *all* of the coefficients $\beta_1 = \beta_2 = \cdots = \beta_p = 0$, or we may test that any *one* coefficient $\beta_k = 0$, $k = 1, ..., p$, or we may test that any *subset* of m coefficients is equal to zero. All of these hypotheses may also be expressed in terms of the independent variables, as follows.

The hypothesis $H_0: \beta_1 = \cdots = \beta_p = 0$ may be viewed as a hypothesis that "the independent variables $X_1, ..., X_p$ do not significantly improve the prediction of Y over $\hat{y} = \bar{y}$." Failure to reject this hypothesis implies that $\bar{y}$ is acceptable as the best predictor of Y. The alternative hypothesis is that not all of the coefficients are equal to zero, that is, that "some of the independent variables significantly improve the prediction of Y over $\hat{y} = \bar{y}$." The test statistic is the F ratio given in the last column of the analysis of variance table (Table 3.2.1), that is,

$$F = \frac{MS_D}{MS_R}. \tag{3.2.10}$$

Under H_0, F has an F distribution with $v_D = p$ and $v_R = n - p - 1$ degrees of freedom. The P value is the area to the right of the computed F under the $F(v_D, v_R)$ frequency curve.

The hypothesis $H_0: \beta_k = 0$ for $k = 1, ..., p$, may be viewed as a hypothesis that "the variable X_k does not significantly improve the prediction of Y over that obtained by regressing Y on the other $p - 1$ variables." One test statistic for the alternative $H_1: \beta_k \neq 0$ is

$$F = \frac{b_k^2}{[\text{se}(b_k)]^2}, \tag{3.2.11}$$

which under H_0 has an F distribution with 1 and $v_R = n - p - 1$ degrees of freedom. The P value is the area to the right of F under the $F(1, v_R)$ frequency curve. Some programs print this value of F for each b_k. Other programs print the equivalent statistic

$$t = \frac{b_k}{se(b_k)}, \tag{3.2.12}$$

which under H_0 has a Student's t distribution with $v_R = n-p-1$ degrees of freedom. The P value is double the area to the right of $|t|$ under the $t(v_R)$ frequency curve.

We may use the t distribution to test H_0: $\beta_k = \beta_k^{(0)}$, $\beta_k^{(0)}$ a constant, versus a one-sided or two-sided alternative. The test statistic is

$$t = \frac{b_k - \beta_k^{(0)}}{se(b_k)}, \tag{3.2.13}$$

and P is obtained from the $t(v_R)$ frequency curve for the particular alternative hypothesis.

The intermediate hypothesis that a subset of m coefficients is zero is more difficult to test. Without loss of generality, let this subset consist of the first m coefficients $\beta_1, ..., \beta_m$. Then testing H_0: $\beta_1 = \cdots = \beta_m = 0$ is equivalent to testing that "the m variables $X_1, ..., X_m$ do not significantly improve the prediction of Y over that obtained by regressing Y on $X_{m+1}, ..., X_p$." We first use the program to calculate the regression of Y on $X_{m+1}, ..., X_p$. From the analysis of variance table we denote the residual sum of squares by SS_R'. We then use the program to calculate the regression of Y on $X_1, ..., X_m, ..., X_p$, and denote the residual sum of squares and mean square by SS_R and MS_R, respectively. Then the test statistic for H_0 is

$$F = \frac{(SS_R' - SS_R)/m}{MS_R}, \tag{3.2.14}$$

which under H_0 has an F distribution with m and $v_R = n-p-1$ degrees of freedom. The P value is the area to the right of F under the $F(m, v_R)$ frequency curve.

COMMENTS 3.2.2   These comments give matrix forms for confidence intervals for the mean of Y and the intercept β_0.

*1.   The variance of $\hat{y}$ at $x_1, ..., x_p$ is

$$V(\hat{y}) = \sigma^2 \left(\frac{1}{n} + d'A^{-1}d \right),$$

where the elements of A are given by Comment 3.2.1.3 and

$$d = (x_1 - \bar{x}_1, ..., x_p - \bar{x}_p)', \quad \text{with} \quad \bar{x}_i = \frac{1}{n} \sum_{k=1}^{n} x_{ik}.$$

A $100(1-\alpha)\%$ confidence interval for the true mean of Y given $x_1, ..., x_p$ is

$$\hat{y} \pm \left[s^2 \left(\frac{1}{n} + d'A^{-1}d \right) \right]^{1/2} t_{1-(\alpha/2)}(n-p-1).$$

A $100(1-\alpha)\%$ *confidence interval for a single new value of Y given* $x_1, ..., x_p$ *is*

$$\hat{y} \pm \left[s^2 \left(1 + \frac{1}{n} + \mathbf{d}'\mathbf{A}^{-1}\mathbf{d} \right) \right]^{\frac{1}{2}} t_{1-(\alpha/2)}(n-p-1).*$$

*2.   Note that b_0 is the predicted value of Y at $x_1 = \cdots = x_p = 0$. Thus, the variance of b_0 is

$$V(b_0) = \sigma^2 \left(\frac{1}{n} + \bar{\mathbf{x}}'\mathbf{A}^{-1}\bar{\mathbf{x}} \right).$$

A $100(1-\alpha)\%$ *confidence interval for* β_0 *is then*

$$b_0 \pm \left[s^2 \left(\frac{1}{n} + \bar{\mathbf{x}}'\mathbf{A}^{-1}\bar{\mathbf{x}} \right) \right]^{\frac{1}{2}} t_{1-(\alpha/2)}(n-p-1).$$

Some programs print the elements of $\mathbf{A}^{-1}$ to facilitate the computations of the above equations.*

3.   If we test the hypothesis $H_0: \beta_k = 0$ for several values of k at the same significance level α, then the joint significance level is not necessarily the nominal α. To avoid this problem we may use multiple confidence intervals for all β_k, $k = 1, ..., p$, such that the joint confidence level is $1 - \alpha$. The multiple $100(1-\alpha)\%$ confidence interval for β_k is

$$b_k \pm \text{se}(b_k)[pF_{1-\alpha}(p, n-p-1)]^{\frac{1}{2}}.$$

The hypothesis $H_0: \beta_k = \beta_k^{(0)}$ is rejected at significant level α if $\beta_k^{(0)}$ is not in this interval.

EXAMPLE 3.2.1 (*continued*)   From the analysis of variance table for this example, we have $F = 54.485$. Comparing F with percentiles of $F(2, 13)$ we reject the hypothesis $H_0: \beta_1 = \beta_2 = 0$ at $P < 0.001$. Thus, the octane count is linearly related to the concentrations of both compounds A and B.

The regression coefficients, standard error of the coefficients, the F statistic of Eq. (3.2.11), and the P values after comparing F with percentiles of $F(1, 13)$ are summarized in the accompanying table. Thus, $H_0: \beta_1 = 0$ is rejected and $H_0: \beta_2 = 0$ is also rejected. That is, X_1 does significantly improve the prediction of Y over that obtained by regressing Y on X_2; and, conversely, X_2 does significantly improve the prediction of Y over that obtained by regressing Y on X_1. We will demonstrate the case of testing that a subset of $m = 2$ coefficients is zero in Example 3.2.3.

Variable	Estimated regression coefficient	Standard error of the coefficient	F	P
X_1	1.833	0.3112	34.67	<0.001
X_2	2.683	0.3112	74.29	<0.001

A 95% confidence interval for β_1 is, by Eq. (3.2.9)

$$1.833 \pm 0.3112(2.160) = (1.161, 2.505),$$

where $t_{0.975}(13) = 2.160$. A confidence interval for β_2 is obtained in the same way.

Finally, to test $H_0: \beta_2 = 3.0$ against $H_1: \beta_2 < 3.0$ at $\alpha = 0.05$, we calculate from Eq. (3.2.13)

$$t = \frac{2.683 - 3.000}{0.3112} = -1.019$$

and compare it with percentiles of the Student's t distribution with $\nu_R = 13$ degrees of freedom. Since the alternative is one tailed, $P > 0.10$, and we accept H_0.

COMMENT 3.2.3   In the regression model, β_i measures the rate of change of Y in terms of X_i when the values of $X_j, j = 1, ..., p, j \neq i$ are held fixed. These coefficients, however, may not be comparable in magnitude due to differences in the units of measurement of $X_1, ..., X_p$. This difficulty may be overcome by using *standardized independent variables*. Thus, for $j = 1, ..., p$, let $Z_j = X_j/s_j$, where $s_j^2 = \sum_{i=1}^{n} (x_{ji} - \bar{x}_j)^2/(n-1)$. The multiple regression model in terms of Z_j is now given by

$$y_i = \gamma_0 + \gamma_1 z_{1i} + \cdots + \gamma_p z_{pi} + e_i, \qquad i = 1, ..., n,$$

where $\gamma_k, k = 0, ..., p$, are unknown parameters, and e_i are independent $N(0, \sigma^2)$ error terms. The least-squares estimators c_k of γ_k and tests of hypotheses about γ_k follow the above theory with z_j and γ_k replacing x_j and β_k, respectively. The advantage of this standardization is that $\gamma_1, ..., \gamma_p$ are now rates of change measured on the same scale. This permits us to make inferences about the contribution of the independent variables $Z_1, ..., Z_p$ (or equivalently, $X_1, ..., X_p$). Hence, a large value of c_j implies a high degree of contribution of Z_j (or X_j), $j = 1, ..., p$.

3.2.3 The Multiple Correlation Coefficient

In this and the next section we discuss the theoretical framework of the multiple linear regression model. The theory assumes that the $p+1$ variables $Y, X_1, ..., X_p$ are all random variables, and, moreover, they have a multivariate normal distribution. It will be shown in this section that the mean of the conditional distribution of Y given $X_1 = x_1, ..., X_p = x_p$ is the multiple linear regression function $\beta_0 + \beta_1 x_1 + \cdots + \beta_p x_p$. This motivates the multiple linear regression model in which the error variance σ^2 is a function of the variance σ_y^2 of Y and a quantity called the multiple correlation coefficient. The reader should review Section I.1.6 (Appendix I) for multivariate concepts.

Consider the multivariate normal distribution of $Y, X_1, ..., X_p$ with means $\mu_y, \mu_1, ..., \mu_p$ and variances $\sigma_y^2, \sigma_1^2, ..., \sigma_p^2$, respectively. Denote the covariance of Y with X_i by σ_{yi}, and the covariance of X_i with X_j by σ_{ij} for $i, j = 1, ..., p$. We also define the population simple correlation coefficients as

$$\rho_{yx_i} = \frac{\sigma_{yi}}{\sigma_y \sigma_i} \quad \text{and} \quad \rho_{x_i x_j} = \frac{\sigma_{ij}}{\sigma_i \sigma_j}.$$

For given values of $X_1 = x_1, ..., X_p = x_p$, there exists a subpopulation of corresponding values of Y. Its distribution, the *conditional distribution of Y given* $X_1 = x_1, ..., X_p = x_p$, is also normal with mean

$$\mu_{y.x_1...x_p} = \mu_y + \beta_1(x_1 - \mu_1) + \cdots + \beta_p(x_p - \mu_p), \tag{3.2.15}$$

called the *conditional expectation of Y given $X_1 = x_1, ..., X_p = x_p$*, or the *regression of Y on $X_1, ..., X_p$*. The quantities $\beta_1, ..., \beta_p$ are called the *(partial) regression coefficients* and are functions of the above variances and covariances. The variance of this conditional distribution is

$$\sigma^2 = \sigma_y^2(1 - \rho_{y.x_1...x_p}^2) \tag{3.2.16}$$

where $\rho_{y.x_1...x_p}$, the positive square root of $\rho_{y.x_1...x_p}^2$, is called the *multiple correlation coefficient* of Y on $X_1, ..., X_p$.

If we define the random variable $e = Y - \mu_{y.x_1...x_p}$, then the conditional distribution of e given $X_1 = x_1, ..., X_p = x_p$ is $N(0, \sigma^2)$. Based on this conditional distribution, we may write

$$Y = \beta_0 + \beta_1 x_1 + \cdots + \beta_p x_p + e, \tag{3.2.17}$$

where

$$\beta_0 = \mu_y - \beta_1 \mu_1 - \cdots - \beta_p \mu_p, \tag{3.2.18}$$

and e is $N(0, \sigma^2)$. Note that Eq. (3.2.17) has the same form as the multiple linear regression model given by Eq. (3.2.3).

COMMENTS 3.2.4   1.   The multiple correlation coefficient $\rho_{y.x_1...x_p}$ is a measure of linear association between Y and the set of variables $\{X_1, ..., X_p\}$. The range of the multiple correlation coefficient is $0 \leqslant \rho_{y.x_1...x_p} \leqslant 1$. A value of zero indicates that Y is independent of the set $\{X_1, ..., X_p\}$, while a value of 1 indicates perfect linear association, that is, Y can be expressed exactly as a linear combination of $X_1, ..., X_p$ with no error.

2.   Solving Eq. (3.2.16) for the multiple correlation coefficient, we obtain

$$\rho_{y.x_1...x_p}^2 = \frac{\sigma_y^2 - \sigma^2}{\sigma_y^2}.$$

Hence, the squared multiple correlation coefficient is equal to the *proportion of the variance of Y "explained" by the regression relationship with* $X_1, ..., X_p$.

3. The multiple correlation coefficient is nonnegative by definition. Hence, in the case of the bivariate normal distribution ($p = 1$), we have

$$\rho_{y.x_1} = \rho_{x_1.y} = |\rho_{x_1y}|,$$

where ρ_{x_1y} is the simple correlation coefficient between X_1 and Y.

4. When $p = 2$, Eq. (3.2.15) can be written as

$$\mu_{y.x_1x_2} = \mu_y + \beta_1(x_1 - \mu_1) + \beta_2(x_2 - \mu_2).$$

The graph of this equation is a plane (called the *regression plane of Y on* X_1 *and* X_2) in the space defined by the coordinate axes x_1, x_2, and $\mu_{y.x_1x_2}$. The graph of Eq. (3.2.15) for $p > 2$ generalizes the regression plane to a hyperplane in the $p + 1$ dimensional space defined by $x_1, ..., x_p$ and $\mu_{y.x_1...x_p}$.

5. The multiple correlation is the maximum simple correlation between Y and any linear combination of $X_1, ..., X_p$. Furthermore, $\mu_{y.x_1...x_p}$ is that linear combination which yields this maximum correlation. The relationship between the multiple correlation coefficient and the regression parameters $\beta_1, ..., \beta_p$ will be discussed in Comment 3.2.6.4.

6. The multiple correlation coefficient is invariant under nonsingular linear transformations of the original variables. In particular, it is invariant to changes in scale or origin of measurement of $Y, X_1, ...,$ or X_p.

3.2.4 The Partial Correlation Coefficient

This section discusses another correlation coefficient called the partial correlation coefficient. It is used to measure the linear association between any two of the $p + 1$ variables $Y, X_1, ..., X_p$ after taking out the "effect" of any nonempty subset of the remaining $p - 1$ variables. In particular, it measures the dependence between Y and an independent variable X_m, say, after removing the linear dependence of Y on a subset of k of the remaining $p - 1$ independent variables X_i, $i = 1, ..., p$, $i \neq m$. This linear dependence of Y on the subset of k variables is the so-called "effect" of the subset mentioned previously. The theory of the partial correlation coefficient involves examining two conditional distributions as follows.

Let l and h stand for any two of the variables $Y, X_1, ..., X_p$, and let c be any nonempty subset of the remaining $p - 1$ variables. Define the quantities $Z_1 = \mu_{l.c}$ as the conditional expectation of l given c, and $Z_2 = \mu_{h.c}$ as the conditional expectation of h given c. Note that Z_1 and Z_2 are random variables, since they are functions of the random variables in c. Then the *partial correlation coefficient between l and h given c* is

$$\rho_{lh.c} = \rho_{z_1z_2}, \tag{3.2.19}$$

where $\rho_{z_1 z_2}$ is the simple correlation coefficient between Z_1 and Z_2. Two special cases appear throughout this section and the next. One case is where $l = Y$, $h = X_m$, $m = 1, \ldots, p$, and c is the full subset of the remaining $p - 1$ independent variables. We will denote this partial correlation by $\rho_{yx_m.c}$. The second case is where $l = Y$, $h = X_m$, and c is the subset of the first k independent variables $\{X_1, X_2, \ldots, X_k\}$, where $1 \leqslant k < m \leqslant p$. The resulting partial will be denoted by $\rho_{yx_m.x_1\ldots x_k}$. In general, if there are k variables in c, then the partial correlation coefficient is said to be of the *kth order*.

COMMENTS 3.2.5   1.   The partial correlation coefficient $\rho_{lh.c}$ is a measure of linear association between l and h when the values of the variables in c are held fixed. The range is from -1 to 1; a value of zero implies that l is independent of h when c is held fixed.

2.   A useful identity between the partial and multiple correlation coefficients for the set of variables $Y, X_1, \ldots, X_{k-1}, X_k, k = 2, \ldots, p$, is

$$1 - \rho_{y.x_1\ldots x_k}^2 = (1 - \rho_{y.x_1\ldots x_{k-1}}^2)(1 - \rho_{yx_k.x_1\ldots x_{k-1}}^2).$$

This implies that

$$V(Y|X_1, \ldots, X_k) = V(Y|X_1, \ldots, X_{k-1})(1 - \rho_{yx_k.x_1\ldots x_{k-1}}^2),$$

where $V(Y|X_1, \ldots, X_i)$ is the conditional variance of Y given $X_1, \ldots, X_i$, $i = 1, \ldots, p$. Hence,

$$\rho_{yx_k.x_1\ldots x_{k-1}}^2 = \frac{V(Y|X_1, \ldots, X_{k-1}) - V(Y|X_1, \ldots, X_k)}{V(Y|X_1, \ldots, X_{k-1})},$$

so that the squared partial correlation coefficient is the *proportion of variance of Y "explained" by the addition of* X_k *to the set* $\{X_1, \ldots, X_{k-1}\}$.

3.   We also have

$$\rho_{yx_m.c} = \beta_m \left[\frac{V(X_m|c)}{V(Y|c)}\right]^{1/2}, \qquad m = 1, \ldots, p,$$

where c is the full subset of the remaining $p - 1$ variables, and $V(X_m|c)$ is the conditional variance of X_m given the variables in c. Thus, testing $\beta_m = 0$ is equivalent to testing $\rho_{yx_m.c} = 0$, as will be seen in the next section.

4.   The partial correlations may be computed from a recursive relationship as follows. If l, h, and d stand for any three distinct variables in the set $\{Y, X_1, \ldots, X_p\}$, then all partial correlations of the first order are given by

$$\rho_{lh.d} = \frac{\rho_{lh} - \rho_{ld}\rho_{hd}}{\sqrt{(1 - \rho_{ld}^2)}\sqrt{(1 - \rho_{hd}^2)}},$$

where the quantities on the right are simple correlations. By successively using the recursive formula

$$\rho_{lh.cd} = \frac{\rho_{lh.c} - \rho_{ld.c}\,\rho_{hd.c}}{\sqrt{(1 - \rho_{ld.c}^2)}\,\sqrt{(1 - \rho_{hd.c}^2)}},$$

where c is any subset of the remaining variables, we may obtain the partial correlation coefficients of any order.

3.2.5 Estimation and Testing Hypotheses about the Multiple and Partial Correlation Coefficients

We now turn to the problem of obtaining and interpreting the sample estimates of these correlation coefficients. Since the theory requires that all $p+1$ variables be random variables, we now assume that our sample $(y_1, x_{11}, ..., x_{p1}), ...,$ $(y_n, x_{1n}, ..., x_{pn})$ was obtained by randomly selecting n individuals from the multivariate normal population with parameters defined in Section 3.2.3. For each individual, all $p+1$ variables were measured simultaneously. The estimators of the means, variances, and covariances of this population are given by the corresponding sample means, variances, and covariances discussed in Chapter 2. These estimates may be obtained from the vector of sample means and sample covariance matrix in either a descriptive program or a multiple linear regression program. In addition, such a program will usually give the matrix of simple correlations between the $p+1$ variables.

From Section 3.2.3 it was seen that the conditional distribution of Y given $X_1 = x_1, ..., X_p = x_p$ leads to the multiple linear regression model. Thus, for our sample we have

$$y_i = \beta_0 + \beta_1 x_{1i} + \cdots + \beta_p x_{pi} + e_i, \qquad i = 1, ..., n, \qquad (3.2.20)$$

with e_i independent $N(0, \sigma^2)$. Since this equation is identical to Eq. (3.2.4), the estimators of $\beta_0, \beta_1, ..., \beta_p$, and σ^2 are the ones discussed in Section 3.2.1, and the tests of hypotheses and confidence intervals discussed in Section 3.2.2 are valid here. The only remaining quantities to be estimated are the multiple and partial correlation coefficients. We first consider the following examples.

EXAMPLE 3.2.2   In this example we continue the experiment discussed in Example 2.4.1. In that situation five arterial pressure measurements (mm Hg) were made on $n = 141$ patients using the intra-arterial catheter and the cuff methods. Thus, all five variables are random. A descriptive program calculated the sample means and covariance and correlation matrices. For convenience, we summarize the means and correlations in the accompanying table and correlation matrix.

Method	Variable	Sample mean	Sample standard deviation
Intra-arterial	X_1 = Systolic pressure	$\bar{x}_{1.}$ = 112.2	s_1 = 28.6
	X_2 = Diastolic pressure	$\bar{x}_{2.}$ = 59.4	s_2 = 17.1
	X_3 = Mean pressure	$\bar{x}_{3.}$ = 76.8	s_3 = 21.0
Cuff	X_4 = Systolic pressure	$\bar{x}_{4.}$ = 107.0	s_4 = 28.9
	X_5 = Diastolic pressure	$\bar{x}_{5.}$ = 66.8	s_5 = 19.3

Correlation matrix

$$\begin{array}{c|ccccc} & X_1 & X_2 & X_3 & X_4 & X_5 \\ \hline X_1 & 1.000 & 0.839 & 0.927 & 0.871 & 0.753 \\ X_2 & & 1.000 & 0.967 & 0.778 & 0.828 \\ X_3 & & & 1.000 & 0.845 & 0.852 \\ X_4 & & & & 1.000 & 0.837 \\ X_5 & & & & & 1.000 \end{array}$$

In Example 2.4.1 we showed that the cuff measurements were significantly inaccurate in estimating the intra-arterial measurements. Since cuff systolic pressure may not be substituted for intra-arterial systolic pressure, it would be useful to compute an equation which estimates $Y \equiv X_1$ = intra-arterial systolic pressure as a linear function of the cuff measurements X_4 = cuff systolic pressure and X_5 = cuff diastolic pressure. Thus, we used a packaged multiple linear regression program to obtain the estimated regression equation $\hat{y} = 21.99 + 0.755x_4 + 0.141x_5$. Note that when $x_4 = 107.0$ and $x_5 = 66.8$—the mean cuff measurements—then $\hat{y} = 112.2$—the mean systolic pressure using the intra-arterial method as indicated in Comment 3.2.1.4. A similar estimation equation could be made for $Y \equiv X_2$ = intra-arterial diastolic pressure.

EXAMPLE 3.2.3   In this example we use the same data from the previous example, but we define the dependent variable now as $Y \equiv X_4$ = cuff systolic pressure, and use the independent variables X_1, X_2, and X_3 = intra-arterial systolic, diastolic, and mean pressures, respectively. With this choice of dependent and independent variables, a multiple linear regression analysis leads to an estimate of the error in the cuff measurement for a given combination of the intra-arterial pressures. Here, the objective is not "estimation," but rather, the examination of how well the cuff measurement is "explained" by the intra-arterial pressures. This gives an indication of the error incurred in using the cuff method. From the packaged program the estimated equation is $\hat{y} = 8.29 + 0.597x_1 - 0.136x_2 + 0.519x_3$. The analysis of variance table is given in the accompanying table. Thus, the unbiased estimate of the error variance σ^2 is $MS_R \equiv s^2 = 197.1$. The standard error of the estimate, $s = \sqrt{197.1} = 14.04$, is a measure of the error induced by the cuff method. Finally, $R^2 = SS_D/SS_T = 89,963/116,971 = 0.769$ implies that 76.9% of the variance of the cuff systolic pressure is explained by the regression relationship.

Source of variation	Sum of squares	Degrees of freedom	Mean square	F ratio
Due regression	89,963.8	3	29,987.9	$F = 152.1$
Residual	27,007.6	137	197.1	
Total	116,971.4	140		

We now use the techniques of Section 3.2.2 to test hypotheses and form confidence intervals. From the analysis of variance table we have $F = 152.1$. Comparing F with percentiles of $F(3,137)$ we reject the hypothesis $H_0: \beta_1 = \beta_2 = \beta_3 = 0$ at $P < 10^{-3}$. Thus, as expected, the cuff systolic measurement is highly dependent on the intra-arterial pressure measurements.

The regression coefficients, standard errors of the coefficients, the F statistics of Eq. (3.2.11), and the P values after comparing F with percentiles of $F(1,137)$ are given in the accompanying table.

Variable	Regression coefficient	Standard error of coefficients	F	P
$X_1 =$ Intra-arterial systolic	0.597	0.136	19.15	<0.001
$X_2 =$ Intra-arterial diastolic	−0.136	0.334	0.17	NS
$X_3 =$ Mean pressure	0.519	0.393	1.74	NS

Thus, $H_0: \beta_1 = 0$ is rejected, while $H_0: \beta_2 = 0$ and $H_0: \beta_3 = 0$ are each accepted. That is, X_1 does significantly improve the prediction of Y over that obtained by regressing Y on X_2 and X_3; X_2 does not significantly improve the prediction of Y over that obtained by regressing Y on X_1 and X_3; and X_3 does not significantly improve the prediction of Y over that obtained by regressing Y on X_1 and X_2.

On the basis of these results it would be incorrect to conclude that X_2 and X_3 together do not significantly improve the prediction of Y over that obtained by regressing Y on X_1. To verify this statistically we test $H_0: \beta_2 = \beta_3 = 0$. Regression of Y on X_1 gives $SS_R' = 28{,}240.4$. From the regression of Y on all three variables, we have $SS_R = 27{,}007.6$ and $MS_R = 197.1$. Thus, from Eq. (3.2.14),

$$F = \frac{(28{,}240.4 - 27{,}007.6)/2}{197.1} = 3.13.$$

When F is compared with percentiles of the $F(2,137)$ distribution, the P value is less than 0.05; thus, X_2 and X_3 together significantly improve the prediction of Y based on X_1 alone.

Finally, a 95% confidence interval for β_1 is

$$0.597 \pm 0.136(1.97) = (0.329, 0.865),$$

where $t_{0.975}(137) = 1.97$. Confidence intervals for β_2 and β_3 would be computed in the same way. This example is continued throughout the chapter.

We now discuss estimation of the multiple and partial correlation coefficients. The estimator of the multiple correlation coefficient, denoted by $r_{y.x_1\ldots x_p}$, is usually designated in the output of a packaged program by the label *multiple R* or *multiple correlation coefficient*. It may also be calculated from the analysis of variance table (Table 3.2.1) by the relationship

$$r_{y.x_1\ldots x_p} = +\sqrt{\frac{SS_D}{SS_T}}. \tag{3.2.21}$$

Equivalently, it may be calculated as the positive square root of the coefficient of determination R^2. This estimate is always nonnegative as is the case with its population analog.

Now what are its uses? First, it is a measure of the linear dependence of Y on all of the independent variables. The closer $r_{y.x_1\ldots x_p}$ is to 1, the stronger the dependence. To test that there is no linear dependence, that is, to test H_0: $\rho_{y.x_1\ldots x_p} = 0$, we may use the F statistic of Eq. (3.2.10), since this hypothesis is equivalent to $H_0: \beta_1 = \cdots = \beta_p = 0$. We may also use the equivalent F statistic

$$F = \frac{n-p-1}{p} \frac{r_{y.x_1\ldots x_p}^2}{1 - r_{y.x_1\ldots x_p}^2}. \tag{3.2.22}$$

The P value is the area to the right of F under the $F(p, n-p-1)$ frequency curve.

The second use of this coefficient is the application of Comment 3.2.4.2. Thus, the square of this estimated coefficient is the "proportion of variance of Y explained by the linear regression of Y on $X_1, \ldots, X_p$."

We now discuss the estimators of the partial correlation coefficients. First, we consider the estimator $r_{yx_m.c}$ of $\rho_{yx_m.c}$, where $m = 1, \ldots, p$, and c is the full subset of the remaining $p-1$ independent variables. This estimate is sometimes included in the output of a packaged multiple linear regression program for each X_m, $m = 1, \ldots, p$. If these estimates are not included in the output, we may readily calculate these coefficients from either one of two test statistics which may be included in the output. If the t statistic for testing $H_0: \beta_m = 0$ [Eq. (3.2.12)] is given in the output, then we compute

$$r_{yx_m.c} = \frac{t}{\sqrt{t^2 + n - p - 1}}, \qquad m = 1, ..., p. \qquad (3.2.23)$$

If on the other hand, the F statistic [Eq. (3.2.11)] for testing this hypothesis is given, then we may compute

$$r_{yx_m.c} = \pm \sqrt{\frac{F}{F + n - p - 1}}, \qquad m = 1, ..., p. \qquad (3.2.24)$$

The sign of the estimate is the same as the estimated regression coefficient b_m.

Estimates of other partial correlations such as: (a) that between Y and X_m holding a subset of k of the remaining $p - 1$ variables fixed $(k < p - 1)$; or (b) that between X_1 and X_2 holding Y fixed, and so forth, may be obtained from a packaged multiple regression program by rearranging the order of the variables, redefining the dependent variable, and redefining p—the number of independent variables.

We now discuss the estimation of any partial correlation coefficient. Let l and h be any pair of the variables $Y, X_1, ..., X_p$, and let c be a nonempty subset of the remaining variables. Denote by $r_{lh.c}$ the estimator of $\rho_{lh.c}$. Then some methods for obtaining any or all estimates are:

1. Use a partial correlation packaged program.
2. Hand calculate or program the recursive relationships given in Comment 3.2.5.4. Initial values are the simple correlations from the output of a multiple linear regression or a descriptive program.
3. In Section 3.3 we discuss a procedure called stepwise regression which computes multiple regression equations in steps. At each step, a subset c of the independent variables appears in the regression equation. These programs calculate at each step the partial correlation between Y and each independent variable not appearing in the regression equation *given* the variables in c. For a given preselected set c, we may *force* its elements in the regression equation (see Comment 3.3.1.4) and obtain all partials between Y and the remaining independent variables given c.

Now, what are the uses of these coefficients? First, the coefficient $r_{yx_m.c}$, where c is the full subset of $p - 1$ independent variables not including X_m, $m = 1, ..., p$, is a measure of the linear association of Y on X_m after removing the effect of the set of variables in c. The closer this quantity is to 1 in absolute value, the stronger the dependence. Testing that there is no significant contribution due to X_m given the variables in c, that is, $H_0: \rho_{yx_m.c} = 0$, is equivalent to testing $H_0: \beta_k = 0$ using either the F test of Eq. (3.2.11) or the t test of Eq. (3.2.12).

Secondly, the coefficient $r_{yx_m.c}$, where c is a subset of $k < p-1$ independent variables, is a measure of the "worth" of X_m in predicting Y after removing the linear effect of the independent variables in c. Thus, if we compared $r_{yx_m.c}$ for all X_m not in c, we may obtain an ordering of the importance of the independent variables in predicting Y relative to c. As will be seen in Section 3.3, this is how a variable is selected for the stepwise procedure.

In general, to test $H_0: \rho_{lh.c} = 0$, we may use

$$t = \frac{r_{lh.c}\sqrt{n-k-2}}{\sqrt{1-r_{lh.c}^2}}, \tag{3.2.25}$$

where k is the number of variables in the set c. Under H_0 this statistic has a Student's t distribution with $n-k-2$ degrees of freedom. The Fisher transformation can be used to test $H_0: \rho_{lh.c} = \rho_0$, where ρ_0 is a constant, by replacing the simple correlations by the partial correlations in Eq. (3.1.37). The variance σ_v^2 is now given by

$$\bullet \qquad \sigma_v^2 = \frac{1}{n-k-3}. \tag{3.2.26}$$

which is substituted in Eq. (3.1.36) to give the test statistic z. The $100(1-\alpha)\%$ *confidence interval for* $\rho_{lh.c}$ can be obtained using the Fisher transformation or the chart in Table 9 (Appendix II) with n replaced by $n-k$.

Finally, for any partial correlation $r_{lh.c}$, the square of this quantity is the estimated proportion of variance of l explained by h after removing the effect of the variables in c.

COMMENTS 3.2.6   1.   The following relationship holds:

$$s^2 = \frac{n-1}{n-p}s_y^2(1-r_{y.x_1...x_p}^2),$$

where s^2 is the residual mean square MS_R from the analysis of variance Table 3.2.1, and s_y^2 is the estimated variance of Y.

2.   The simple correlation between the observed y_i and the predicted $\hat{y}_i$, $i = 1, ..., n$, is equivalent to the sample multiple correlation coefficient $r_{y.x_1...x_p}$.

3.   Since the multiple correlation coefficient is invariant under nonsingular transformations, the sample estimates obtained from the original model, from the "centered model," and from the "standardized model," will be equal.

*4.   A theoretical discussion of the above concepts utilizing matrix notation is given as follows. Let $\mathbf{Z} = (Y, X_1, ..., X_p)'$ be the $(p+1) \times 1$ random vector of variables. This vector is assumed to have a multivariate normal distribution with the $(p+1) \times 1$ mean vector $E(\mathbf{Z}) = (\mu_y, \mu_1, ..., \mu_p)'$ and the $(p+1) \times (p+1)$ covariance matrix

$$\Sigma_z = \begin{bmatrix} \sigma_y^2 & \sigma_{y1} & \sigma_{y2} & \cdots & \sigma_{yp} \\ \sigma_{y1} & \sigma_1^2 & \sigma_{12} & \cdots & \sigma_{1p} \\ \vdots & \vdots & \vdots & \vdots & \vdots \\ \sigma_{yp} & \sigma_{1p} & \sigma_{2p} & \cdots & \sigma_p^2 \end{bmatrix}.$$

To obtain the equations for the regression coefficients $\beta_1, \ldots, \beta_p$ and for the multiple correlation coefficient $\rho_{y.x_1\ldots x_p}$, we write $\mathbf{Z}$ as the partitioned vector $\mathbf{Z} = (Y, \mathbf{X}')'$, where the $p \times 1$ vector $\mathbf{X} = (X_1, \ldots, X_p)'$. The mean vector and covariance matrix are correspondingly partitioned as $E(\mathbf{Z}) = (\mu_y, \mathbf{\mu}_x')'$ where $\mathbf{\mu}_x = (\mu_1, \ldots, \mu_p)'$, and

$$\Sigma_z = \begin{pmatrix} \sigma_y^2 & \Sigma_{yx} \\ \Sigma_{xy} & \Sigma_{xx} \end{pmatrix}.$$

The submatrices Σ_{yx}, Σ_{xy}, and Σ_{xx} have dimensions $1 \times p$, $p \times 1$, and $p \times p$, respectively. Note $\Sigma_{yx}' = \Sigma_{xy}$. Thus, the conditional distribution of Y given $\mathbf{X} = \mathbf{x}$ is normal with mean

$$\mu_{y.x_1\ldots x_p} = \mu_y + \tilde{\mathbf{\beta}}'(x - \mathbf{\mu}_x), \qquad \text{where} \qquad \tilde{\mathbf{\beta}} = \Sigma_{xx}^{-1} \Sigma_{xy},$$

and variance $\sigma^2 = \sigma_y^2 - \Sigma_{yx} \Sigma_{xx}^{-1} \Sigma_{xy}$. It then follows that

$$\rho_{y.x_1\ldots x_p} = \frac{\sqrt{\Sigma_{yx} \Sigma_{xx}^{-1} \Sigma_{xy}}}{\sigma_y},$$

or equivalently,

$$\rho_{y.x_1\ldots x_p} = \frac{\sqrt{\tilde{\mathbf{\beta}}' \Sigma_{xx} \tilde{\mathbf{\beta}}}}{\sigma_y}.$$

This last equation defines the relationship between the regression coefficients and the multiple correlation coefficient.

To derive the partial correlation coefficient $\rho_{lh.c}$, renumber the X_i so that $c = \{X_1, \ldots, X_k\}$. Define the 2×1 random vector $\mathbf{W}_1 = (l, h)'$ and the $k \times 1$ random vector $\mathbf{W}_2 = (X_1, \ldots, X_k)'$. The distribution of $\mathbf{W}_1$ is bivariate normal with mean vector $E(\mathbf{W}_1) = (\mu_l, \mu_h)'$ and covariance matrix

$$\Sigma_{w_1 w_1} = \begin{bmatrix} \sigma_l^2 & \sigma_{lh} \\ \sigma_{lh} & \sigma_h^2 \end{bmatrix}.$$

Similarly, the distribution of $\mathbf{W}_2$ is k-variate normal with mean vector $E(\mathbf{W}_2) = (\mu_1, \ldots, \mu_k)'$ and covariance matrix

$$\Sigma_{w_2 w_2} = \begin{bmatrix} \sigma_1^2 & \sigma_{12} & \cdots & \sigma_{1k} \\ \vdots & \vdots & \vdots & \vdots \\ \sigma_{1k} & \sigma_{2k} & \cdots & \sigma_k^2 \end{bmatrix}.$$

Define the $2 \times k$ covariance matrix between $\mathbf{W}_1$ and $\mathbf{W}_2$ as

$$\Sigma_{w_1 w_2} = \begin{bmatrix} \sigma_{l1} & \sigma_{l2} & \cdots & \sigma_{lk} \\ \sigma_{h1} & \sigma_{h2} & \cdots & \sigma_{hk} \end{bmatrix}$$

with $\Sigma_{w_2 w_1} = \Sigma'_{w_1 w_2}$.

Then, the conditional distribution of $\mathbf{W}_1$ given that the elements of $\mathbf{W}_2$ are fixed, say at $\mathbf{w}_2$, is bivariate normal with mean vector

$$E(\mathbf{W}_1) + \Sigma_{w_1 w_2} \Sigma_{w_2 w_2}^{-1} (\mathbf{w}_2 - E(\mathbf{W}_2)),$$

called the *conditional expectation of* $\mathbf{W}_1$ *given* $\mathbf{W}_2 = \mathbf{w}_2$, and covariance matrix

$$\Sigma_{w_1 w_1} - \Sigma_{w_1 w_2} \Sigma_{w_2 w_2}^{-1} \Sigma_{w_2 w_1} = \begin{bmatrix} \sigma_{l.c}^2 & \sigma_{lh.c} \\ \sigma_{lh.c} & \sigma_{h.c}^2 \end{bmatrix}.$$

Thus, the partial correlation coefficient is

$$\rho_{lh.c} = \frac{\sigma_{lh.c}}{\sigma_{l.c} \sigma_{h.c}}.$$

To obtain estimators of the multiple and partial correlation coefficients, we select a random sample $\mathbf{z}_i = (y_i, x_{1i}, \ldots, x_{pi})'$, $i = 1, \ldots, n$. The maximum likelihood estimator of $E(\mathbf{Z})$ is

$$\bar{\mathbf{z}} = \frac{1}{n} \sum_{i=1}^{n} \mathbf{z}_i,$$

and an unbiased estimator of Σ_z is

$$\mathbf{S}_z = \frac{1}{n-1} \sum_{i=1}^{n} (\mathbf{z}_i - \bar{\mathbf{z}})(\mathbf{z}_i - \bar{\mathbf{z}})'$$

$$= \begin{bmatrix} s_y^2 & s_{y1} & s_{y2} & \cdots & s_{yp} \\ s_{y1} & s_1^2 & s_{12} & \cdots & s_{1p} \\ \vdots & \vdots & \vdots & \vdots & \vdots \\ s_{yp} & s_{1p} & s_{2p} & \cdots & s_p^2 \end{bmatrix}.$$

Thus, unbiased estimators of Σ_{xx}, Σ_{xy}, and Σ_{yx} are obtained by partitioning $\mathbf{S}_z$ in the same way as Σ_z. Denote these estimators by $\mathbf{S}_{xx}$, $\mathbf{S}_{xy}$, and $\mathbf{S}_{yx}$, respectively. Note that

$$\mathbf{S}_{xx} = \frac{1}{n-1} \mathbf{A},$$

where $\mathbf{A}$ is given in Comment 3.2.1.3. Hence,

$$\tilde{\mathbf{b}} = \mathbf{S}_{xx}^{-1} \mathbf{S}_{xy},$$

and the sample multiple correlation coefficient is

$$r_{y.x_1...x_p} = \frac{\sqrt{S_{yx}S_{xx}^{-1}S_{xy}}}{s_y}.$$

To obtain the estimator of the partial correlation coefficient, we define $S_{w_1w_1}$, $S_{w_1w_2}$, and $S_{w_2w_2}$ in the same way as their population analogs with sample variances and covariances substituted for the corresponding parameters. Thus, the estimator of the partial correlation coefficient is given by

$$r_{lh.c} = \frac{S_{lh.c}}{S_{l.c}\,S_{h.c}}.*$$

EXAMPLE 3.2.3 (*continued*)   Since the second sampling procedure was used to obtain the measurements of arterial pressure, it is appropriate here to obtain estimates of the multiple and partial correlation coefficients of the variables $Y =$ cuff systolic pressure, and X_1, X_2, and $X_3 =$ intra-arterial systolic, diastolic, and mean pressures respectively.

For the estimate of the multiple correlation coefficient, we used the analysis of variance table and obtained $r^2_{y.x_1x_2x_3} = SS_D/SS_T = 89,963/116,971 = 0.769$. Thus, $r_{y.x_1x_2x_3} = \sqrt{0.769} = 0.877$: the three intra-arterial pressures X_1, X_2, and X_3 explain 76.9% of the variance of the cuff systolic pressure Y.

A test of the hypothesis $H_0: \rho_{y.x_1x_2x_3} = 0$ was given above by the equivalent test $H_0: \beta_1 = \beta_2 = \beta_3 = 0$. (This hypothesis was rejected since $F = 152$.) Equivalently, we can use Eq. (3.2.22) with $n = 141$ and $p = 3$ to obtain

$$F = \frac{141-3-1}{3}\frac{0.769}{1-0.769} = 152.$$

The program printed the estimate of the partial correlation coefficient between Y and X_2 given X_1 and X_3 as $r_{yx_2.x_1x_3} = -0.035$. Since $r^2_{yx_2.x_1x_3} = 0.00123$, $X_2 =$ intra-arterial diastolic pressure explains less than 1% of the variance of $Y =$ cuff systolic pressure, when the intra-arterial systolic and mean arterial pressures X_1 and X_3 are held fixed. A test of the hypothesis that X_2 contributes significantly to the prediction of Y was given previously by the equivalent test $H_0: \beta_2 = 0$. Since $F = 0.17$, H_0 was accepted. Alternately, we can use Eq. (3.2.25) with $l = Y$, $h = X_2$, and $c = \{X_1, X_3\}$ to obtain

$$t = \frac{-0.035\sqrt{141-2-2}}{\sqrt{1-(-0.035)^2}} = -0.41.$$

Again we accept H_0.

Referring to Comment 3.2.5.4, we may obtain $r_{yx_2.x_1x_3}$ from the simple correlations as follows. Letting $l = Y$, $h = X_2$, $c = X_1$, and $d = X_3$, we have

$$r_{yx_3.x_1} = \frac{0.845 - 0.871(0.927)}{\sqrt{1-(0.871)^2}\sqrt{1-(0.927)^2}} = 0.204,$$

$$r_{yx_2.x_1} = \frac{0.778 - 0.871(0.839)}{\sqrt{1-(0.871)^2}\sqrt{1-(0.839)^2}} = 0.177,$$

$$r_{x_2x_3.x_1} = \frac{0.967 - 0.927(0.839)}{\sqrt{1-(0.927)^2}\sqrt{1-(0.839)^2}} = 0.927,$$

where the simple correlations,

$$r_{yx_1} = 0.871, \qquad r_{yx_2} = 0.778, \qquad r_{yx_3} = 0.845,$$

$$r_{x_1x_2} = 0.839, \qquad r_{x_1x_3} = 0.927, \qquad r_{x_2x_3} = 0.967,$$

are taken from the correlation matrix given in Example 3.2.2 (with $Y \equiv X_4$). Hence, within slide rule accuracy, we have

$$r_{yx_2.x_1x_3} = \frac{r_{yx_2.x_1} - r_{yx_3.x_1}\, r_{x_2x_3.x_1}}{\sqrt{1-r_{yx_3.x_1}^2}\sqrt{1-r_{x_2x_3.x_1}^2}}$$

$$= \frac{0.177 - 0.204(0.927)}{\sqrt{1-(0.204)^2}\sqrt{1-(0.927)^2}} = -0.033.$$

A 95% confidence interval for $\rho_{yx_2.x_1x_3}$ may be obtained from the chart given in Table 9 (Appendix II) with n replaced by $n-k = 139$. This interval is $(-0.19, 0.08)$.

It should be noted from the above partial correlations that $Y =$ cuff systolic pressure and $X_3 =$ mean arterial pressure are not highly correlated when the value $X_1 =$ intra-arterial systolic pressure is held fixed. A similar conclusion is obtained when X_3 is replaced by $X_2 =$ intra-arterial diastolic pressure. On the other hand, X_2 is highly correlated with X_3 when X_1 is held fixed. All these findings are in agreement with known clinical facts.

3.3 Stepwise Regression

In many regression situations the experimenter does not have sufficient information about the order of importance of the independent variables $X_1, X_2, \ldots, X_p$ in predicting the dependent variable Y. Testing $H_0: \beta_i = 0$ for each variable X_i, $i = 1, \ldots, p$, does not reveal this ordering. For instance, in Example 3.2.3 rejecting the hypothesis that $\beta_1 = 0$ while accepting the hypothesis that $\beta_2 = 0$ and the hypothesis $\beta_3 = 0$ could have led to the false conclusion that X_1 was the only variable of importance in predicting Y.

Since the statistic for determining the effectiveness of a set of independent variables as predictors is the multiple correlation coefficient, one solution to the above problem is to regress Y on all possible subsets of the independent variables and then to select a best subset according to the following procedure. For each subset of size k, $k = 1, ..., p$, select the subset S_k yielding the largest multiple correlation coefficient. For the regression of Y on the subset S_1, test the hypothesis that inclusion of the remaining $p - 1$ variables does not improve the prediction of Y. Use the test statistic given by Eq. (3.2.14). If the hypothesis is rejected, then for the regression of Y on the subset S_2, test whether the inclusion of the remaining $p - 2$ variables does not improve the prediction of Y. Repeat this test successively until for some subset S_m, $1 \leqslant m \leqslant p$, the hypothesis that the remaining $p - m$ variables do not improve the prediction of Y is accepted. Then S_m is the *best subset of variables in predicting* Y since: (a) it yields the largest multiple correlation among all subsets of size m; and (b) inclusion of the remaining $p - m$ variables does not significantly improve the prediction of Y. If this solution is not unique, the nature of the problem may suggest a more appropriate subset.

When the number of independent variables is large it becomes impractical, even with the availability of high speed computers, to determine the best subset using this procedure. For example, when $p = 5$, there are $5 + 10 + 10 + 5 + 1 = 31$ regression equations, and when $p = 10$, there are $2(10 + 45 + 120 + 210) + 252 + 1 = 1023$ regression equations. In general, there are $2^p - 1$ regression equations needed. The computer time and expense involved necessitates finding another approach to the problem.

One solution is the technique of *stepwise regression* which selects a best subset according to the following procedure: The first step selects the single variable which best predicts Y. The second step finds the variable which best predicts Y given the first variable entered. In the steps that follow, either: (a) a variable is entered which best improved the prediction of Y given all the variables entered from the previous steps; or (b) a variable is removed from the set of predictors if its predictive ability falls below a given level. The process is terminated when no further variable improves the prediction of Y. Details are as follows.

3.3.1 Procedures for Stepwise Regression

Assume that we have a random sample of size n. A typical stepwise regression program first prints the sample means of each variable $Y, X_1, ..., X_p$ and the $(p + 1) \times (p + 1)$ sample covariance matrix of these variables. Then,

Step 0 The simple correlation coefficient r_{yx_i} and the F to enter,

$$F_{yx_i} = \frac{r_{yx_i}^2 (n - 2)}{1 - r_{yx_i}^2}, \qquad (3.3.1)$$

are calculated for $i = 1, ..., p$. This statistic is the square of t given in Comment 3.1.5.3, has 1 and $n-2$ degrees of freedom, and tests $H_0: \rho_{yx_i} = 0, i = 1, ..., p$. If all the F to enter are less than a prescribed inclusion level, called the F to *include*, the process is terminated, and we conclude that Y is estimated by $\bar{y}$ for any values of $X_1, ..., X_p$.

Step 1   The variable X_{i_1} having the largest F to enter (or equivalently, the largest squared correlation with Y) is selected as the best predictor of Y. The least-squares equation, the analysis of variance table, and the multiple correlation coefficient $r_{y.x_{i_1}} = |r_{yx_{i_1}}|$ are calculated. Also, the F to *remove* for X_{i_1} which is equal to the F to enter for X_{i_1} is calculated. Then the partial correlation coefficient $r_{yx_i.x_{i_1}}$ and the F to enter

$$F_{yx_i.x_{i_1}} = \frac{r_{yx_i.x_{i_1}}^2 (n-3)}{1 - r_{yx_i.x_{i_1}}^2} \qquad (3.3.2)$$

are calculated for $i = 1, ..., p, i \neq i_1$, that is, for each variable not included in the regression equation. This statistic is the square of Eq. (3.2.25), has 1 and $n-3$ degrees of freedom, and tests $H_0: \rho_{yx_i.x_{i_1}} = 0$ for $i = 1, ..., p, i \neq i_1$. If all the F to enter are less than the F to include, then Step S is executed. Otherwise, we go to Step 2.

Step 2   The variable X_{i_2} having the largest F to enter (or equivalently, the largest squared partial correlation with Y given X_{i_1}) is selected as the best predictor of Y given that X_{i_1} has already been selected. The least-squares equation, the analysis of variance table, and the multiple correlation coefficient $r_{y.x_{i_1}x_{i_2}}$ are calculated. Also, the F to remove $F_{yx_{i_1}.x_{i_2}}$ and $F_{yx_{i_2}.x_{i_1}}$ are calculated. These statistics which are the square of Eq. (3.2.25) with 1 and $n-3$ degrees of freedom are given by

$$F_{yx_{i_1}.x_{i_2}} = \frac{r_{yx_{i_1}.x_{i_2}}^2 (n-3)}{1 - r_{yx_{i_1}.x_{i_2}}^2} \quad \text{and} \quad F_{yx_{i_2}.x_{i_1}} = \frac{r_{yx_{i_2}.x_{i_1}}^2 (n-3)}{1 - r_{yx_{i_2}.x_{i_1}}^2}. \qquad (3.3.3)$$

They test the hypotheses that $H_0: \rho_{yx_{i_1}.x_{i_2}} = 0$ and $H_0: \rho_{yx_{i_1}.x_{i_2}} = 0$, respectively. Finally, the partial correlation coefficient $r_{yx_i.x_{i_1}x_{i_2}}$ and the F to enter

$$F_{yx_i.x_{i_1}x_{i_2}} = \frac{r_{yx_i.x_{i_1}x_{i_2}}^2 (n-4)}{1 - r_{yx_i.x_{i_1}x_{i_2}}^2} \qquad (3.3.4)$$

testing $H_0: \rho_{yx_i.x_{i_1}x_{i_2}} = 0$ with 1 and $n-4$ degrees of freedom for $i = 1, ..., p$, $i \neq i_1, i \neq i_2$ are calculated. If all F to enter are less than F to include, then Step S is executed. Otherwise we go to Step 3.

Step 3   (a) Let L denote the set of l independent variables which have been entered into the regression equation. If any of the F to remove for the variables in L are less than a prescribed deletion level, called the F *to exclude*, then the variable with the smallest F to remove is deleted from L and Step (3b) is executed with l replaced by $l-1$. If all of the F to enter for the variables not in L are less than F to include, then Step S is executed. Otherwise, the variable with the largest F to enter is chosen and is added to L so that l is replaced by $l+1$. (b) The least-squares equation, the analysis of variance table, and the multiple correlation coefficient $r_{y.l}$ are calculated for Y and the variables in L. Also, the F to remove $F_{yx_{i_j}.(l-1)}$ is calculated between Y and X_{i_j} in L given the $l-1$ remaining variables in L. This is the square of Eq. (3.2.25) with 1 and $n-l-1$ degrees of freedom and tests $H_0: \rho_{yx_i.(l-1)} = 0$. Finally, the partial correlation coefficient $r_{yx_i.l}$ and the F to enter $F_{yx_i.l}$ are calculated for Y and X_i not in L given the variables in L. This statistic is the square of Eq. (3.2.25) with 1 and $n-l-2$ degrees of freedom and tests $H_0: \rho_{yx_i.l} = 0$ for X_i not in L, $i = 1, ..., p$.

Steps 4, 5, ...   Repeat Step 3 recursively. Step S is executed (a) when the F to enter for all variables not in L is less than the F to include, (b) when all the variables have been entered and the F to remove for entered variables is greater than F to exclude, or (c) when the number of variables entered is $n-2$.

Step S   A summary table is usually available upon request by the user. For each step the quantities printed include the step number, the variable entered or removed, the F to enter or F to remove, and the multiple correlation coefficient of Y with the entered variables.

COMMENTS 3.3.1   1.   The summary table is extremely useful for selecting which variables should be used in the final regression equation. For a prescribed significance level α, a procedure for selecting a set H of predictors is as follows:

(a) At Step 1 in the summary table, a variable X_{i_1} was entered. If the corresponding F to enter is nonsignificant, that is, the F to enter $< F_{1-\alpha}(1, n-2)$, then the regression is meaningless, and the user should seek other methods of analyzing his data. Otherwise, $H = \{X_{i_1}\}$.

(b) At Step 2, a variable X_{i_2} was entered. If its F to enter $< F_{1-\alpha}(1, n-3)$, then H consists only of the variable X_{i_1}, and the best regression is given at Step 1. Otherwise, $H = \{X_{i_1}, X_{i_2}\}$.

(c) For each Step q, $q = 3, 4, ...$, if a variable is removed then delete it from the set H and go to Step $q+1$. On the other hand, if a variable is entered, compare its F to enter with $F_{1-\alpha}(1, n-h-2)$, where h is the number of variables in H at step $q-1$. If the F to enter is significant, augment H by the variable and go to Step $q+1$. If it is not significant, stop amending H and take as the best regression equation the one given at Step $q-1$.

2.   An alternative procedure requires that we carefully select the F to include and the F to exclude. The F to include may be chosen so that a variable which is potentially useful in predicting Y would be included. For example, we may decide to give each variable approximately a 50–50 chance of being included. Thus, we would choose the F to include to be $F_{0.50}(1, n-2)$. On the other hand, the F to exclude may be chosen so that there is a small chance of removing any variable once it has been entered. Thus, we would choose the F to exclude to be small, for example, 0.01. Then, attention is restricted to those variables which have been entered in the summary table where q' is the last step. Let L be this set of l variables, $l \leqslant p$, and let $r_{y.l}$ be the multiple correlation between Y and all the variables in L. (Note that if one or more variables have been removed, then it is necessary to compute $r_{y.l}$ separately.) Let H be the set of h variables in the regression equation at Step q, $q = 1, ..., q'-1$. The procedure consists in testing $H_0: \rho_{y.h} = \rho_{y.l}$ using

$$F = \frac{n-l-1}{l-h} \frac{r_{y.l}^2 - r_{y.h}^2}{1 - r_{y.l}^2}. \tag{3.3.5}$$

Under H_0, F has an F distribution with $l-h$ and $n-l-1$ degrees of freedom. We perform this test successively for $q = 1, 2, ..., q'-1$ until the first nonsignificant F is obtained. Suppose this occurs at Step q, where Y is regressed on a set H of h variables:

If at Step $q+1$ a variable is entered, we stop amending H and take as our best regression equation the one given at Step q. If, on the other hand, a variable is removed, we perform the above test for Step $q+1$. If the test is significant, then H is the set of variables at Step q and we take as our best regression the one given at this step. If the test is nonsignificant, we define H to be the set of $h-1$ variables at Step $q+1$, we define q to be $q+1$, and we repeat this process. If no nonsignificant F is obtained, then the regression equation at Step q is the best.

3.   The user is warned against concluding that the chosen set H is the best set of h variables for predicting Y. Indeed, to obtain such a set, all possible combinations of h variables taken from the p original variables must be examined.

4.   Since the stepwise procedure does not take into account the interpretation of the variables, the set of variables selected for the final regression equation may not be the most meaningful. For this reason, the user is given the option of *forcing* certain variables to be included in the regression equation. The stepwise procedure is then applied to the remaining variables.

5.   An advantage of the stepwise procedure is that it allows the user to include transformations of his independent variables. This is very useful in the exploratory phases of his research.

EXAMPLE 3.3.1   We performed a stepwise regression analysis on the $n = 141$ observations of Example 3.2.3 with $Y =$ cuff systolic pressure and $X_1, X_2, X_3 =$ intra-arterial, systolic, diastolic, and mean pressures (mm Hg), respectively. We performed the calculations with F to include $= 0.01$ and F to exclude $= 0.005$. Paralleling the steps outlined above, we have:

Step 0   The simple correlation coefficients are $r_{yx_1} = 0.871$, $r_{yx_2} = 0.778$, and $r_{yx_3} = 0.845$, and the corresponding F to enter are $F_{yx_1} = 436.8$, $F_{yx_2} = 213.2$, and $F_{yx_3} = 347.0$, with 1 and 139 degrees of freedom. Since all F to enter are greater than F to include, Step 1 is executed.

Step 1   Since X_1 has the largest F to enter, it is selected as the best predictor of Y. The least-squares equation is $\hat{y} = 8.08 + 0.88x_1$, and the analysis of variance table is given in the accompanying table. The multiple correlation coefficient $r_{y.x_1} = 0.8710$, and the F to remove is 436.8, with 1 and 139 degrees of freedom. The partial correlation coefficients are $r_{yx_2.x_1} = 0.178$ and $r_{yx_3.x_1} = 0.206$, and the corresponding F to enter are $F_{yx_2.x_1} = 4.49$ and $F_{yx_3.x_1} = 6.12$, with 1 and 138 degrees of freedom. Since both F to enter are greater than F to include, Step 2 is executed.

Source of variation	Sum of squares	Degrees of freedom	Mean square	F ratio
Due regression	88731.0	1	88731.0	436.8
Residual	28240.4	139	203.2	
Total	116971.4	140		

Step 2   Since X_3 has the largest F to enter, it is selected as the best predictor of Y given X_1. The least-squares equation is $\hat{y} = 7.93 + 0.63x_1 + 0.37x_3$, and the analysis of variance table is given in the accompanying table. The multiple correlation coefficient is $r_{y.x_1x_3} = 0.8768$, and the F to remove are $F_{yx_1.x_3} = 32.6$ and $F_{yx_3.x_1} = 6.12$, with 1 and 138 degrees of freedom. The partial correlation coefficient is $r_{yx_2.x_1x_3} = -0.035$ with an F to enter $F_{yx_2.x_1x_3} = 0.17$.

Source of variation	Sum of squares	Degrees of freedom	Mean square	F ratio
Due regression	89931.1	2	44965.6	229.5
Residual	27040.3	138	195.9	
Total	116971.4	140		

Step 3   (a) The set $L = \{X_1, X_3\}$ with $l = 2$. Since the F to remove for X_1 and X_3 are greater than F to exclude, and since F to enter for X_2 is greater than F to include, we augment L. Hence, $L = \{X_1, X_2, X_3\}$ with $l = 3$.

(b) The least-squares equation is $\hat{y} = 8.29 + 0.60x_1 - 0.14x_2 + 0.52x_3$, and the analysis of variance table is given in the accompanying table. The multiple correlation coefficient is $r_{y.x_1x_2x_3} = 0.8770$, and the F to remove are $F_{yx_1.x_2x_3} = 19.2$, $F_{yx_2.x_1x_3} = 0.17$, and $F_{yx_3.x_1x_2} = 1.7$. Since all F to remove are greater than F to exclude, and since there are no further variables to enter, Step S is executed.

Source of variation	Sum of squares	Degrees of freedom	Mean square	F ratio
Due regression	89963.8	3	29987.9	152.1
Residual	27007.6	137	197.1	
Total	116971.4	140		

Step S   The summary table is shown. Applying Comment 3.3.1.1 to the summary table with $\alpha = 0.05$, we have at Step 1, the F to enter $> F_{0.95}(1, 139) \simeq 3.92$. At Step 2, the F to enter $> F_{0.95}(1, 138) \simeq 3.92$, but at Step 3, the F to enter $< F_{0.95}(1, 138) \simeq 3.92$. Hence, $H = \{X_1, X_3\}$ so that the regression of Y on X_1 and X_3 is the best equation for predicting Y based on the stepwise procedure.

Step number	Variable Entered	Removed	F to Enter	Remove	Multiple correlation
1	X_1	—	436.74	—	0.8710
2	X_3	—	6.12	—	0.8768
3	X_2	—	0.17	—	0.8770

Alternatively, we can apply Comment 3.3.1.2, in which case, $L = \{X_1, X_2, X_3\}$ and $r_{y.l} = 0.8770$. Applying Eq. (3.3.5) with $q = 1$, we have

$$F = \frac{(141 - 3 - 1)}{(3 - 1)} \frac{(0.7691 - 0.7586)}{(1 - 0.7691)} = 3.11,$$

which is significant, since $F_{0.95}(2, 137) \simeq 3.07$. For $q = 2$, we have

$$F = \frac{(141 - 3 - 2)}{(3 - 2)} \frac{(0.7691 - 0.7688)}{(1 - 0.7691)} = 0.17,$$

which is nonsignificant, since $F_{0.95}(1, 137) \simeq 3.92$. Hence $q = 2$, $H = \{X_1, X_3\}$, and $h = 2$. Since at Step $q + 1$ a new variable is entered, the process is terminated, and we use the regression equation at Step 2. Either procedure yields the best regression equation as

$$\hat{y} = 7.93 + 0.63x_1 + 0.37x_3.$$

EXAMPLE 3.3.2   The purpose of this example is to demonstrate that the two procedures of selecting the best regression equation from the summary table may give different results. A stepwise regression analysis was performed on $p = 10$ independent variables from a fictitious sample of size $n = 200$. The summary table is given in the accompanying table.

Step number	Variable Entered	Variable Removed	F to Enter	F to Remove	Multiple correlation	Squared multiple correlation
1	5	—	8.1	—	0.1982	0.0393
2	8	—	6.2	—	0.2619	0.0686
3	4	—	5.8	—	0.3089	0.0954
4	—	8	—	0.002	0.3087	0.0953
5	3	—	2.4	—	0.3256	0.1062
6	—	5	—	...	0.3302	0.1089
7	2	—	...	—	0.3311	0.1097
8	9	—	...	—	0.3319	0.1102

Applying Comment 3.3.1.1 with $\alpha = 0.05$ to the summary table, we see that variables X_5, X_8, and X_4 all enter with a significant F to enter, since $F_{0.95}(1, 198) \simeq F_{0.95}(1, 197) \simeq F_{0.95}(1, 196) \simeq 3.8$. Hence, at Step $q = 3$, $H = \{X_5, X_8, X_4\}$ and $h = 3$. At Step $q = 4$, variable X_8 is removed, since its F to remove is less than F to exclude $= 0.01$. Hence $H = \{X_5, X_4\}$ and $h = 2$. At Step $q = 5$, variable X_3 is entered with an F to enter less than $F_{0.95}(1, 196) \simeq 3.8$. Hence, we stop amending H and take as our best regression the one given at Step $q = 4$, namely the regression of Y on X_5 and X_4.

Applying Comment 3.3.1.2 to the summary table, we have $q' = 8$, $L = \{X_2, X_3, X_4, X_5, X_8, X_9\}$, and $l = 6$. Since X_5 and X_8 have been removed, we reran the stepwise program forcing all six variables in L to enter the regression equation. This yielded a multiple correlation coefficient of $r_{y.l} = 0.3324$. Using Eq. (3.3.5) with $q = 1$, we have $F = 3.09$, which is greater than $F_{0.95}(5, 193) \simeq 2.25$. At Step $q = 2$, $F = 2.28$, which is less than $F_{0.95}(4, 193) \simeq 2.40$. Hence, the first non-significant F occurs at Step $q = 2$, with $H = \{X_5, X_8\}$ and $h = 2$. Since at Step $q + 1 = 3$, a variable is entered, we terminate the process. The best regression equation is the one given at Step 2, namely the regression of Y on X_5 and X_8.

Note that the difference in the two procedures becomes apparent at Step 2. The method of Comment 3.3.1.1 asks whether any single variable not included at Step 2 improves the prediction of Y given the two variables already entered, while the method of Comment 3.3.1.2 asks whether the two variables entered at Step 2 give the same prediction accuracy as that attained by all six variables in L.

3.4 Nonlinear Regression

In the previous sections we have been concerned with models which are linear in the parameters, namely of the form

$$y_i = \beta_0 + \beta_1 x_{1i} + \cdots + \beta_p x_{pi} + e_i, \qquad i = 1, ..., n. \tag{3.4.1}$$

As noted in the beginning of the chapter, for many problems the linear model may be appropriate at least as a first approximation to the true underlying model. Furthermore, it was noted in Example 3.1.3 that for other problems transformations of the variables may induce a linear relationship among the parameters. There are many situations, however, for which a linear model is not appropriate, for example, where the underlying model is a sum of exponential and/or trigonometric functions. In this case, the assumed model can neither be easily transformed into, nor appropriately approximated by, a linear model.

Any model not of the form given by Eq. (3.4.1) is called a *nonlinear regression model* and can be written as

$$y_i = f(x_{1i}, ..., x_{pi}; \theta_1, ..., \theta_m) + e_i, \qquad i = 1, ..., n, \tag{3.4.2}$$

where $f(\ )$ is a nonlinear function in the parameters $\theta_1, ..., \theta_m$, and e_i are uncorrelated error terms. Two examples of nonlinear functions are

$$f(x_i; \theta_1, \theta_2, \theta_3) = \theta_1 + \theta_2 e^{\theta_3 x_i}, \qquad \text{and}$$

$$f(x_{1i} x_{2i}; \theta_1, \theta_2, \theta_3) = \theta_1 + \theta_2 \sin(x_{1i} + \theta_3 \cos x_{2i}), \qquad i = 1, ..., n.$$

When the underlying model is assumed to be linear, then the least-squares estimators of the parameters are optimal since they are minimum variance unbiased estimators. However, when the model is nonlinear, there are no such best estimators of the parameters. The method of *maximum likelihood*, though, does produce estimators $\hat{\theta}_1, \hat{\theta}_2, ..., \hat{\theta}_m$ which have the desirable properties of being *consistent* and *asymptotically efficient* under some general conditions. Furthermore, if the error terms e_i are independent $N(0, \sigma^2)$, then these maximum likelihood estimators are also the least-squares estimators. As in the previous sections, the least-squares estimators are the values of $\theta_1, \theta_2, ..., \theta_m$ which minimize the *sum of squares of deviations*

$$S = \sum_{i=1}^{n} (y_i - f(x_{1i}, ..., x_{pi}; \theta_1, ..., \theta_m))^2. \tag{3.4.3}$$

Recall that in the case of linear regression models, the least-squares estimators were obtained by solving a set of linear equations. In the case of nonlinear models, the set of equations to be solved are unfortunately also nonlinear, and hence, the least-squares solutions are not obtainable in closed form. For this reason, it becomes imperative to use one of various iterative procedures to obtain them numerically.

3.4.1 Iterative Methods for Obtaining Numerical Least-Squares Estimates

All packaged programs for obtaining least-squares estimates $\hat{\theta}_1, ..., \hat{\theta}_m$ successively calculate numerical approximations $\theta_1^{(j)}, ..., \theta_m^{(j)}, j = 1, 2, ...$. Most programs require the user to specify initial guesses $\theta_1^{(0)}, ..., \theta_m^{(0)}$. Some programs terminate the approximation process when the increments in the estimates are negligible, that is, when

$$\left| \frac{\theta_i^{(j+1)} - \theta_i^{(j)}}{\theta_i^{(j)}} \right| < \delta \qquad (3.4.4)$$

for $i = 1, ..., m$, and for some preselected small number δ. Other programs terminate when the computed error sum of squares converges.

The technical details underlying the numerical approximations are not discussed in this book but may be found in Draper and Smith (1968, Ch. 10) or Ralston and Wilf (1960). The most commonly used techniques are *linearization* (Hartley, 1961), *scoring* (Rao, 1965, p. 302) *steepest descent* (Davies, 1954) and *Marquardt's compromise* (Marquardt, 1963).

In addition to (a) the initial guesses for $\theta_1, ..., \theta_m$, the user of a packaged nonlinear regression program is usually required to supply: (b) upper and lower limits on the values of the parameters, and (c) a subroutine specifying the form of the function $f(\ )$, and the first, and sometimes, the second partial derivatives of $f(\ )$ with respect to $\theta_1, ..., \theta_m$. For users not familiar with calculus and/or elementary programming, professional assistance may be needed.

The output of a typical nonlinear program includes (a) the final (and sometimes, intermediate) estimates of the parameters, (b) the final (and sometimes, intermediate) sum of squares of deviations S as a measure of goodness-of-fit, and (c) estimates of the asymptotic variances $V(\hat{\theta}_i)$ and covariances $\text{cov}(\hat{\theta}_i, \hat{\theta}_j)$ of $\hat{\theta}_i$ and $\hat{\theta}_j$, $i \neq j = 1, ..., m$. The quantity

$$s^2 = \frac{S}{(n-m)} \qquad (3.4.5)$$

(sometimes called the *error mean square*) serves as an estimate of the error variance σ^2. The estimates of the asymptotic variances can be used to perform approximate tests of hypotheses and to obtain approximate $100(1-\alpha)\%$ confidence intervals for the parameters. Additional output may include the predicted value $\hat{y}_i$, the corresponding standard deviation of $\hat{y}_i$, and the residual $y_i - \hat{y}_i$, for $i = 1, ..., n$.

3.4.2 Approximate Tests of Hypotheses and Confidence Intervals

To *test the hypothesis* $H_0: \theta_i = \theta_{i0}$, $i = 1, \ldots, m$, we use the statistic

$$z = \frac{\hat{\theta}_i - \theta_{i0}}{[\hat{V}(\hat{\theta}_i)]^{1/2}}, \tag{3.4.6}$$

where $\hat{\theta}_i$ is the numerical least-squares estimate of θ_i, and $\hat{V}(\hat{\theta}_i)$ is the estimate of the asymptotic variance of $\hat{\theta}_i$. Under H_0, this quantity is approximately $N(0, 1)$ for large n.

An *approximate* $100(1 - \alpha)\%$ *confidence interval for* θ_i, $i = 1, \ldots, m$, is given by

$$\hat{\theta}_i \pm z_{1-(\alpha/2)} [\hat{V}(\hat{\theta}_i)]^{1/2}. \tag{3.4.7}$$

An *approximate* $100(1 - \alpha)\%$ *confidence interval for the mean value of Y at the sample values* $x_{1i}, x_{2i}, \ldots, x_{pi}$, $i = 1, \ldots, n$, is given by

$$\hat{y}_i \pm z_{1-(\alpha/2)} [\mathrm{sd}(\hat{y}_i)], \tag{3.4.8}$$

where $\hat{y}_i$ is the estimated value of Y at $x_{1i}, x_{2i}, \ldots, x_{pi}$, and $\mathrm{sd}(\hat{y}_i)$ is the corresponding standard deviation of $\hat{y}_i$. It is not easy to obtain confidence intervals for the mean of Y at values of $X_1, \ldots, X_p$ not in the sample.

The list of residuals $y_i - \hat{y}_i$, $i = 1, \ldots, n$ may be used in the manner described in Section 3.1.5.

COMMENTS 3.4.1   1.   The choice of initial values for $\theta_1^{(0)}, \ldots, \theta_m^{(0)}$ is very important, since a poor guess may result in slow convergence or even divergence. Previous experience may indicate reasonable initial guesses, or in the case of a single independent variable X, a plot of the data points along with $f(x; \theta_1, \ldots, \theta_m)$ for various choices of $\theta_1, \ldots, \theta_m$ may be useful.

2.   Since some programs do not permit the numerical estimates to exceed the upper and lower limits, the user should be careful in not making these limits too restrictive.

3.   A useful quantity (but one not usually contained in the output) is the correlation between the true observations y_i and the predicted values $\hat{y}_i$. If alternative models are considered, then the best model is the one which maximizes this correlation.

4.   Some programs written for special nonlinear functions (such as exponential or trigonometric functions) do not require initial guesses or the user-written subroutine.

EXAMPLE 3.4.1   Data on $Y = $ cardiac index (liters/min/m^2) and $X = $ mean circulation time (sec) were collected on 107 critically ill patients. The scattergram of the data (Fig. 3.1.3) exhibits a relationship of the form

$$y_i = \theta_1 + \theta_2 e^{\theta_3 x_i} + e_i.$$

A packaged nonlinear regression program (BMDX85) was used to fit the data. The initial guesses as well as the upper and lower limits were obtained by plotting the data and the function

$$f(x; \theta_1, \theta_2, \theta_3) = \theta_1 + \theta_2 e^{\theta_3 x}$$

for various choices of θ_1, θ_2, and θ_3. From these plots we chose $\theta_1^{(0)} = 1.0$, $\theta_2^{(0)} = 1.0$, and $\theta_3^{(0)} = -0.2$ with limits $0.1 < \theta_1 < 5.0$, $0.1 < \theta_2 < 100.0$, and $-0.5 < \theta_3 < 0.0$.

The results of the iterative process (which used the linearization method) are given in Table 3.4.1. The final equation is given by

$$\hat{y} = 1.3707 + 1.8925 e^{-0.1580x}$$

and the estimate of σ^2 is $s^2 = 0.7304$.

TABLE 3.4.1. *Numerical Iterations to Obtain a Solution for the Model* $y = \theta_1 + \theta_2 e^{\theta_3 x}$

Iteration	Error mean square	$\hat{\theta}_1$	$\hat{\theta}_2$	$\hat{\theta}_3$
0	3.8168	1.0000	1.0000	−0.2000
1	2.4508	1.0400	2.5492	−0.1000
2	2.0765	1.4036	1.3727	−0.2562
3	0.8968	1.4393	8.8124	−0.1281
4	0.7857	1.3574	1.6380	−0.1618
5	0.7305	1.3744	1.8426	−0.1580
6	0.7304	1.3709	1.8303	−0.1580
7	0.7304	1.3707	1.8296	−0.1580
8	0.7304	1.3707	1.8295	−0.1580
9	0.7304	1.3707	1.8295	−0.1580
10	0.7304	1.3707	1.8295	−0.1580
11	0.7304	1.3707	1.8295	−0.1580

The estimates of the asymptotic standard deviations are $[\hat{V}(\hat{\theta}_1)]^{1/2} = 0.1774$, $[\hat{V}(\hat{\theta}_2)]^{1/2} = 5.728$, and $[\hat{V}(\hat{\theta}_3)]^{1/2} = 0.02822$. Testing $H_0: \theta_1 = 0$ versus $H_1: \theta_1 \neq 0$ was achieved by computing

$$z = \frac{1.3707 - 0}{0.1774} = 7.726,$$

which is highly significant ($P < 10^{-6}$). To test $H_0: \theta_2 = 2$ versus $H_1: \theta_2 < 2$, we calculate

$$z = \frac{1.8295 - 2.0}{5.728} = -0.030,$$

which is not significant.

An approximate 95% confidence interval for θ_3 is

$$-0.1580 \pm 1.96(0.02822) = (-0.213, -0.103).$$

Since this interval does not include 0, the hypothesis $H_0: \theta_3 = 0$ is rejected at the $\alpha = 0.05$ level.

An approximate 95% confidence interval for the mean value of Y at $x_{23} = 20.5$ is

$$2.089 \pm 1.96(0.1106) = (1.872, 2.306),$$

where 0.1106 is the standard deviation of $\hat{y}_{23} = 2.089$.

The simple correlation between Y and X was calculated to be $r_L = -0.659$. This is a measure of the linear association between these variables. The association using the nonlinear model was estimated by computing the simple correlation coefficient r_{NL} between y_i and $\hat{y}_i$. Since $r_{NL} = 0.771 > |r_L|$, we conclude from Comment 3.4.1.3 that the nonlinear model may be more appropriate than the linear model in fitting the data.

Problems

Note: Data Set A is the data in Example 1.4.1, Tables 1.4.1 and 1.4.2.
Data Set B is the data in Example 1.4.2, Tables 1.4.3 and 1.4.4.

Section 3.1

3.1.1. (Data Set B)   Plot a scattergram of $Y =$ systolic pressure (1962) against $X =$ systolic pressure (1950) and note that a linear relationship is a reasonable approximation. Estimate the means, variances, covariance, and correlation coefficient between X and Y.

3.1.2. (Data Set B)   For the data in Problem 3.1.1, test the hypothesis that X and Y are independent and compute a 95% confidence interval for the correlation coefficient.

3.1.3. (Data Set B)   (a) Repeat Problem 3.1.1 using diastolic pressure (DP) instead of systolic pressure.

(b) Compute the estimated regression line of Y on X and plot it on the scattergram.

(c) Compute a 95% confidence interval for the mean DP in 1962 for individuals whose DP in 1950 was 75.

(d) Compute a 90% interval estimate of DP (1962) for an individual whose DP (1950) was 80.

(e) Draw approximate parallel 95% confidence bands for the line drawn in (b).

Section 3.2

3.2.1. (Data Set A) (a) Using the initial data on all patients, estimate the multiple linear regression of $Y = $ PVI on $X_1 = $ RCI, $X_2 = $ Hgb, and $X_3 = $ Hct.
   (b) Test the hypothesis that Y is independent of X_1, X_2, and X_3.

3.2.2. (Data Set A) For the data of Problem 3.2.1:
   (a) Obtain separate 95% confidence intervals for β_1, β_2, and β_3.
   (b) Compute simultaneous 95% confidence intervals for β_1, β_2, and β_3.
   (c) Compare (a) and (b).

3.2.3. (Data Set A) For the data of Problem 3.2.1:
   (a) Compute a 90% confidence interval for the mean of Y at $X_1 = 21$, $X_2 = 12$, $X_3 = 32$.
   (b) Compute a 90% interval estimate of Y for an individual whose X's are the same as in (a).
   (c) Compare (a) and (b).

3.2.4. (Data Set A) For the data of Problem 3.2.1:
   (a) Estimate the multiple correlation coefficient of Y on X_1, X_2, and X_3.
   (b) Estimate the multiple correlation coefficient of Y on X_1.
   (c) Test whether the addition of X_2 and X_3 to X_1, as independent variables, improves the prediction of Y.

3.2.5. (Data Set A) For the data of Problem 3.2.1:
   (a) Estimate the partial correlation coefficient between Y and each X_i given the other two X variables, $i = 1, 2, 3$.
   (b) Compute simultaneous 95% confidence intervals for the three population partial correlation coefficients estimated in (a). [Hint: use the intervals computed in Problem 3.2.2(b) and the relationship given in Comment 3.2.5.3.]

Section 3.3

3.3.1. (Data Set A) (a) Using the initial data on all patients, perform a step-wise regression of $Y = $ CI on $X_1 = $ SP, $X_2 = $ MAP, and $X_3 = $ DP.
   (b) Obtain the summary table and use it to choose a "best" set of predictors according to both Comments 3.3.1.1 and 3.3.1.2, with $\alpha = 0.05$.
   (c) Interpret these results.

Section 3.4

3.4.1. (Data Set A) (a) Using initial data on all patients, plot a scattergram of $Y = $ CI against $X = $ AT. Obtain the estimated regression line of Y on X using the two models:

   (1) $y_i = \alpha + \beta_1 x_i + \beta_2 x_i^2 + e_i, \quad i = 1, ..., n;$
   (2) $y_i = \theta_1 + \theta_2 e^{\theta_3 x_i} + e_i, \quad i = 1, ..., n.$

(b) Plot these estimated lines and decide which seems to fit the data better (visually).

(c) Answer (b) on the basis of the error mean square for (1) and (2).

(d) Answer (b) using Comment 3.4.1.3.

(e) What is your final answer to (b)?

3.4.2. (Data Set A)   Using initial data on all patients, define $Y = CI$, $X_1 = AT$, $X_2 = MCT$. Assume the model

$$y_i = \theta_1 + \theta_2 e^{\theta_3 x_{1i}} + \theta_4 e^{\theta_5 x_{2i}} + e_i, \qquad i = 1, \ldots, n.$$

(a) Using a nonlinear regression program, compute the estimated regression equation.

(b) Test the null hypothesis $\theta_1 = 0$.

(c) Obtain an approximate confidence interval for θ_5. [Hint: In many programs the choice of the initial value is crucial. The reader should first try plotting Y versus X_1 and Y versus X_2 and approximating the individual functions. These values may be of help in determining the initial values.]

4 THE ANALYSIS OF VARIANCE

The subject of analysis of variance (*anova*) was introduced for the first time in Section 2.4 as a technique for comparing $k \geqslant 2$ means from k strata or subpopulations. The strata or subpopulations may be viewed as different *levels* of some basis for classification—called a *factor*. The general anova problem involves several factors, each with more than one level. Underlying each problem is an *experimental design*, that is, a specification of the assignment of each experimental unit to a particular combination of the levels of the factors under consideration, and an anova *model*, that is, the mathematical equation expressing each observation as a sum of a mean and an error term. The mean of each observation is expressed as the sum of an overall mean and an "effect" for each factor and for every combination of factors. The statistical problems involved are to estimate these effects and test hypotheses concerning them.

The subject of the analysis of variance was developed primarily with the work of Fisher (1918, 1925, 1935) (see also Scheffé, 1956 for a historical summary). The anova defined by Fisher was later put into the different theoretical framework of the *general linear model*. This elegant theory, which represents the modern approach to the subject, is discussed in Section 4.1. In addition, the estimators and tests of hypotheses for the simple and multiple linear regression models are shown to be applications of this theory. The reader interested in specific analysis of variance problems may read only the summary at the end of that section and proceed to Section 4.2.

In Sections 4.2 and 4.3 we deviate from our policy of interweaving computer and theoretical discussions. Rather, we discuss various experimental designs, theoretical models, and applications. Beginning with Section 4.4 we become computer oriented and show how various packaged computer programs may be used to perform analyses of variance. The reader already familiar with the theory of anova may wish to commence with Section 4.4.

In Section 4.2–4.3 we consider some of the standard anova problems. Section 4.2 reconsiders the one-way anova model of Section 2.4 and distinguishes between the so-called *Model I* (or *fixed effects*) and *Model II* (or *random effects*) interpretations. Section 4.3 discusses problems involving two factors arising from the *crossed* and the *nested* designs. Distinction is made between the Model I, Model II, and the so-called *mixed* model interpretations. In both sections we assume that computations are performed using available packaged programs.

In Section 4.4 we discuss the *factorial* program which is available in many statistical packages. We show how the models discussed to that point, as well as some other models, may be analyzed by a factorial program. In Section 4.5 we show how a multiple linear regression program may be used for solving an analysis of variance problem. In Section 4.6 we discuss a very general program which may be used for solving most anova problems. Finally, in Section 4.7, the *one-way analysis of covariance* is presented.

It should be noted that although anova assumes a "designed" experiment, many anova problems arise from "non-designed" experiments, especially in the life and social sciences. For example, an interviewer, collecting attitudinal measurements on residents of a given town, may wish to examine the effect of the factors "ethnic group" and "socio-economic status" on the responses. He then formulates this problem as a two-way anova problem (Section 4.3) and proceeds with the standard analysis. The reader should be aware that the interpretations of the results of an analysis of variance depend on whether the data arose from a designed experiment or a survey.

4.1 Basic Theory of the General Linear Model

The material of this section is more theoretical in nature than that in the remainder of this chapter. The reader interested in specific analyses may be satisfied by reading the summary of this section and then proceeding with the next section.

We assume that we have n observations $y_1, \ldots, y_n$ which are realizations of the random variables $Y_1, \ldots, Y_n$. Furthermore, we assume that the average value of each Y_i is a linear function of p unknown parameters $\theta_1, \ldots, \theta_p$, that is,

$$E(Y_i) = \theta_1 x_{1i} + \cdots + \theta_p x_{pi}, \qquad i = 1, \ldots, n, \qquad (4.1.1)$$

where $x_{1i}, \ldots, x_{pi}$ are known constants. Thus, each observation y_i can be written as

$$y_i = \theta_1 x_{1i} + \cdots + \theta_p x_{pi} + e_i, \qquad i = 1, \ldots, n, \qquad (4.1.2)$$

where $e_1, \ldots, e_n$ are error variables. The *general linear model* consists of this equation along with the assumptions that

$$E(e_i) = 0, \qquad V(e_i) = \sigma^2,$$

and

$$\text{cov}(e_i, e_j) = 0, \qquad i, j = 1, \ldots, n, \quad i \neq j. \qquad (4.1.3)$$

Thus, the errors are assumed to be uncorrelated with zero means and equal variances σ^2.

We have seen this model in Section 3.2 when we discussed multiple linear regression, for if we define $x_{1i} \equiv 1$ and renumber the constants and parameters, we can write Eq. (4.1.2) in the form of the multiple linear regression model, that is

$$y_i = \theta_0 + \theta_1 x_{1i} + \cdots + \theta_q x_{qi} + e_i, \qquad \text{where} \quad q = p - 1.$$

In this section we wish to find point estimators of $\theta_1, \ldots, \theta_p$, and to describe methods for obtaining confidence intervals and for testing hypotheses concerning the parameters. For point estimation we do not require any additional assumptions about the model; however, for obtaining confidence intervals and for testing hypotheses, we assume that the errors are normally distributed.

EXAMPLE 4.1.1   To obtain the simple linear regression model [see Eq. (3.1.2)] from Eq. (4.1.2), we let $\theta_1 = \beta_0$, $\theta_2 = \beta_1$, $x_{1i} = 1$, and $x_{2i} = x_i$, so that we have $y_i = \beta_0 + \beta_1 x_i + e_i$, $i = 1, \ldots, n$.

EXAMPLE 4.1.2   Suppose we have a sample $z_{11}, \ldots, z_{1n_1}$ from $N(\mu_1, \sigma^2)$ and a sample $z_{21}, \ldots, z_{2n_2}$ from $N(\mu_2, \sigma^2)$. Let

$$y_1 = z_{11}, \ldots, y_{n_1} = z_{1n_1}, \qquad y_{n_1+1} = z_{21}, \ldots, y_n = z_{2n_2}, \qquad \text{where} \quad n = n_1 + n_2.$$

Thus, we can write

$$y_i = \begin{cases} \mu_1 + e_i, & i = 1, \ldots, n_1, \\ \mu_2 + e_i, & i = n_1 + 1, \ldots, n, \end{cases}$$

where e_i is $N(0, \sigma^2)$. This can be rewritten in the form of Eq. (4.1.2) as

$$y_i = \mu_1 x_{1i} + \mu_2 x_{2i} + e_i, \qquad i = 1, \ldots, n,$$

where

$$x_{1i} = \begin{cases} 1, & i = 1, \ldots, n_1, \\ 0, & i = n_1 + 1, \ldots, n, \end{cases} \qquad \text{and} \qquad x_{2i} = \begin{cases} 0, & i = 1, \ldots, n_1, \\ 1, & i = n_1 + 1, \ldots, n. \end{cases}$$

This is the one-way anova model with $k = 2$ subpopulations. It is also called the *two-sample problem* (see Section 2.3.2).

Although Dr. Gesund may think that this complicates a simple situation, it will be shown that there are advantages in expressing such problems in the framework of the general linear model.

4.1.1 Point Estimation

The usual procedure to obtain estimators of $\theta_1, \ldots, \theta_p$, is the *least-squares method*. The *least-squares estimators* are defined to be the values $\hat{\theta}_1, \ldots, \hat{\theta}_p$ of $\theta_1, \ldots, \theta_p$ which minimize the sum of squares

$$S = \sum_{i=1}^{n} (y_i - \theta_1 x_{1i} - \cdots - \theta_p x_{pi})^2 \tag{4.1.4}$$

with respect to $\theta_1, \ldots, \theta_p$. The solutions $\hat{\theta}_1, \ldots, \hat{\theta}_p$, which are linear functions of the observations, may be unique (as in the case of multiple linear regression), or there may be an infinite number of solutions which minimize S. This non-uniqueness may occur in the analysis of variance situation. To obtain unique solutions, the customary procedure is to impose certain restrictions, called *side conditions*, on the parameters and their estimators. These side conditions will be introduced in the following sections when needed.

For the remainder of this section we assume that we have obtained unique estimators $\hat{\theta}_1, \ldots, \hat{\theta}_p$ either directly or after imposing some side conditions. Then the least-squares estimator of any linear function of the parameters is the same linear function of the least-squares estimators of the parameters, that is,

$$\widehat{\sum_{i=1}^{p} c_i \theta_i} = \sum_{i=1}^{p} c_i \hat{\theta}_i, \tag{4.1.5}$$

where c_i are known constants, $i = 1, \ldots, p$. Since $\hat{\theta}_i$ are linear in $y_1, \ldots, y_n$, $\sum_{i=1}^{p} c_i \hat{\theta}_i$ is also linear in the observations. The importance of the least-squares method is that the least-squares estimators are unbiased and have minimum variance among all unbiased estimators which are linear in the observations. This is the statement of the Gauss–Markov theorem.

Gauss–Markov Theorem.[†]   Under the general linear model [Eq. (4.1.2)] with assumptions [Eq. (4.1.3)], the least-squares estimator $\sum_{i=1}^{p} c_i \hat{\theta}_i$ of $\sum_{i=1}^{p} c_i \theta_i$ ($\hat{\theta}_i$ is the unique least-squares estimator of θ_i and c_i's are constants) is unbiased

---

† This is the most general form of this theorem when unique estimators are assumed. A more general form may be found in Scheffé (1959).

and has minimum variance among all unbiased estimators of $\sum_{i=1}^{p} c_i \theta_i$ which are linear in the observations $y_1, ..., y_n$.

Having obtained estimators of $\theta_1, ..., \theta_p$ we need to obtain an estimator of the error variance σ^2. The usual estimator of σ^2, called the *residual* (or *error*) *mean square*, is

$$\mathrm{MS_R} \equiv s^2 = \frac{\mathrm{SS_R}}{\nu_R}, \qquad (4.1.6)$$

where $\mathrm{SS_R}$, the *residual* (or *error*) *sum of squares*, is

$$\mathrm{SS_R} = \sum_{i=1}^{n} (y_i - \hat\theta_1 x_{1i} - \cdots - \hat\theta_p x_{pi})^2 \qquad (4.1.7)$$

and ν_R is the *residual* (or *error*) *degrees of freedom*. This last quantity ν_R, which will be specified in each situation that we consider (see Comment 4.1.1.2), is chosen so that s^2 is unbiased. The quantities $\mathrm{SS_R}$, ν_R, and $\mathrm{MS_R}$ appear in *anova tables* such as Tables 2.4.1, 3.1.1, and 3.2.1 for the cases of one-way analysis of variance, simple linear regression, and multiple linear regression, respectively. Note that $\mathrm{SS_R}$ is a measure of how well the estimated model fits the data; the smaller $\mathrm{SS_R}$ is, the better the fit.

COMMENTS 4.1.1   *1.   The general linear model takes on a simple form in matrix notation. Let

$$\mathbf{y}^{n \times 1} = (y_1, ..., y_n)', \qquad \boldsymbol{\theta}^{p \times 1} = (\theta_1, ..., \theta_p)', \qquad \mathbf{e}^{n \times 1} = (e_1, ..., e_n)',$$

and

$$(\mathbf{X}')^{n \times p} = \begin{bmatrix} x_{11} & \cdots & x_{p1} \\ \vdots & \vdots & \vdots \\ x_{1n} & \cdots & x_{pn} \end{bmatrix}.$$

Then Eq. (4.1.2) becomes

$$\mathbf{y} = \mathbf{X}'\boldsymbol{\theta} + \mathbf{e},$$

and Eq. (4.1.3) becomes

$$E(\mathbf{e}) = \mathbf{0} \qquad \text{and} \qquad \mathrm{cov}(\mathbf{e}) = \sigma^2 \mathbf{I},$$

where $\mathbf{0}$ is the zero vector and $\mathbf{I}$ is the identity matrix. The sum of squares to be minimized is now

$$S = (\mathbf{y} - \mathbf{X}'\boldsymbol{\theta})'(\mathbf{y} - \mathbf{X}'\boldsymbol{\theta}),$$

and the least-squares estimators are the solutions to the equations (called the *normal equations*)

$$(\mathbf{X}\mathbf{X}')\boldsymbol{\theta} = \mathbf{X}\mathbf{y}.$$

If rank $\mathbf{X}' = p$, then $\mathbf{XX}'$ is nonsingular, and there exists a unique solution

$$\hat{\boldsymbol{\theta}} = (\mathbf{XX}')^{-1}(\mathbf{Xy})$$

for $\boldsymbol{\theta}$. This is called the *full-rank case*. The covariance matrix of $\hat{\boldsymbol{\theta}}$ is

$$\mathrm{cov}(\hat{\boldsymbol{\theta}}) = \sigma^2 (\mathbf{XX}')^{-1}.$$

If rank $\mathbf{X}' = r < p$, this is called the *less-than-full-rank case* and the estimators are not unique.

2. The residual degrees of freedom $v_R = n - r$, where $r = $ rank $\mathbf{X}'$. Hence,

$$\mathrm{SS}_R = \frac{(\mathbf{y} - \mathbf{X}'\hat{\boldsymbol{\theta}})'(\mathbf{y} - \mathbf{X}'\hat{\boldsymbol{\theta}})}{(n-r)} .*$$

EXAMPLE 4.1.1 (*continued*)   The least-squares estimators of β_0 and β_1 are obtained by minimizing

$$S = \sum_{i=1}^{n} (y_i - \beta_0 - \beta_1 x_i)^2$$

with respect to β_0 and β_1. These estimators are uniquely given by [see Eqs. (3.1.5) and (3.1.6)]

$$\hat{\beta}_0 = \bar{y} - \hat{\beta}_1 \bar{x} \quad \text{and} \quad \hat{\beta}_1 = \frac{\sum_{i=1}^{n}(x_i - \bar{x})y_i}{\sum_{i=1}^{n}(x_i - \bar{x})^2}.$$

The unbiased estimator of σ^2 is

$$s^2 = \frac{\sum_{i=1}^{n}(y_i - \hat{\beta}_0 - \hat{\beta}_1 x_i)^2}{(n-2)},$$

where $v_R = n - 2$ in this case. As an application of the Gauss–Markov theorem, note that the least-squares estimator of $\beta_0 + 2\beta_1$ is

$$\widehat{\beta_0 + 2\beta_1} = \hat{\beta}_0 + 2\hat{\beta}_1.$$

This estimator has minimum variance among all unbiased estimators of $\beta_0 + 2\beta_1$ which are linear in $y_1, \ldots, y_n$.

EXAMPLE 4.1.2 (*continued*)   The least-squares estimators of μ_1 and μ_2 are obtained by minimizing

$$S = \sum_{i=1}^{n} (y_i - \mu_1 x_{1i} - \mu_2 x_{2i})^2$$

$$= \sum_{i=1}^{n_1} (y_i - \mu_1 \cdot 1)^2 + \sum_{i=n_1+1}^{n} (y_i - \mu_2 \cdot 1)^2$$

with respect to μ_1 and μ_2. These estimators are uniquely given by

$$\hat{\mu}_1 = \sum_{i=1}^{n_1} \frac{y_i}{n_1} = \bar{y}_1 \quad \text{and} \quad \sum_{i=n_1+1}^{n} \frac{y_i}{n_2} = \bar{y}_2.$$

The residual sum of squares SS_R is

$$SS_R = \left[\sum_{i=1}^{n_1} (y_i - \bar{y}_1)^2 + \sum_{i=n_1+1}^{n_2} (y_i - \bar{y}_2)^2 \right]$$

and $v_R = n_1 + n_2 - 2$. It may be noted that the mean square $s^2 \equiv MS_R = SS_R/v_R$ is the pooled variance s_p^2 discussed in Eq. (2.3.6).

4.1.2 Confidence Intervals

In order to derive confidence intervals for the parameters or for functions of the parameters, we need to make an assumption about the distribution of the errors. The standard procedure is to assume that the errors are normally distributed. Thus, in the remainder of this chapter we assume that

$$e_1, \ldots, e_n \quad \text{are independent} \quad N(0, \sigma^2). \tag{4.1.8}$$

Then, it may be shown that for any linear combination of the parameters $\psi = \sum_{i=1}^{p} c_i \theta_i$, a $100(1-\alpha)\%$ confidence interval for ψ is

$$\hat{\psi} \pm t_{1-(\alpha/2)}(v_R) \sqrt{\hat{V}(\hat{\psi})}, \tag{4.1.9}$$

where $\hat{\psi} = \sum_{i=1}^{p} c_i \hat{\theta}_i$ and $\hat{V}(\hat{\psi})$ is the estimate of the variance of $\hat{\psi}$. Since $\hat{\psi}$ is a linear function of the observations, we may express $\hat{\psi}$ in the form $\hat{\psi} = \sum_{i=1}^{n} a_i y_i$, where a_i are known constants. Hence,

$$V(\hat{\psi}) = \sigma^2 \sum_{i=1}^{n} a_i^2 \tag{4.1.10}$$

and

$$\hat{V}(\hat{\psi}) = s^2 \sum_{i=1}^{n} a_i^2, \tag{4.1.11}$$

where s^2 is the unbiased estimator of σ^2 with v_R degrees of freedom.

If it is desired to obtain $100(1-\alpha)\%$ confidence intervals for more than one linear combination of the parameters, we may apply Eq. (4.1.9) for each. However, the overall confidence level is no longer the nominal $1-\alpha$. It should be recalled that in the one-way analysis of variance discussed in Section 2.4, the solution to this problem was to compute multiple confidence intervals. For the general linear model, there also exist methods for calculating multiple confidence intervals, but they are beyond the scope of this book. Interested readers will find this subject discussed in Scheffé (1959, p. 68). The principle underlying this general method is given in the following starred comment.

COMMENT 4.1.2   *Let $\psi_1 = \mathbf{a}_1'\mathbf{\theta}, ..., \psi_q = \mathbf{a}_q'\mathbf{\theta}$ be q linear functions of the parameters $\mathbf{\theta} = (\theta_1, ..., \theta_p)'$, where $\{\mathbf{a}_1, ..., \mathbf{a}_q\}$ is a linearly independent set of vectors. Also, let $\hat{\psi}_1, ..., \hat{\psi}_q$ be the least-squares estimators for $\psi_1, ..., \psi_q$ and s^2 be the usual unbiased estimator of σ^2 with v_R degrees of freedom. Then if

$$\mathbf{\psi}^{q \times 1} = (\psi_1, ..., \psi_q)' \qquad \text{and} \qquad \hat{\mathbf{\psi}}^{q \times 1} = (\hat{\psi}_1, ..., \hat{\psi}_q)',$$

a $100(1 - \alpha)\%$ *confidence set* for $\mathbf{\psi}$ is given by

$$(\mathbf{\psi} - \hat{\mathbf{\psi}})' \mathbf{B}^{-1} (\mathbf{\psi} - \hat{\mathbf{\psi}}) \leqslant qs^2 F_{1-\alpha}(q, v_R),$$

where

$$\text{Cov}(\hat{\mathbf{\psi}}) = \sigma^2 \mathbf{B}.*$$

EXAMPLE 4.1.1 (*continued*)   Assuming $e_1, ..., e_n$ are independent $N(0, \sigma^2)$, we let $\psi = 0\beta_0 + 1\beta_1 = \beta_1$. The estimator for ψ is

$$\hat{\psi} = \hat{\beta}_1 = \sum_{i=1}^n a_i y_i, \qquad \text{with} \quad a_i = \frac{(x_i - \bar{x})}{\sum_{j=1}^n (x_j - \bar{x})^2}.$$

Hence,

$$\hat{V}(\hat{\psi}) = s^2 \frac{\sum_{i=1}^n (x_i - \bar{x})^2}{[\sum_{j=1}^n (x_j - \bar{x})^2]^2} = \frac{s^2}{\sum_{j=1}^n (x_j - \bar{x})^2}.$$

Thus, we get the $100(1 - \alpha)\%$ confidence interval for β_1 given in Eq. (3.1.14).

EXAMPLE 4.1.2 (*continued*)   Assuming $e_1, ..., e_n$ are independent $N(0, \sigma^2)$, we let $\psi = \mu_1 - \mu_2$. The estimator for ψ is $\hat{\psi} = \hat{\mu}_1 - \hat{\mu}_2 = \bar{y}_1 - \bar{y}_2$. Hence, $a_i = 1/n_1$ for $i = 1, ..., n_1$, and $a_i = 1/n_2$ for $i = n_1 + 1, ..., n$. The estimate of $V(\hat{\psi})$ is $\hat{V}(\hat{\psi}) = s_p^2(1/n_1 + 1/n_2)$. Thus, we obtain the $100(1 - \alpha)\%$ confidence interval for $\mu_1 - \mu_2$ [see Eq. (2.3.7)],

$$(\bar{y}_1 - \bar{y}_2) \pm t_{1-(\alpha/2)}(n_1 + n_2 - 2)\sqrt{s_p^2 \left(\frac{1}{n_1} + \frac{1}{n_2}\right)}.$$

4.1.3 Tests of Hypotheses

In most analysis of variance problems the hypotheses of interest may be written in the form $H_0: \theta_{i_1} = \cdots = \theta_{i_m} = 0$, that is, that m of the p parameters are equal to zero. Without loss of generality, we let these parameters be the last m parameters. Thus, the null hypothesis is $H_0: \theta_{p-m+1} = \cdots = \theta_p = 0$. To obtain the likelihood ratio test of this hypothesis, we first write the *restricted model*

$$y_i = \theta_1 x_{1i} + \cdots + \theta_{p-m} x_{p-m,i} + e_i, \qquad i = 1, ..., n. \qquad (4.1.12)$$

This is the general model after imposing the restrictions of H_0. We then obtain least-squares estimators $\tilde{\theta}_1, \ldots, \tilde{\theta}_{p-m}$ of $\theta_1, \ldots, \theta_{p-m}$. The residual sum of squares under H_0 is

$$\mathrm{SS_R}' = \sum_{i=1}^{n} (y_i - \tilde{\theta}_1 x_{1i} - \cdots - \tilde{\theta}_{p-m} x_{p-m,i})^2 \qquad (4.1.13)$$

and v_R' is its degrees of freedom. Then, the likelihood ratio test statistic is

$$F = \frac{(\mathrm{SS_R}' - \mathrm{SS_R})/(v_R' - v_R)}{\mathrm{SS_R}/v_R} \qquad (4.1.14)$$

and the P value is the area to the right of F under the $F(v_R' - v_R, v_R)$ frequency curve. The quantity $\mathrm{SS_H} = \mathrm{SS_R}' - \mathrm{SS_R}$ is called the *hypothesis sum of squares* and $v_H = v_R' - v_R$ is called the *hypothesis degrees of freedom*. As before, $\mathrm{SS_R}$ is the *residual* (or *error*) *sum of squares* and v_R is the *residual* (or *error*) *degrees of freedom*. Note that the quantity $\mathrm{SS_R}'$ is a measure of how well the restricted model fits the observations. Since $\mathrm{SS_R}' \geqslant \mathrm{SS_R}$, $\mathrm{SS_H}$ is a measure of how much worse the restricted model fits the observations, and F is a measure of how much is lost by assuming H_0 relative to the goodness-of-fit of the original model. The larger F is, the less appropriate the restricted model is. Hence, large values of F dictate rejection of H_0.

In addition to $\mathrm{SS_R}$, v_R, and $\mathrm{MS_R}$, the test statistic F, called the *F ratio*, may be given in the anova table. Such tables will be given in appropriate places throughout the remainder of this chapter.

COMMENTS 4.1.3   1. The fact that Eq. (4.1.14) follows an F distribution comes from the result known as *Cochran's theorem*. A form of this theorem is the following.

Cochran's Theorem   Assume the general linear model of Eq. (4.1.2) with assumptions given by Eqs. (4.1.3) and (4.1.8). Let S_i be a sum of squares with v_i degrees of freedom,[†] $i = 1, \ldots, q$. If $S = \sum_{i=1}^{q} S_i$ is distributed as $\sigma^2 \chi^2(v)$ and if $v_1 + v_2 + \cdots + v_q = v$, then

(a) S_i is distributed as $\sigma^2 \chi^2(v_i)$, $i = 1, \ldots, q$; and
(b) $S_1, \ldots, S_q$ are independent.

---

† *A sum of squares S_i can be written as $\mathbf{y}'\mathbf{A}\mathbf{y}$ where $\mathbf{y}^{n \times 1}$ is an observation vector and $\mathbf{A}^{n \times n}$ is a matrix of known constants. Then, the number of degrees of freedom of S_i is defined to be the rank of $\mathbf{A}$.*

From Section I.2.8 we know that if S_i/σ^2 is $\chi^2(v_i)$, S_j/σ^2 is $\chi^2(v_j)$, and S_i and S_j are independent, then

$$F = \frac{S_i/v_i}{S_j/v_j}$$

has an $F(v_i, v_j)$ distribution.

2. It may be shown that SS_R' is distributed as $\sigma^2\chi^2(v_R')$. Since $SS_R' = SS_R + SS_H$ with $v_R' = v_R + v_H$, we have

(a) SS_R is $\sigma^2\chi^2(v_R)$ and SS_H is $\sigma^2\chi^2(v_H)$;
(b) SS_R and SS_H are independent.

Thus,

$$F = \frac{SS_H/v_H}{SS_R/v_R}$$

is distributed as $F(v_H, v_R)$.

EXAMPLE 4.1.1 (*continued*)   For the simple linear regression model, we have

$$SS_R = \sum_{i=1}^n (y_i - \hat{\beta}_0 - \hat{\beta}_1 x_i)^2 = \sum_{i=1}^n (y_i - \bar{y})^2 - \hat{\beta}_1^2 \sum_{i=1}^n (x_i - \bar{x})^2.$$

Under $H_0: \beta_1 = 0$, the restricted model is $y_i = \beta_0 + e_i$, so that $\tilde{\beta}_0 = \bar{y}$ and

$$SS_R' = \sum_{i=1}^n (y_i - \tilde{\beta}_0)^2 = \sum_{i=1}^n (y_i - \bar{y})^2.$$

The degrees of freedom are $v_R = n - 2$ and $v_R' = n - 1$. Hence, the hypothesis sum of squares is

$$SS_H = (SS_R' - SS_R) = \hat{\beta}_1^2 \sum_{i=1}^n (x_i - \bar{x})^2.$$

Thus,

$$F = \frac{\hat{\beta}_1^2 \sum_{i=1}^n (x_i - \bar{x})^2/1}{\sum_{i=1}^n (y_i - \hat{y}_i)^2/(n-2)}, \qquad \text{where} \quad \hat{y}_i = \hat{\beta}_0 + \hat{\beta}_1 x_i.$$

This is the same test statistic given in Eq. (3.1.11).

EXAMPLE 4.1.2 (*continued*)   For this example, we have

$$SS_R = \sum_{i=1}^{n_1} (y_i - \bar{y}_1)^2 + \sum_{i=n_1+1}^n (y_i - \bar{y}_2)^2.$$

Under $H_0{}^\dagger$: $\mu_1 = \mu_2 = \mu$, the restricted model is $y_i = \mu + e_i$, $i = 1, \ldots, n$. We have $\tilde{\mu} = \bar{y} = (1/n) \sum_{i=1}^{n} y_i$, so that

$$\mathrm{SS_R}' = \sum_{i=1}^{n} (y_i - \bar{y})^2 = \sum_{i=1}^{n_1} (y_i - \bar{y}_1)^2 + \sum_{i=n_1+1}^{n} (y_i - \bar{y}_2)^2 + \frac{n_1 n_2}{n_1 + n_2} (\bar{y}_1 - \bar{y}_2)^2.$$

The degrees of freedom are $v_R = n_1 + n_2 - 2$ and $v_R{}' = n_1 + n_2 - 1$. Hence the hypothesis sum of squares is

$$\mathrm{SS_H} = (\mathrm{SS_R}' - \mathrm{SS_R}) = \frac{n_1 n_2}{n_1 + n_2} (\bar{y}_1 - \bar{y}_2)^2.$$

Thus,

$$F = \frac{\dfrac{n_1 n_2}{n_1 + n_2} (\bar{y}_1 - \bar{y}_2)^2}{s_p{}^2}$$

$$= \frac{(\bar{y}_1 - \bar{y}_2)^2}{s_p{}^2 (1/n_1 + 1/n_2)} = t^2,$$

where t is the test statistic given in Eq. (2.3.9) with $\delta = 0$.

Summary

In this section we have introduced the general linear model as the theoretical framework of the analysis of variance as well as regression analysis. Since all of the anova models introduced in the remainder of this chapter can be written in this form, we can apply the theory of the general linear model to any specific anova model. Thus:

1.   The least-squares procedure is the optimal procedure for estimating the parameters of the anova model. The resulting estimators are unbiased and have minimum variance among all other unbiased estimators which are linear in the observations.

2.   For any anova model the usual unbiased estimator of the error variance σ^2 is given in the anova table by the residual mean square $\mathrm{MS_R}$. This quantity is sometimes denoted by s^2—its more familiar notation. This estimator of the variance is used in forming a confidence interval [Eq. (4.1.9)] for any linear combination $\sum_{i=1}^{p} c_i \theta_i$ of the parameters. The number of degrees of freedom of the t distribution involved in this interval is the same as the residual degrees of freedom v_R.

3.   Each test of a hypothesis concerning the parameters of an anova model reduces to an F test. Each F test is a ratio of mean squares usually summarized in the anova table.

† This hypothesis is not of the form $H_0: \theta_i's = 0$. However, if we write $\mu_1 = \theta_1$, $\mu_2 = \theta_1 + \theta_2$, then we are testing $H_0: \theta_2 = 0$.

4.2 One-Way Analysis of Variance

We now discuss the simplest analysis of variance model—the *one-way analysis of variance* (also called the *one-way layout* or the *one-way classification*). This model, previously discussed in Section 2.4, is reconsidered here to develop some of the concepts of the analysis of variance. Recall from that section that we were given I subpopulations which could also be viewed as I strata from a larger population. Denoting the mean of the ith subpopulation by μ_i, $i = 1, ..., I$, the major concern of that section was to estimate these means from random samples from the I subpopulations and then to test hypotheses about them. To do this, we imposed the conditions that the distribution of each subpopulation was normal and that the I variances were equal. Thus, we have I normal subpopulations $N(\mu_1, \sigma^2), ..., N(\mu_I, \sigma^2)$, and we express the above conditions by the equation

$$y_{ij} = \mu_i + e_{ij}, \qquad j = 1, ..., J_i, \quad i = 1, ..., I, \qquad (4.2.1)$$

where y_{ij} is the jth observation from the ith subpopulation, and e_{ij} are independent $N(0, \sigma^2)$ error variables. This equation is one form of the *one-way anova model*.

In many situations it is desirable to express the ith mean μ_i as a sum of an *overall mean* μ and a *differential* (or *main*) *effect* α_i specific to the ith subpopulation. This may be achieved by defining

$$\mu = \frac{1}{n} \sum_{i=1}^{I} J_i \mu_i, \qquad (4.2.2)$$

where $n = \sum_{i=1}^{I} J_i$, and

$$\alpha_i = \mu_i - \mu. \qquad (4.2.3)$$

Thus, we can write the one-way anova model as

$$y_{ij} = \mu + \alpha_i + e_{ij}, \qquad j = 1, ..., J_i, \quad i = 1, ..., I, \qquad (4.2.4)$$

where e_{ij} are independent $N(0, \sigma^2)$ error terms. This equation is another form of the *one-way anova model* and is the form that we use in this chapter.

In the terminology of the analysis of variance, this one-way anova model arises from an experimental design involving a factor. Roughly speaking, a *factor A* is a basis for classification of a population of experimental units. Let Y be a random variable defined for this population, and let μ be its mean. The population is then divided into I subpopulations such that each subpopulation is represented by a *level i* of the factor, $i = 1, ..., I$. If $\mu_i = \mu + \alpha_i$ is the mean of Y for the ith subpopulation, then α_i is the differential effect of the ith level of the factor, $i = 1, ..., I$. For each level i, we observe Y for J_i randomly selected experimental units. Thus, we have a random sample $y_{i1}, ..., y_{iJ_i}$ from the ith subpopulation where y_{ij} is the observation made on the jth experimental unit "subjected" to the ith level of the factor, $j = 1, ..., J_i$, $i = 1, ..., I$. Thus, if each subpopulation is assumed to be normal with variance σ^2, then the above one-way anova model describes this situation. Each observation y_{ij} is equal to the sum of an overall mean μ (common to all I levels of the factor), a differential effect α_i specific to the ith level of the factor, and a random error variable e_{ij}. Two examples will clarify these concepts.

EXAMPLE 4.2.1   In an x-ray experiment, rats are given doses of 0, 100, 200, and 300 roentgens. The factor A is radiation, there are $I = 4$ levels of the factor corresponding to 0, 100, 200, and 300 roentgens, and the ith subpopulation consists of those rats (experimental units) receiving the ith dose, $i = 1, ..., 4$. The doses are indexed by 1 to 4, so that 0 roentgens is the first level, ..., and 300 roentgens is the fourth level of the factor. Let Y be the area of burned skin in the rat after radiation. Then μ_i is the mean burned area for the ith level of radiation, and α_i is the differential effect of the ith level of radiation, $i = 1, ..., 4$.

EXAMPLE 4.2.2   In a diet experiment, men all weighing approximately 200 pounds and all having similar builds are given one of three diets. The factor A is diet, there are $I = 3$ levels of the factor corresponding to the particular diet, and the ith subpopulation consists of those men (experimental units) following the ith diet. The diets are arbitrarily indexed from 1 to 3. Let Y be the weight loss after following the diet for one month. Then μ_i is the mean weight loss for the ith diet and α_i is the differential effect of the ith diet on weight loss, $i = 1, 2, 3$.

In the first example the levels have some physical significance, while in the second example the levels correspond to categories. In either case if a factor has I levels, we represent these levels by the integers $1, 2, ..., I$.

For the remainder of this chapter we will view analysis of variance models as arising from a design which involves one or more factors. Furthermore, each factor will be interpreted either as *Model I* or as *Model II*. A factor is said to be Model I if the particular subpopulations represented by the levels of the factor are those of interest to the experimenter. That is, if the experiment were repeated, random samples from these same subpopulations would be analyzed. In Example 4.2.1, radiation is a Model I factor if the experimenter is interested in the reaction of rats to the particular doses 0, 100, 200, and 300 roentgens. Similarly, the diet factor of Example 4.2.2 is Model I if the three diets are the ones of interest. In contrast to this, a factor is said to be Model II if the subpopulation represented by the levels of the factor are selected randomly from a very large (infinite) number of subpopulations. In other words, if the experiment were repeated, it would be most likely that random samples from different subpopulations would be analyzed. The radiation factor of Example 4.2.1 is Model II if 4 dose levels are picked at random and the rats are subjected to these levels.

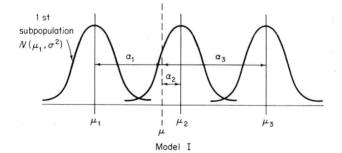

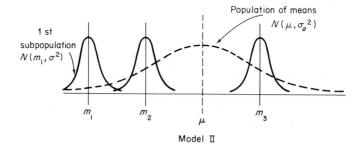

FIG. 4.2.1. *Comparison of a Model I and Model II factor. Model I: test H_0: $\mu_1 = \mu_2 = \mu_3 = \mu$, or $\alpha_1 = \alpha_2 = \alpha_3 = 0$; μ_1, μ_2, and μ_3 are selected by design. Model II: test H_0: $\sigma_a^2 = 0$; m_1, m_2, and m_3 are randomly selected.*

Figure 4.2.1 is a representation of a Model I factor and a Model II factor each with $I = 3$ levels. For the Model II case, m_i is the mean of the randomly chosen subpopulation corresponding to the ith level, $i = 1, 2, 3$ and σ_a^2 is the variance of the population of means corresponding to the levels of the factor. The next two sections discuss these two cases, and the reader is urged to refer to this figure for clarification. Section 4.2.1 discusses the *fixed effects* one-way anova model which is the case where factor A is Model I. Section 4.2.2 then discusses the *random effects* one-way anova model which is the case where factor A is Model II.

4.2.1 The Fixed Effects Model

The *fixed effects* (or *Model I*) *one-way analysis of variance model* is given either by Eq. (4.2.1) or (4.2.4). In the latter form, we see why the name "fixed effects" is used to describe the model. In this case each observation is expressed as a sum of an overall mean μ and a differential "effect" α_i which is "fixed" in the sense that the subpopulation with mean $\mu_i = \mu + \alpha_i$ is fixed by the experimenter. Since either form of the model may be put into the form of the general linear model [Eq. (4.1.1)], we may use the theory of Section 4.1 to derive estimators of the parameters and to test hypotheses about them. Using Eq. (4.2.1) as the model, we obtain the least-squares estimator $\hat{\mu}_i$ for μ_i, $i = 1, ..., I$. The least-squares estimators for the parameters μ and α_i of Eq. (4.2.4) are $\hat{\mu} = (1/n) \sum_{i=1}^{I} J_i \hat{\mu}_i$ and $\hat{\alpha}_i = \hat{\mu}_i - \hat{\mu}$, $i = 1, ..., I$.

Using Eq. (4.2.4) as the model, we obtain the least-squares estimators $\hat{\mu}$ and $\hat{\alpha}_i$ for μ and α_i, $i = 1, ..., I$. The least-squares estimator for the parameter μ_i of Eq. (4.2.1) is then $\hat{\mu}_i = \hat{\mu} + \hat{\alpha}_i$, $i = 1, ..., I$. Thus, it is seen that the estimators of the parameters of one form of the model will give the estimators of the parameters of the other form of the model. Since the anova table is the same for either form of the one-way anova model, the estimator for σ^2 is the same, and the F ratio tests $H_0: \mu_1 = \cdots = \mu_I = \mu$ for Eq. (4.2.1) or $H_0: \alpha_1 = \cdots = \alpha_I = 0$ for Eq. (4.2.4). Note that one hypothesis implies the other hypothesis. Thus, working with one form of the model is equivalent to working with the other form. We choose to work with Eq. (4.2.4).

To obtain unique least-squares estimators we need to impose one side condition on the parameters $\alpha_1, ..., \alpha_I$. It is customary to require that the weighted sum of differential effects is zero, that is,

$$\sum_{i=1}^{I} J_i \alpha_i = 0. \tag{4.2.5}$$

Thus, we minimize the sum of squares

$$S = \sum_{i=1}^{I} \sum_{j=1}^{J_i} (y_{ij} - \mu - \alpha_i)^2 \tag{4.2.6}$$

with respect to μ and α_i, $i = 1, ..., I$ subject to the condition given by Eq. (4.2.5). The unique least-squares estimators are then

$$\hat{\mu} = \bar{y}_{..} = \frac{1}{n} \sum_{i=1}^{I} \sum_{j=1}^{J_i} y_{ij} \tag{4.2.7}$$

and

$$\hat{\alpha}_i = \bar{y}_{i.} - \bar{y}_{..} = \frac{1}{J_i} \sum_{j=1}^{J_i} y_{ij} - \bar{y}_{..}, \qquad i = 1, ..., I. \tag{4.2.8}$$

The "dot" notation implies averaging over the subscript that the dot replaces. The usual unbiased estimator for σ^2 is then given by

$$MS_R \equiv s^2 = \sum_{i=1}^{I} \sum_{j=1}^{J_i} \frac{(y_{ij} - \bar{y}_{i.})^2}{(n - I)}. \tag{4.2.9}$$

A typical packaged one-way anova program prints the least-squares estimates $\hat{\mu}_i$ and $\hat{\mu}$. From these quantities the user obtains the estimates of α_i by

$$\hat{\alpha}_i = \hat{\mu}_i - \hat{\mu}. \tag{4.2.10}$$

In addition, the program calculates and prints an anova table similar to that shown in Table 4.2.1. The entries in the table typically consist of the *sums of squares* (SS), *degrees of freedom* (v), and *mean squares* (MS) for two sources of variation—the *between levels* (or *groups*) and the *within levels* (or *groups*). The latter quantity is also called the *residual* (or *error*). Sometimes the sum of squares and degrees of freedom are printed for the *total* variation. (These quantities are the sums of the between and within quantities.) From the table we can obtain s^2; the numerator in Eq. (4.2.9) is the residual sum of squares SS_R, the denominator is the residual degrees of freedom v_R, and s^2 is the residual mean square MS_R.

TABLE 4.2.1. *One-Way Analysis of Variance Table*

Source of variation	Sum of squares	Degrees of freedom	Mean square	F ratio
Between levels (groups)	$SS_B = \sum_{i=1}^{I} J_i(\bar{y}_{i.} - \bar{y}_{..})^2$	$v_B = I - 1$	$MS_B = \dfrac{SS_B}{v_B}$	$F = \dfrac{MS_B}{MS_R}$
Within levels (groups)	$SS_R = \sum_{i=1}^{I}\sum_{j=1}^{J_i}(y_{ij} - \bar{y}_{i.})^2$	$v_R = n - I$	$MS_R = \dfrac{SS_R}{v_R}$	
Total	$SS_T = \sum_{i=1}^{I}\sum_{j=1}^{J_i}(y_{ij} - \bar{y}_{..})^2$	$v_T = n - 1$		

To test the hypothesis $H_0: \alpha_1 = \cdots = \alpha_I = 0,$[†] that is, that the I differential effects are all equal to zero, we use the theory of Section 4.1.3. Thus, it can be shown that we test H_0 by calculating the F ratio which is the ratio of the between mean square to the within mean square. The P value is the area to the right of F under the $F(I-1, n-I)$ frequency curve. Acceptance of this hypothesis implies $H_0: \mu_1 = \cdots = \mu_I = \mu$, that is, all I subpopulation means are equal to the overall mean.

EXAMPLE 4.2.2 (*continued*)   In the diet experiment, $J_1 = 9$ men followed the first diet, $J_2 = 6$ men followed the second diet, and $J_3 = 8$ men followed the third diet. The weight losses after one month are given in Table A.

The least-squares estimates of the means are $\hat{\mu}_1 = 10.23$, $\hat{\mu}_2 = 18.83$, $\hat{\mu}_3 = 16.69$, and $\hat{\mu} = 14.71$. Hence, the estimated differential effects of the diets are $\hat{\alpha}_1 = -4.48$, $\hat{\alpha}_2 = 4.12$, and $\hat{\alpha}_3 = 1.98$. The analysis of variance table is given in Table B. From this table we obtain $MS_R = 11.79$ as an estimate of σ^2. To test $H_0: \alpha_1 = \alpha_2 = \alpha_3 = 0$, we compare $F = 13.31$ with the percentiles of $F(2, 20)$. We reject this hypothesis at $P < 0.001$.

[†] In the remainder of this chapter, hypotheses of this form will be written simply as "H_0: all $\alpha_i = 0$."

TABLE A

| | Diet factor | |
Diet 1	Diet 2	Diet 3
9.40	22.84	17.35
9.48	15.32	16.36
7.56	11.04	15.88
11.52	17.92	14.28
11.56	19.68	18.60
12.12	26.20	19.32
11.36		14.20
4.60		17.52
14.48		

TABLE B

Source of variation	Sum of squares	Degrees of freedom	Mean square	F ratio
Between levels	313.87	2	156.93	13.31
Within levels	235.83	20	11.79	
Total	549.70	22		

4.2.2 The Random Effects Model

For the second interpretation of the one-way layout, *the random effects model* (*components of variance model* or *Model II*), we assume that we randomly select I subpopulations from an infinite population of possible subpopulations. Each subpopulation is indexed from 1 to I, the ith subpopulation corresponding to the ith level of the factor. From the ith subpopulation we randomly select J_i experimental units and observe $y_{i1}, ..., y_{iJ_i}$, $i = 1, ..., I$. These observations are assumed to be normally distributed about a mean m_i with variance σ^2, $i = 1, ..., I$ (see Fig. 4.2.1). It is assumed that $m_1, ..., m_I$ represents a random sample from a population normally distributed about a mean μ with variance σ_a^2. We define the ith *differential* (or *main*) *effect* of the factor to be $a_i = m_i - \mu$. Note that this effect (as contrasted to its counterpart α_i in the fixed effects model) is now a *random variable* normally distributed about a mean 0 and variance σ_a^2. Thus, the *random effects one-way anova model* is

$$y_{ij} = \mu + a_i + e_{ij}, \qquad j = 1, ..., J_i, \quad i = 1, ..., I, \qquad (4.2.11)$$

where a_i are independent $N(0, \sigma_a^2)$, e_{ij} are independent $N(0, \sigma^2)$, and a_i and e_{ij} are independent, $j = 1, ..., J_i$, $i = 1, ..., I$.

In the Model I case we are interested in estimating the differential effect α_i of the ith level of the factor and in testing the hypothesis that these effects are all zero. In the Model II case, we are not interested in the specific effect a_i, but rather, the variance σ_a^2 of the population of differential effects. Thus, we are interested in estimating μ and the two *components of variance* σ^2 and σ_a^2, and in testing the hypothesis that $H_0: \sigma_a^2 = 0$, that is, that there is no variation due to the factor.

To estimate σ_a^2 and to test $H_0: \sigma_a^2 = 0$, we need additional quantities called the *expected mean squares* (abbreviated as EMS). These quantities are derived for each source of variation (excluding the *total*) and are, by definition, the expected values of the mean squares under the original model. The EMS are also derived for the fixed effects model, but they are not needed for the determination of the tests for that model.

The EMS are usually not computed by a packaged program, so that a user must refer to this or another book to determine the form of the EMS. These quantities are usually presented in a table such as Table 4.2.2. This table lists the

TABLE 4.2.2. *The EMS for the One-Way Anova Model (Model I and Model II)*

Source of variation	EMS Model I	EMS Model II
Between levels (groups)	$\sigma^2 + \dfrac{\sum_{i=1}^{I} J_i \alpha_i^2}{I-1}$	$\sigma^2 + k\sigma_a^2$ [see Eq. (4.2.12)]
Within levels (groups) (residual)	σ^2	σ^2

EMS for the between and within levels for both the fixed and the random effects models. Note that the residual EMS is σ^2 in both cases. (This will be true for all anova models.) Note also that if $\alpha_1 = \cdots = \alpha_I = 0$ in the Model I case, or if $\sigma_a^2 = 0$ in the Model II case, then both the between and within levels mean squares are estimators of σ^2. For the Model II case, the EMS for between levels is

$$\sigma^2 + k\sigma_a^2,$$

where

$$k = \frac{1}{I-1}\left[\sum_{i=1}^{I} J_i - \sum_{i=1}^{I} J_i^2 \bigg/ \sum_{i=1}^{I} J_i\right]. \tag{4.2.12}$$

When $J_1 = \cdots = J_I = J$, then $k = J$.

From these quantities we obtain an unbiased estimator for σ_a^2. Since the between EMS minus the within EMS is equal to $k\sigma_a^2$, the unbiased estimator for σ_a^2 is

$$\hat{\sigma}_a^2 = \frac{MS_B - MS_R}{k}, \qquad (4.2.13)$$

where MS_B and MS_R are the between and the within mean squares, respectively (see Table 4.2.1).

To test the hypothesis $H_0: \sigma_a^2 = 0$, it may be shown that we can use the same F ratio as in the fixed effects case. For more complicated models the EMS determine the appropriate F ratio for a given test of hypothesis (see Comments 4.2.1).

EXAMPLE 4.2.2 (*continued*)   We now assume for the sake of illustration, that the three diets were chosen at random from a very large number of diets. Thus, the model is given by Eq. (4.2.11). The least-squares estimate of μ is $\hat{\mu} = 14.71$, and the unbiased estimate for σ^2 is $MS_R = 11.79$. The unbiased estimate for σ_a^2, the variance of the population of differential effects of the diet factor, is

$$\hat{\sigma}_a^2 = \frac{156.93 - 11.79}{k} = 19.2, \qquad \text{where} \quad k = \frac{1}{2}\left[23 - \frac{181}{23}\right] = 7.57.$$

To test $H_0: \sigma_a^2 = 0$, we compare, as before, $F = 13.31$ with the percentiles of $F(2, 20)$. We reject this hypothesis at $P < 0.001$.

COMMENTS 4.2.1   1.   For any anova model with $m \geqslant 1$ factors, if all of the m factors are Model I, then the anova model is called a *fixed effects model* or *Model I*. Similarly, if all of the factors are Model II, then the model is a *random effects model* or *Model II*. When some factors are Model I, while others are Model II, then we have a *mixed model*.

2.   For an anova model with $m \geqslant 1$ factors, each level of the factor is represented by a term in the anova model. If a given factor is Model I, then the corresponding term is a constant parameter; otherwise, it is a random variable.

3.   The formulas for the sums of squares and degrees of freedom (and hence, the mean squares) which appear in the anova table are the same regardless of whether the model is Model I, II, or mixed.

4.   The quantities to be estimated are different depending on the type of factors in the anova design. For a Model I factor, we estimate the constant parameters in the model corresponding to its levels. For a Model II factor, we estimate the overall mean μ and the variance of the random variables in the model representing its levels.

5.   For a random effects or mixed model with $m \geqslant 1$ factors, we view the anova model as a fixed effects model to obtain least-squares estimators of the constant parameters and to test hypotheses about them. To obtain estimators of the components of variance and to test hypotheses about them, we refer to this or some other book to obtain the EMS for each source of variation. As will be seen in the next sections, this determines the formula for the estimates of the components of variance, as well as the correct F ratios to test hypotheses about them.

6.   For a fixed effects model with $m \geqslant 1$ factors, the denominator for the F ratio is *always* the residual mean square $\mathrm{MS_R}$.

4.3 *Two-Way Analysis of Variance*

In this section we consider various models for analyzing the differential effects of two factors A and B. We will be interested in two types of relations among the factors, usually called *crossing* and *nesting*. Two factors A and B are said to be *crossed* (denoted by $A \times B$) if all possible combinations of levels of the factors are present in the experimental design. That is, if there are I levels of A and J levels of B, then there is at least one observation for each of the IJ combinations of levels of the two factors. If i is a level of A and j is a level of B, then the combination ij will sometimes be called the ijth *cell*, $i = 1, ..., I, j = 1, ..., J$. In each cell we randomly assign K_{ij} experimental units and observe a random variable Y. Equivalently, we may assign a single experimental unit to the ijth cell and make K_{ij} measurements on the random variable Y. In either case, for the ijth cell we have a random sample

$$y_{ij1}, y_{ij2}, ..., y_{ijK_{ij}}, \qquad i = 1, ..., I, \quad j = 1, ..., J.$$

EXAMPLE 4.3.1   In an experiment, 10 rats were given one of three doses of radiation—0, 100, or 300 roentgens—and one of two diets—high protein or low protein. Here, factor A is radiation with $I = 3$ levels; $i = 1$ refers to 0 roentgens, ..., $i = 3$ refers to 300 roentgens. Factor B is diet with $J = 2$ levels; $j = 1$ refers to high protein, $j = 2$ refers to low protein. Thus, $K_{11} = 10$ rats (experimental units) were given no radiation and a high protein diet, $K_{12} = 10$ rats were given no radiation and a low protein diet, ..., $K_{32} = 10$ rats were given 300 roentgens and a low protein diet. For each cell, the percent increase in weight Y was measured. Thus, the cells of this experiment are denoted by the pairs of subscripts 11, 12, 21, 22, 31, and 32.

A model with two crossed factors is called a *two-way layout*, a *two-way classification*, or a *factorial* model with two factors. Section 4.3.1 discusses this design for the fixed effects interpretation (that is, both factors are Model I) and the random effects case (that is, both factors are Model II). Section 4.3.2 presents the case when all $K_{ij} = 1$. Section 4.3.3 discusses a mixed model (one factor is Model I, the other is Model II) which is the model for a *randomized blocks* design.

The other relation among two factors, that is, nesting, is defined as follows. A factor B is said to be *nested* within A if each level of B occurs with at most one level of A. We denote this by $B(A)$ and we say that A nests B, A is the *nesting* factor, and B is the *nested* factor. If there are I levels of A and J levels of B, then there are less than IJ combinations of levels for which observations are made. In those cells for which the combinations of the factors are defined, we randomly assign K_{ij} experimental units and observe a random variable Y. Equivalently, we may assign a single experimental unit and make K_{ij} measurements on Y. Thus, in either case if the jth level of B is nested within the ith level of A we have the random sample $y_{(i)j1}, ..., y_{(i)jK_{ij}}$. Parentheses are sometimes used to denote the nesting relationship—the index of the nesting factor appears in parentheses. When the nesting relationship is understood, we will drop the parentheses.

EXAMPLE 4.3.2   In an experiment, 12 female mosquito pupae are divided at random among three cages, and the wing length is measured twice on each mosquito. In this case factor A is the cages with $I = 3$ levels—$i = 1$ refers to the first cage, ..., $i = 3$ refers to the third cage. Factor B is the pupae (experimental unit) with $J = 12$ levels—j refers to the jth pupae. Finally, $K_{ij} = 2$ determinations of $Y =$ wing length are made on each pupae. Suppose that the random assignment results in assigning pupae 2, 3, 7, and 10 to cage 1; 1, 4, 8, and 12 to cage 2; and 5, 6, 9, and 11 to cage 3. Then in anova terminology the 1,1th cell represents the first cage and the second pupae; the 1,2th cell represents the first cage and the third pupae; ...; the 3,4th cell represents the third cage and the eleventh pupae. Similarly, y_{111} is the first measurement in the 1,1th cell; y_{232} is the second determination in the 2,3th cell; and so forth. An important aspect of the nesting design is that there is no correspondence between the same level of the nested factor for different levels of the nesting factor. Thus, for example, there is no correspondence between the pupae in the 1,2th and 2,2th cells. This is in contrast to the crossed relationship in which 2 represents the same level in the 1,2th and 2,2th cells.

A model with one factor nested within another factor is called a *two-way nested* or *two-way hierarchical* model. Section 4.3.4 discusses this design for the fixed effects interpretation (that is, both factors are Model I) and the random effects case (both factors are Model II). Section 4.3.5 compares all of the models of this chapter in a hypothetical experimental situation.

We assume that the anova tables are generated using a packaged anova program. Most of these programs, called (*m-way*) *factorial anova* programs, require constant K, that is $K_{ij} = K$ for all i, j. Furthermore, some programs require that $K = 1$, while others permit $K \geqslant 1$. The details of these programs are discussed in Section 4.4. Section 4.5 discusses a method of generating the anova table when the K_{ij} are not all equal. In the present section we assume constant $K \geqslant 1$. When $K > 1$, we say the experiment has been *replicated* K times.

4.3.1 The Replicated Two-Way Layout—Fixed and Random Effects

In this section we assume that we have a factor A at I levels and a factor B at J levels. Each combination of levels is replicated exactly $K > 1$ times. The case of unequal replicates is discussed in Section 4.5. Denote by y_{ijk} the value of Y for the kth replication in the ijth cell, $i = 1, ..., I$, $j = 1, ..., J$, $k = 1, ..., K$. If both factors are Model I, then the *fixed effects two-way layout* is

$$y_{ijk} = \mu + \alpha_i + \beta_j + (\alpha\beta)_{ij} + e_{ijk}, \qquad i = 1, ..., I, \quad j = 1, ..., J, \quad k = 1, ..., K,$$

(4.3.1)

where μ is an overall mean, α_i is the ith *differential* (or *main*) *effect* of factor A, and β_j is the jth differential effect of factor B. The quantity $(\alpha\beta)_{ij}$ is called the *(two-way) interaction* of the ith level of A and the jth level of B. This quantity is defined to be the differential effect of the combination of the ith level of A and the jth level of B not accounted for by the sum $\mu + \alpha_i + \beta_j$. If $(\alpha\beta)_{ij} = 0$ for all i and j, then this model is said to be *additive*. As will be seen we may test for additivity. Finally, the e_{ijk} are independent $N(0, \sigma^2)$ error variables.

We now discuss the estimation of the parameters, that is, the differential effects and interactions. Since the least-squares estimators are not unique for this model, it is customary to impose side conditions on all of the differential effects. Thus, we assume

$$\sum_{i=1}^{I} \alpha_i = 0, \qquad \sum_{j=1}^{J} \beta_j = 0,$$

$$\sum_{i=1}^{I} (\alpha\beta)_{ij} = 0 \quad \text{for } j = 1, ..., J; \qquad \sum_{j=1}^{J} (\alpha\beta)_{ij} = 0 \quad \text{for } i = 1, ..., I.$$

(4.3.2)

The least-squares estimators for the parameters are then uniquely given by

$$\hat{\mu} = \bar{y}_{...}, \qquad \hat{\alpha}_i = \bar{y}_{i..} - \bar{y}_{...}, \qquad \hat{\beta}_j = \bar{y}_{.j.} - \bar{y}_{...},$$

$$\widehat{(\alpha\beta)}_{ij} = \bar{y}_{ij.} - \bar{y}_{i..} - \bar{y}_{.j.} + \bar{y}_{...} \quad \text{for } i = 1, ..., I \text{ and } j = 1, ..., J.$$

(4.3.3)

These quantities are obtained from a packaged factorial anova program. The $\bar{y}_{i..}$ are sometimes called *row means*, the $\bar{y}_{.j.}$ *column means*, and $\bar{y}_{...}$ the *grand mean.*

TABLE 4.3.1. *Replicated Two-Way Anova Table*

Source of variation	Sum of squares	Degrees of freedom	Mean square
Factor A	$SS_A = JK \sum_{i=1}^{I} (\bar{y}_{i..} - \bar{y}_{...})^2$	$v_A = I - 1$	$MS_A = \dfrac{SS_A}{v_A}$
Factor B	$SS_B = IK \sum_{j=1}^{J} (\bar{y}_{.j.} - \bar{y}_{...})^2$	$v_B = J - 1$	$MS_B = \dfrac{SS_B}{v_B}$
Interaction AB	$SS_{AB} = K \sum_{i=1}^{I} \sum_{j=1}^{J} (\bar{y}_{ij.} - \bar{y}_{i..} - \bar{y}_{.j.} + \bar{y}_{...})^2$	$v_{AB} = (I-1)(J-1)$	$MS_{AB} = \dfrac{SS_{AB}}{v_{AB}}$
Residual R (error)	$SS_R = \sum_{i=1}^{I} \sum_{j=1}^{J} \sum_{k=1}^{K} (y_{ijk} - \bar{y}_{ij.})^2$	$v_R = IJ(K-1)$	$MS_R = \dfrac{SS_R}{v_R}$
Total	$SS_T = \sum_{i=1}^{I} \sum_{j=1}^{J} \sum_{k=1}^{K} (y_{ijk} - \bar{y}_{...})^2$	$v_T = IJK - 1$	

In addition to these quantities, the program computes and prints an anova table similar to that of Table 4.3.1. This table includes the sums of squares, degrees of freedom, and mean squares for sources of variation due to the residual, the factors A and B, and the interaction of A and B. The interaction source is usually denoted by AB (or $A \times B$). From Table 4.3.1 we estimate the error variance σ^2 by the residual mean square MS_R. We may also test hypotheses concerning the differential effects. These tests, the appropriate test statistics and the degrees of freedom are summarized in Table 4.3.2. The P value is the area to the right of F under the $F(v_1, v_2)$ frequency curve. It should be noted that the sums of squares SS_{AB} and SS_R may be added together (that is, *pooled*) in certain cases. We discuss pooling in Comments 4.3.1.

TABLE 4.3.2. *Tests of Hypotheses for the Replicated Fixed Effects Two-Way Layout*

H_0: all $(\alpha\beta)_{ij} = 0$ (no interaction effects)	H_0: all $\alpha_i = 0$ (no factor A main effects)	H_0: all $\beta_j = 0$ (no factor B main effects)
$F = \dfrac{MS_{AB}}{MS_R}$	$F = \dfrac{MS_A}{MS_R}$	$F = \dfrac{MS_B}{MS_R}$
$v_1 = (I-1)(J-1)$	$v_1 = I - 1$	$v_1 = J - 1$
$v_2 = IJ(K-1)$	$v_2 = IJ(K-1)$	$v_2 = IJ(K-1)$

COMMENTS 4.3.1   1. It is customary to test H_0: all $(\alpha\beta)_{ij} = 0$ first. If this test is nonsignificant, then the investigator may: (a) proceed with testing main effects as in columns 2 and 3 of Table 4.3.2; or (b) "pool" the interaction and residual sums of squares and degrees of freedom to obtain another estimate of σ^2. In this case, the pooled sum of squares is $SS_P = SS_R + SS_{AB}$, and the pooled number of degrees of freedom is $v_P = v_R + v_{AB}$. Then $MS_P = SS_P/v_P$ is the estimate of σ^2, and H_0: all $\alpha_i = 0$ is tested by $F = MS_A/MS_P$ and H_0: all $\beta_j = 0$ is tested by $F = MS_B/MS_P$. For either test the P value is the area to the right of F under the $F(v_1, v_P)$ frequency curve.

2. Investigators tend to fall into three categories—big-time poolers, small-time poolers and nonpoolers. Nonpoolers never pool regardless of the outcome of the test for interactions. Big-time poolers always pool when the test for interactions is nonsignificant. Small-time poolers pool when the P value for that hypothesis is large, $P > 0.5$, say. Since there is no decisive rule to dictate a unique decision, it is up to the experimenter to decide his "pool state." (Dr. Gesund, being a "nonextremist," is a small-time pooler.)

EXAMPLE 4.3.3   In an experiment, it was desired to evaluate the effect of $I = 4$ drugs (factor A) crossed with $J = 3$ experimentally induced diseases (factor B). Each disease–drug combination was applied to $K = 6$ randomly selected dogs (experimental units). The measurement Y to be analyzed was the increase in systolic pressure (mm Hg) due to the treatment. The data collected are given in Table A. A packaged factorial analysis of variance program was used to obtain the analysis of variance table, Table B.

TABLE A

Drug	Disease		
	1	2	3
1	42, 44 36, 13 19, 22	33, 40 26, 34 33, 21	31, −3 19, 25 25, 24
2	28, 40 23, 34 42, 13	31, 34 33, 31 33, 36	3, 26 28, 32 4, 16
3	28, 21 1, 29 6, 19	−4, 11 9, 7 1, −6	21, 1 2, 9 3, 9
4	24, 19 9, 22 −2, 15	27, 12 12, −5 16, 15	22, 7 25, 5 12, 7

TABLE B

Source of variation	Sum of squares	Degrees of freedom	Mean square
Factor A (drugs)	4649.6	3	1549.9
Factor B (disease)	746.1	2	373.0
Interaction AB	1053.6	6	175.6
Residual	5915.5	60	98.6
Total	12364.8	71	

Using Table 4.3.2, the tests of hypotheses, F statistics, and P values are given in Table C. Therefore, at the $\alpha = 0.05$ level, the experimenter concluded that there are no significant drug–disease interactions, but that there are significant drug and disease differential effects. Since the P value for interactions is only

TABLE C

H_0: all $(\alpha\beta)_{ij} = 0$ (no drug–disease interactions)	H_0: all $\alpha_i = 0$ (no drug differential effects)	H_0: all $\beta_j = 0$ (no disease differential effects)
$F = 1.8$	$F = 15.8$	$F = 3.8$
$v_1 = 6$	$v_1 = 3$	$v_1 = 2$
$v_2 = 60$	$v_2 = 60$	$v_2 = 60$
$P \simeq 0.10$	$P < 0.001$	$P < 0.05$

$P \simeq 0.10$, a small-time pooler would decide not to pool. The big-time pooler, however, would compute (according to Comment 4.3.1.1), $SS_P = 1053.6 + 5915.5 = 6969.1$, $v_P = 6 + 60 = 66$ and hence, $MS_P = 105.6$. The F values for testing main effects are $F = 1549.9/105.6 = 14.7$ and $F = 373.0/105.6 = 3.53$, respectively, which are again significant. Finally, note that the estimate of the variance σ^2 is $MS_R = 98.6$ if we do not pool and $MS_P = 105.6$ if we pool.

When factors A and B are both Model II, then the appropriate model is the *random effects two-way layout* given by

$$y_{ijk} = \mu + a_i + b_j + (ab)_{ij} + e_{ijk}, \quad i = 1, ..., I, \quad j = 1, ..., J, \quad k = 1, ..., K,$$

$$(4.3.4)$$

where μ is an overall mean, a_i are independent $N(0, \sigma_a^2)$, b_j are independent $N(0, \sigma_b^2)$, $(ab)_{ij}$ are independent $N(0, \sigma_{ab}^2)$, and e_{ijk} are independent $N(0, \sigma^2)$. The random variables a_i, b_j, $(ab)_{ij}$, and e_{ijk} are all assumed to be independent of each other.

For this model we have four components of variance—$\sigma_a{}^2$, $\sigma_b{}^2$, σ_{ab}^2, and σ^2—attributable to the sources of variation A, B, AB, and R, respectively. To determine unbiased estimators for the first three components of variance as well as the appropriate test statistics for testing that these components are zero, we need the EMS for each source of variation. Table 4.3.3 lists these quantities for both the fixed (Model I) and random (Model II) effects cases.

TABLE 4.3.3. *The EMS for the Replicated Fixed and Random Effects Two-Way Layout*

Source of variation	EMS Model I	EMS Model II
A	$\sigma^2 + \dfrac{JK\sum_i \alpha_i^2}{I-1}$	$\sigma^2 + K\sigma_{ab}^2 + JK\sigma_a^2$
B	$\sigma^2 + \dfrac{IK\sum_j \beta_j^2}{J-1}$	$\sigma^2 + K\sigma_{ab}^2 + IK\sigma_b^2$
AB	$\sigma^2 + \dfrac{K\sum_i \sum_j (\alpha\beta)_{ij}^2}{(I-1)(J-1)}$	$\sigma^2 + K\sigma_{ab}^2$
R	σ^2	σ^2

The unbiased estimator for a component of variance is obtained by first finding a linear combination of EMS (under Model II) which is equal to this component of variance, and then by taking the same linear combination of the corresponding mean squares. Thus, we have the estimators

$$\hat{\sigma}_a{}^2 = \frac{MS_A - MS_{AB}}{JK}, \qquad \hat{\sigma}_b{}^2 = \frac{MS_B - MS_{AB}}{IK}, \qquad \hat{\sigma}_{ab}^2 = \frac{MS_{AB} - MS_R}{K}.$$

$$(4.3.5)$$

The F ratio for a hypothesis H_0 concerning a component of variance is determined from this table as follows. The numerator of the F ratio is the MS of the source of variation corresponding to the component of variance. The denominator of the F ratio is the MS of a source of variation whose EMS is the same as the EMS of the numerator source of variation under H_0. For example, to test $H_0 : \sigma_a{}^2 = 0$, the numerator is MS_A. Under H_0, $EMS_A = EMS_{AB}$; thus the denominator is MS_{AB}; that is, $F = MS_A/MS_{AB}$. Thus, we obtain the test statistics summarized in Table 4.3.4. The P value for each test is the area to the right of F under the $F(v_1, v_2)$ frequency curve. Note that the F ratios are different from those for the fixed effects model.

TABLE 4.3.4. *Tests of Hypotheses for the Replicated Two-Way Layout—Random Effects*

$H_0: \sigma_{ab}^2 = 0$	$H_0: \sigma_a^2 = 0$	$H_0: \sigma_b^2 = 0$
$F = \dfrac{MS_{AB}}{MS_R}$	$F = \dfrac{MS_A}{MS_{AB}}$	$F = \dfrac{MS_B}{MS_{AB}}$
$\nu_1 = (I-1)(J-1)$	$\nu_1 = I-1$	$\nu_1 = J-1$
$\nu_2 = IJ(K-1)$	$\nu_2 = (I-1)(J-1)$	$\nu_2 = (I-1)(J-1)$

COMMENT 4.3.2   The question of pooling is more critical for this model than for the fixed effects model, since in this case, pooling may considerably increase the denominator degrees of freedom, and hence, the power of the test for main effects. For example, let $I = J = 3$ and $K = 10$, and suppose that we accept the hypothesis of additivity [$H_0: (\alpha\beta)_{ij} = 0$ for Model I, or $H_0: \sigma_{ab}^2 = 0$ for Model II]. The accompanying table shows the increase in the number of degrees of freedom for the denominator of the F ratio when the experimenter pools. In the case of Model I he gains only 4 degrees of freedom, while in the case of Model II he gains 81 degrees of freedom.

	Model I	Model II
Not pool	$\nu_2 = 81$	$\nu_2 = 4$
Pool	$\nu_P = 85$	$\nu_P = 85$

EXAMPLE 4.3.3 (*continued*)   Suppose that this data were the result of a Model II experiment—this would be the case if the drugs were randomly chosen from a large (conceivably infinite) number of drugs, and similarly, if the three diseases were also chosen randomly. Although this interpretation seems unrealistic, the point is to illustrate the above concepts. The analysis of variance table would be the same as above, and the estimates of the components of variance are

MS_R = error variance = 98.6 from the anova table,

$$\hat{\sigma}_a^2 = \text{drug variance} = \frac{1549.9 - 175.6}{18} = 76.3,$$

$$\hat{\sigma}_b^2 = \text{disease variance} = \frac{373.0 - 175.6}{24} = 8.2, \quad \text{and}$$

$$\hat{\sigma}_{ab}^2 = \text{interaction variance} = \frac{175.6 - 98.6}{6} = 12.8 \text{ from Eq. (4.3.5).}$$

The tests of hypotheses and the computed F values from Table 4.3.4 are given in the accompanying table. It is interesting to note that in addition to nonsignificance of the interaction variance σ_{ab}^2, we also have a nonsignificant disease variance σ_b^2. Therefore, a small-time pooler might be enticed to pool, since the denominator degrees of freedom for testing main effects are so small. Pooling, he obtains $F = 14.7$ and $F = 3.53$ for testing $H_0: \sigma_a^2 = 0$ and $H_0: \sigma_b^2 = 0$, respectively. As before, these are significant tests.

$H_0: \sigma_{ab}^2 = 0$	$H_0: \sigma_a^2 = 0$	$H_0: \sigma_b^2 = 0$
$F = 1.8$	$F = 8.8$	$F = 2.1$
$\nu_1 = 6$	$\nu_1 = 3$	$\nu_1 = 2$
$\nu_2 = 60$	$\nu_2 = 6$	$\nu_2 = 6$
$P \simeq 0.10$	$P < 0.025$	$P > 0.10$

4.3.2 The Unreplicated Two-Way Layout— Fixed and Random Effects

In this section we assume that we have two factors A and B with I and J levels, respectively, but for every cell we observe Y only $K = 1$ time. Thus, the experiment is *unreplicated*, and as seen from Table 4.3.1, the number of residual degrees of freedom $\nu_R = IJ(K-1)$ becomes zero. In order to derive test statistics for the fixed effects case, we assume that all differential effects due to the interaction of A and B are zero. We then use the interaction sum of squares and degrees of freedom from the replicated case as the residual sum of squares and degrees of freedom for this case. Thus, the *unreplicated fixed effects two-way layout* is

$$y_{ij} = \mu + \alpha_i + \beta_j + e_{ij}, \qquad i = 1, ..., I, \quad j = 1, ..., J, \qquad (4.3.6)$$

where μ is an overall mean, α_i is the ith differential effect of factor A, β_j is the jth differential effect of factor B, and e_{ij} are independent $N(0, \sigma^2)$ error variables.

To obtain unique least-squares estimators for the parameters, we impose the side conditions

$$\sum_{i=1}^{I} \alpha_i = 0, \qquad \sum_{j=1}^{J} \beta_j = 0, \qquad (4.3.7)$$

and obtain

$$\hat{\mu} = \bar{y}_{..}, \qquad \hat{\alpha}_i = \bar{y}_{i.} - \bar{y}_{..}, \qquad \text{and}$$

$$\hat{\beta}_j = \bar{y}_{.j} - \bar{y}_{..} \qquad \text{for} \quad i = 1, ..., I, \quad j = 1, ..., J. \qquad (4.3.8)$$

The *unreplicated random effects two-way layout* is

$$y_{ij} = \mu + a_i + b_j + e_{ij}, \qquad i = 1, ..., I, \quad j = 1, ..., J, \qquad (4.3.9)$$

where μ is an overall mean, a_i are $N(0, \sigma_a{}^2)$, b_j are $N(0, \sigma_b{}^2)$, e_{ij} are $N(0, \sigma^2 + \sigma_{ab}^2)$, and all random variables are jointly independent. Note that the error variance in this case is the sum of two variances σ_{ab}^2 and σ^2.

TABLE 4.3.5. *Unreplicated Two-Way Anova Table*

Source of variation	Sum of squares	Degrees of freedom	Mean square
Factor A	$SS_A = J \sum_{i=1}^{I} (\bar{y}_{i.} - \bar{y}_{..})^2$	$\nu_A = I - 1$	$MS_A = \dfrac{SS_A}{\nu_A}$
Factor B	$SS_B = I \sum_{j=1}^{J} (\bar{y}_{.j} - \bar{y}_{..})^2$	$\nu_B = J - 1$	$MS_B = \dfrac{SS_B}{\nu_B}$
Residual R (error)	$SS_R = \sum_{i=1}^{I} \sum_{j=1}^{J} (y_{ij} - \bar{y}_{i.} - \bar{y}_{.j} + \bar{y}_{..})^2$	$\nu_R = (I-1)(J-1)$	$MS_R = \dfrac{SS_R}{\nu_R}$
Total	$SS_T = \sum_{i=1}^{I} \sum_{j=1}^{J} (y_{ij} - \bar{y}_{..})^2$	$\nu_T = IJ - 1$	

TABLE 4.3.6. *The EMS for the Unreplicated Fixed and Random Effects Two-Way Layout*

Source of variation	EMS Model I	EMS Model II
A	$\sigma^2 + \dfrac{J \sum_i \alpha_i^2}{I-1}$	$\sigma^2 + \sigma_{ab}^2 + J\sigma_a^2$
B	$\sigma^2 + \dfrac{I \sum_j \beta_j^2}{J-1}$	$\sigma^2 + \sigma_{ab}^2 + I\sigma_b^2$
R	σ^2	$\sigma^2 + \sigma_{ab}^2$

For either model the anova table is given in Table 4.3.5, and the EMS for Model I and Model II are given in Table 4.3.6. Thus, MS_R is an unbiased estimator for σ^2 in the Model I case, but unless we assume $\sigma_{ab}^2 = 0$ in the Model II case, we are unable to estimate σ^2. Estimators for $\sigma_a{}^2$ and $\sigma_b{}^2$ are

$$\hat{\sigma}_a{}^2 = \frac{MS_A - MS_R}{J} \quad \text{and} \quad \hat{\sigma}_b{}^2 = \frac{MS_B - MS_R}{I}. \qquad (4.3.10)$$

Finally, tests of hypothesis and their F ratios are given in Table 4.3.7. The P value is the area to the right of F under the $F(v_1, v_2)$ frequency curve. The question of pooling is not relevant in this situation, since there are no terms to pool. However, in the Model I case if the assumption of no interactions is unrealistic, then the tests for main effects may be conservative, that is, the test would reject H_0 less often than expected, thus reducing the power of the test.

TABLE 4.3.7. *Tests for the Unreplicated Two-Way Layout*

Model I: H_0: all $\alpha_i = 0$	H_0: all $\beta_j = 0$
Model II: H_0: $\sigma_a^2 = 0$	H_0: $\sigma_b^2 = 0$

$$F = \frac{MS_A}{MS_R} \qquad F = \frac{MS_B}{MS_R}$$

$$v_1 = I - 1 \qquad v_1 = J - 1$$

$$v_2 = (I-1)(J-1) \qquad v_2 = (I-1)(J-1)$$

COMMENT 4.3.3   For the unreplicated two-way layout, we made the assumption that the model is additive. If the investigator doubts the validity of this assumption, he may assume a nonadditive model and then test the hypothesis that there are no interactions (see Tukey, 1949). To perform this test we calculate the quantities

$$SS_G = \frac{[\sum_{i=1}^{I} \sum_{j=1}^{J} (\bar{y}_{i.} - \bar{y}_{..})(\bar{y}_{.j} - \bar{y}_{..}) y_{ij}]^2}{\sum_{i=1}^{I} (\bar{y}_{i.} - \bar{y}_{..})^2 \sum_{j=1}^{J} (\bar{y}_{.j} - \bar{y}_{..})^2},$$

$$SS_{AB} = \sum_{i=1}^{I} \sum_{j=1}^{J} (y_{ij} - \bar{y}_{i.} - \bar{y}_{.j} + \bar{y}_{..})^2, \qquad \text{and}$$

$$SS_R = SS_{AB} - SS_G.$$

The test statistics for the hypothesis of no interactions is

$$F = \frac{(IJ - I - J) SS_G}{SS_R}.$$

The P value is the area to the right of F under the $F(1, IJ - I - J)$ frequency curve.

EXAMPLE 4.3.4   An experimenter performed a similar experiment to that of Example 4.3.3 except that he used $K = 1$ dog for each drug–disease combination. The data obtained are given in Table A. The analysis of variance table for this

TABLE A

Drug	Disease			Row means $\bar{y}_{i.}$
	1	2	3	
1	25	47	12	28.0
2	34	23	25	27.3
3	9	14	11	11.3
4	−1	25	8	10.7
Column means $\bar{y}_{.j}$	16.8	27.3	14.0	$\bar{y}_{..} = 19.3$

TABLE B

Source of variation	Sum of squares	Degrees of freedom	Mean square
Factor A (drugs)	834.7	3	278.2
Factor B (diseases)	391.2	2	195.6
R (residual)	664.8	6	110.8
Total	1890.7	11	

experiment appears as Table B. The tests of hypotheses and their F values are given in Table C. It is interesting to note that this experimenter does not obtain any significant results. This may be explained by the small number of denominator degrees of freedom resulting in low power for his tests.

TABLE C

H_0: all $\alpha_i = 0$ (no drug differential effects)	H_0: all $\beta_j = 0$ (no disease differential effects)
$F = 2.5$	$F = 1.8$
$v_1 = 3$	$v_1 = 2$
$v_2 = 6$	$v_2 = 6$
$P > 0.10$	$P > 0.10$

Note that the parameter estimates can be easily obtained from the row, column and grand means [see Eq. (4.3.8)]. These are

$$\hat{\mu} = 19.3, \quad \hat{\alpha}_1 = 28.0 - 19.3 = 8.7, \quad \dots, \quad \hat{\beta}_1 = 16.8 - 19.3 = -2.5, \dots.$$

Also, the estimate of σ^2 is 110.8 which compares with the estimate of 98.6 of Example 4.3.3.

4.3.3 Mixed Models—Randomized Blocks Designs

In the analysis of variance a *mixed model* is one in which some of the factors are Model I while the others are Model II. When we have two factors, then there are two possible mixed models. In this section we assume, without loss of generality, that factor A is Model I with I levels, and factor B is Model II with J levels. Furthermore, we assume that we do not replicate the experiment, that is, $K = 1$. Then we write the *mixed two-way layout* as

$$y_{ij} = \mu + \alpha_i + b_j + e_{ij}, \qquad i = 1, ..., I, \quad j = 1, ..., J, \qquad (4.3.11)$$

where μ is an overall mean; α_i is the ith differential effect of A; b_j are independent $N(0, \sigma_b^2)$; and e_{ij} are independent $N(0, \sigma^2)$. Furthermore, we assume that b_j and e_{ij} are independent, and that there are no interactions between A and B. Note that fixed effects are represented by Greek letters and random effects by Latin letters. This device will be used in the remainder of the chapter to distinguish between Model I and Model II factors.

To obtain unique least-squares estimators of the parameters μ and α_i, $i = 1, ..., I$, we impose the customary side condition that $\sum_i \alpha_i = 0$. Thus, we obtain $\hat{\mu} = \bar{y}_{..}$ and $\hat{\alpha}_i = \bar{y}_{i.} - \bar{y}_{..}$, $i = 1, ..., I$. The anova table is the same as that in Table 4.3.5, and the estimator of σ^2 is MS_R. The EMS are generated from Table 4.3.6, that is, the EMS for factor A is that given in the Model I column, while the EMS for factor B and the residual are those in the Model II column, assuming that the interaction variance $\sigma_{ab}^2 = 0$. These quantities are summarized in Table 4.3.8. Thus, an unbiased estimator of σ_b^2 is $\hat{\sigma}_b^2 = (MS_B - MS_R)/I$; the F ratio to test H_0: all $\alpha_i = 0$ is $F = MS_A/MS_R$ with $\nu_1 = I - 1$ and $\nu_2 = (I-1)(J-1)$ degrees of freedom; and the F ratio to test H_0: $\sigma_b^2 = 0$ is $F = MS_B/MS_R$ with $\nu_1 = J - 1$ and $\nu_2 = (I-1)(J-1)$ degrees of freedom.

TABLE 4.3.8. *Expected Mean Squares for the Mixed Two-Way Model*

Source of variation	EMS mixed model
A: Model I (between treatments)	$\sigma^2 + \dfrac{J \sum_i \alpha_i^2}{I-1}$
B: Model II (between blocks)	$\sigma^2 + I\sigma_b^2$
R	σ^2

This model describes an experimental design called the *randomized blocks design*. In this situation an investigator is interested in comparing the differential effects α_i of I treatments (factor A). He randomly assigns these treatments to I experimental units which are homogeneous with respect to some characteristic that affects the measurement of Y. This set of I experimental units is called a *block* and each unit is called a *plot*. If the experimenter replicates the experiment J times, that is, he assigns all of the I treatments randomly to each of J blocks (factor B), then the above model and anova table is appropriate for this situation. Thus, factor A is the Model I treatment factor; α_i is the differential effect due to the ith treatment; factor B is the Model II block factor; and $\sigma_b{}^2$ is the variation between the blocks. Since each treatment is applied to only one plot within a block, a block–treatment interaction cannot be estimated. Therefore, it is assumed that there is no interaction between A and B.

EXAMPLE 4.3.5   An experimenter is interested in estimating and comparing the differential effects of I varieties of corn on the variable $Y =$ yield. Since agricultural fields may vary in fertility thereby affecting the yield of corn, the experimenter divides his field into J blocks such that each block is internally homogeneous with respect to fertility. Each block is then subdivided into I plots and each plot is sown with one variety of corn. If the varieties are assigned at random to the plots within each block, then we have the randomized blocks design.

If each block is subdivided into KI plots so that each variety of corn is randomly assigned to K plots within a block, then we have the *replicated randomized blocks design*. In this case the experimenter is able to measure a block–treatment interaction.

The model describing the replicated randomized blocks design is the *replicated mixed two-way layout*:

$$y_{ijk} = \mu + \alpha_i + b_j + (\alpha b)_{ij} + e_{ijk}, \qquad i = 1, \ldots, I, \quad j = 1, \ldots, J, \quad k = 1, \ldots, K,$$

$$(4.3.12)$$

where α_i is the ith differential effect of A, b_j are independent $N(0, \sigma_b{}^2)$, the interactions $(\alpha b)_{ij}$ are $N(0, \sigma_{ab}^2)$, and e_{ij} are $N(0, \sigma^2)$ error variables. All random variables are assumed jointly independent. Note that the interaction term is a combination of a Greek and a Latin letter.

To obtain unique least-squares estimators for μ, α_i, and $(\alpha b)_{ij}$, we assume $\sum_{i=1}^{I} \alpha_i = 0$ and $\sum_{i=1}^{I} (\alpha b)_{ij} = 0$ for $j = 1, \ldots, J$. The anova table is given by Table 4.3.1. The EMS are given by Table 4.3.9, and the test of hypotheses by Table 4.3.10.

TABLE 4.3.9. *Expected Mean Squares for the Replicated Mixed Two-Way Layout*

Source of variation	EMS mixed models
A: Model I	$\sigma^2 + K\sigma_{ab}^2 + \dfrac{JK\sum_i \alpha_i^2}{I-1}$
B: Model II	$\sigma^2 + IK\sigma_b^2$
AB: Mixed	$\sigma^2 + K\sigma_{ab}^2$
R	σ^2

TABLE 4.3.10. *Tests of Hypotheses for the Replicated Mixed Two-Way Layout*

H_0: $\sigma_{ab}^2 = 0$	H_0: $\sigma_b^2 = 0$	H_0: all $\alpha_i = 0$
$F = \dfrac{MS_{AB}}{MS_R}$	$F = \dfrac{MS_B}{MS_R}$	$F = \dfrac{MS_A}{MS_{AB}}$
$v_1 = (I-1)(J-1)$	$v_1 = J-1$	$v_1 = I-1$
$v_2 = IJ(K-1)$	$v_2 = IJ(K-1)$	$v_2 = (I-1)(J-1)$

EXAMPLE 4.3.3 (*continued*)   For this example, we view the disease factor A as being Model I and the drug factor B as being Model II. Thus, the $I = 3$ levels of factor A are fixed, whereas the $J = 4$ levels of factor B are chosen at random from a very large (infinite) number of drugs. Thus, we have the replicated mixed two-way layout. The anova table is given in that example, but now the tests of hypotheses follow those given in Table 4.3.10. These are summarized in the accompanying table.

H_0: $\sigma_{ab}^2 = 0$	H_0: $\sigma_b^2 = 0$	H_0: all $\alpha_i = 0$
$F = 1.8$	$F = 3.8$	$F = 8.8$
$v_1 = 6$	$v_1 = 2$	$v_1 = 3$
$v_2 = 60$	$v_2 = 60$	$v_2 = 6$
$P \simeq 0.10$	$P < 0.05$	$P < 0.025$

4.3.4 Two-Way Nested Analysis of Variance

Recall from the introduction to this chapter that a factor B (with J levels) is *nested* within a factor A (with I levels) if each level of B occurs with at most one level of A. In most situations the nested factor B is regarded as Model II, while the nesting factor A may be either Model I or Model II. If A is assumed to be Model II then we have the *random effects two-way nested (hierarchical) layout* given by

$$y_{ijk} = \mu + a_i + b_{j(i)} + e_{ijk}, \qquad i = 1, \dots, I, \quad j = 1, \dots, J, \quad k = 1, \dots, K, \tag{4.3.13}$$

where μ is an overall mean, a_i are independent $N(0, \sigma_a^2)$, $b_{j(i)}$ are independent $N(0, \sigma_{b(a)}^2)$, and e_{ijk} are $N(0, \sigma^2)$ error variables. It is also assumed that these random variables are jointly independent. The parentheses around the subscript i for $b_{j(i)}$ indicate the nesting relationship. The same applies to the component of variance $\sigma_{b(a)}^2$.

If, on the other hand, factor A is assumed to be Model I, then we have the *mixed two-way nested (hierarchical) layout* given by

$$y_{ijk} = \mu + \alpha_i + b_{j(i)} + e_{ijk}, \qquad i = 1, \dots, I, \quad j = 1, \dots, J, \quad k = 1, \dots, K, \tag{4.3.14}$$

where α_i is the differential effect due to the ith level of A. To obtain unique least-squares estimators for the parameters $\mu, \alpha_1, \dots, \alpha_I$, we impose the customary side condition that $\sum_i^I \alpha_i = 0$ and obtain the estimates $\hat{\mu} = \bar{y}_{\dots}$ and $\hat{\alpha}_i = \bar{y}_{i\dots} - \bar{y}_{\dots}$, $i = 1, \dots, I$.

TABLE 4.3.11. *Two-Way Nested Layout Anova Table*

Source of variation	Sum of squares	Degrees of freedom	Mean squares
Factor A	$SS_A = JK \sum\limits_{i=1}^{I} (\bar{y}_{i\dots} - \bar{y}_{\dots})^2$	$v_A = I - 1$	$MS_A = \dfrac{SS_A}{v_A}$
Factor B within A	$SS_{B(A)} = K \sum\limits_{i=1}^{I} \sum\limits_{j=1}^{J} (\bar{y}_{ij\cdot} - \bar{y}_{i\cdot\cdot})^2$	$v_{B(A)} = I(J-1)$	$MS_{B(A)} = \dfrac{SS_{B(A)}}{v_{B(A)}}$
Residual R (error)	$SS_R = \sum\limits_{i=1}^{I} \sum\limits_{j=1}^{J} \sum\limits_{k=1}^{K} (y_{ijk} - \bar{y}_{ij\cdot})^2$	$v_R = IJ(K-1)$	$MS_R = \dfrac{SS_R}{v_R}$
Total	$SS_T = \sum\limits_{i=1}^{I} \sum\limits_{j=1}^{J} \sum\limits_{k=1}^{K} (y_{ijk} - \bar{y}_{\dots})^2$	$v_T = IJK - 1$	

The analysis of variance table for either model is given in Table 4.3.11. Note that the nesting relationship is denoted by $B(A)$ for the sum of squares, degrees of freedom, and mean square. To estimate the components of variance σ_a^2 and $\sigma_{b(a)}^2$ and to test hypotheses about them, the EMS for both models are given in Table 4.3.12. Thus, the unbiased estimators for σ_a^2 and $\sigma_{b(a)}^2$ are

$$\hat{\sigma}_a^2 = \frac{MS_A - MS_{B(A)}}{KJ} \quad \text{and} \quad \hat{\sigma}_{b(a)}^2 = \frac{MS_{B(A)} - MS_R}{K}. \quad (4.3.15)$$

TABLE 4.3.12. *EMS for the Two-Way Nested Layout*

Source of variation	EMS Model II	EMS mixed model
A	$\sigma^2 + K\sigma_{b(a)}^2 + KJ\sigma_a^2$	$\sigma^2 + K\sigma_{b(a)}^2 + \dfrac{JK\sum_i \alpha_i^2}{I-1}$
$B(A)$	$\sigma^2 + K\sigma_{b(a)}^2$	$\sigma^2 + K\sigma_{b(a)}^2$
R	σ^2	σ^2

Finally, the appropriate tests and test statistics are summarized in Table 4.3.13.

TABLE 4.3.13. *Tests of Hypotheses for the Two-Way Nested Layout*

$H_0: \sigma_{b(a)}^2 = 0$	$H_0:$ all $\alpha_i = 0$ (mixed) $H_0: \sigma_a^2 = 0$ (Model II)
$F = \dfrac{MS_{B(A)}}{MS_R}$	$F = \dfrac{MS_A}{MS_{B(A)}}$
$\nu_1 = I(J-1)$	$\nu_1 = I-1$
$\nu_2 = IJ(K-1)$	$\nu_2 = I(J-1)$

Most packaged programs do not include a program that specifically solves the anova problem arising from a nested design. However, the m-way factorial program discussed in detail in Section 4.4, or the regression program discussed in Section 4.5, may be used to generate the anova table given in Table 4.3.11. We postpone an example of this design until those sections (see Example 4.4.2).

4.3.5 Comparison of Models

This section compares the various models to point out some of the advantages and disadvantages of each. To do this, we propose a hypothetical experiment which might arise in some form in a real-life situation. The experiment is to

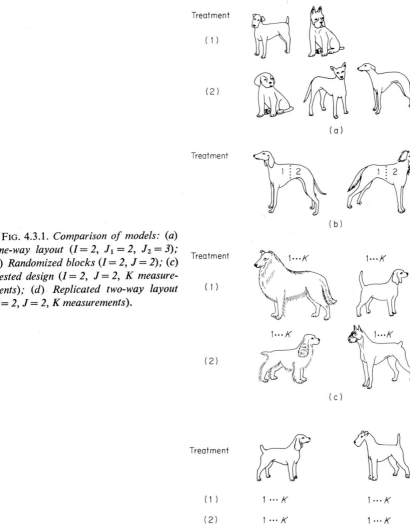

FIG. 4.3.1. *Comparison of models:* (a) *One-way layout* ($I = 2$, $J_1 = 2$, $J_2 = 3$); (b) *Randomized blocks* ($I = 2$, $J = 2$); (c) *Nested design* ($I = 2$, $J = 2$, K *measurements*); (d) *Replicated two-way layout* ($I = 2$, $J = 2$, K *measurements*).

compare the effects of I treatments on a blood characteristic in dogs. Hence, factor A is a Model I treatment factor with I levels, the variable Y is the blood characteristic, and the experimental unit is the dog. The question is: which design should we use? (See Figure 4.3.1.)

Answer 1: One-way layout   In this case we apply the ith treatment to J_i dogs, $i = 1, ..., I$, and we measure the blood characteristic on each dog (see Fig. 4.3.1a). This has the advantage of being the simplest design but the disadvantage that the error variance includes measurement variation as well as variation between dogs. Hence, this may be an inefficient design for determining treatment differences. One improvement may be to select dogs which are as "homogeneous" as possible, thereby reducing the variation between dogs.

Answer 2: Randomized blocks   Here, we regard the treatment as factor A and "dogs" as a random blocks factor B. We select J dogs and each dog (block) receives the I treatments in a randomly selected order (see Fig. 4.3.1b). We measure the blood characteristic after each treatment. This design has the advantage that the effects of the I treatments are measured on a homogeneous block (namely, the same dog), but it has the obvious disadvantage that the observations for any block are dependent. This dependence may be partially eliminated by waiting a sufficient length of time between subsequent treatment applications.

Answer 3: Nested design   In this design we nest factor B (dogs) within factor A (treatments). That is, a random sample of J dogs is subjected to the first treatment, another random sample of J dogs is subjected to the second treatment, and so forth, making a total of IJ dogs in the experiment (see Fig. 4.3.1c). Thus far this is the one-way layout with $J_1 = \cdots = J_I = J$. To make this a nested design, we make K independent determinations of Y. This design may be the most appropriate for such an experiment, since it enables us to estimate measurement error and the variation of dogs within treatments, as well as to test for differences among treatments.

Answer 4: Replicated two-way layout   Here, factor B (dogs) is crossed with factor A (treatments). In this case we take J dogs and apply each of the I treatments to each dog in a random order. We make K determinations of Y for each dog–treatment combination (see Fig. 4.3.1d). This enables us to estimate measurement error and the variation of dogs *across* treatments, as well as to test for differences among treatments. Note that this is the replicated randomized blocks design. This design has the same disadvantage as the unreplicated randomized blocks design.

4.4 The General Factorial Design Program

In this section we return to being "computer oriented" and focus our attention on a standard packaged analysis of variance program. Many packages include only one anova program which computes the anova table for a *factorial design*. Section 4.4.1 defines this design, distinguishes between programs which do and do not permit replicates, and explains how a program of the latter category may be used to analyze replicated factorial designs.

Factorial programs can also be used to analyze special designs such as the randomized blocks, replicated randomized blocks, and nested designs. Since most packages do not include programs for these special designs, Section 4.4.2 discusses methods of utilizing a standard factorial program for analyzing these designs as well as two other designs—the *split plot* and *Latin square*.

4.4.1 The Analysis of Variance for Factorial Designs

We assume that we have m factors $A_1, ..., A_m$, where each factor A_i has $I_i \geqslant 2$ levels, $i = 1, ..., m$. Furthermore, we assume that every combination of levels is replicated exactly N times. Thus every factor is crossed with every other factor, resulting in a completely crossed design. If all m factors are Model I, then we have a *fixed effects* (or *Model I*) *m-factorial* (or *m-way*) *layout*. On the other hand, if all the factors are Model II, we have a *random effects* (or *Model II*) *m-factorial layout*. *Mixed models* arise when some factors are Model I and the remaining factors are Model II. In this section we discuss only the fixed effects model. General programs which handle all three models are presented in Section 4.6.

The fixed effects model assumes that for each combination of levels, $i_1 i_2 \cdots i_m$ of the factors $A_1, A_2, ..., A_m$, an observation is the sum of a mean specific to the combination of levels and an error term. Thus,

$$y_{i_1 i_2 \cdots i_m n} = \mu_{i_1 i_2 \cdots i_m} + e_{i_1 i_2 \cdots i_m n},$$

$$i_1 = 1, ..., I_1, \quad ..., \quad i_m = 1, ..., I_m, \quad n = 1, ..., N. \quad (4.4.1)$$

For arbitrary subscripts $j, k, l = 1, ..., m$, the mean is assumed to be the sum of:

(a) an overall mean μ;

(b) a differential effect due to each factor A_j, which is denoted by $(\alpha_j)_{i_j}$;

(c) a differential effect due to the (*two-way*) *interaction* of each distinct pair of factors $A_j A_k$, which is denoted by $(\alpha_j \alpha_k)_{i_j i_k}, j < k$;

(d) a differential effect due to the (*three-way*) *interaction* of each distinct combination of three factors $A_j A_k A_l$, which is denoted by

$$(\alpha_j \alpha_k \alpha_l)_{i_j i_k i_l}, \qquad j < k < l,$$
$$\vdots$$

(e) a differential effect due to the (*m-way*) *interaction* of all m factors $A_1 A_2 \cdots A_m$, which is denoted by $(\alpha_1 \alpha_2 \cdots \alpha_m)_{i_1 i_2 \cdots i_m}$.

As before, the error terms are assumed to be $N(0, \sigma^2)$. To obtain unique estimators of all these parameters, side conditions on these parameters and their estimators must be assumed. The usual side conditions are that the differential effects of each factor sum to zero, and that the differential effects of each k-way interaction sum to zero over *every* subscript for *all* values of the remaining subscripts $k = 2, \ldots, m$. For example,

$$\sum_{i_j} (\alpha_j \alpha_k)_{i_j i_k} = 0 \qquad \text{for all } i_k \qquad \text{and}$$

$$\sum_{i_k} (\alpha_j \alpha_k)_{i_j i_k} = 0 \qquad \text{for all } i_j.$$

Since the calculations for the various sums of squares becomes difficult when m is large, it becomes especially helpful to use a packaged factorial program. These programs generally fall into one of two categories: those which permit replicates (that is, $N \geqslant 1$) and those which do not (that is, $N = 1$). In the first category, the program is usually called "analysis of variance for a factorial design," "*m*-way analysis of variance," or "analysis of variance for m crossed factors." In the second category, the program is usually designated as simply "analysis of variance." We now consider both of these cases separately.

1. *Programs permitting replicates*   In this case, the output from a typical factorial program lists the sums of squares, degrees of freedom, and mean squares for the residual (or error) component R and for each other source of variation $A_1, A_2, \ldots, A_m, A_1 A_2, \ldots, A_{m-1} A_m, \ldots$, and $A_1 \cdots A_m$, that is, the variation due to each factor, due to the interaction of each distinct pair of factors, ..., and due to the interaction of all m factors. For each source of variation (other than the residual R), the user can test the null hypothesis H_0 that all of the corresponding differential effects are equal to zero. For example, for the source $A_1 A_2 A_3$, the hypothesis is $H_0: (\alpha_1 \alpha_2 \alpha_3)_{i_1 i_2 i_3} = 0$ for all $i_1, i_2, i_3, i_1 = 1, \ldots, I_1, i_2 = 1, \ldots, I_2, i_3 = 1, \ldots, I_3$. The EMS for each source of variation is equal to the sum of the error variance σ^2 and a quantity which vanishes when H_0 is true. Since the residual mean square MS_R is an unbiased estimator of σ^2, we test H_0 by dividing the hypothesis mean square MS_H corresponding to the source of variation by the residual mean square MS_R. Let ν_H and ν_R be the degrees of freedom corresponding to these two mean squares. Thus, we test H_0 by

$$F = \frac{MS_H}{MS_R}. \tag{4.4.2}$$

Under H_0, this statistic has an F distribution with v_H and v_R degrees of freedom. The P value is the area to the right of F under the $F(v_H, v_R)$ frequency curve. We reject H_0 if $P < \alpha$.

When the number of replicates in the experiment is $N = 1$, the model assumes that the differential effects of the m-way interaction are all zero. Thus, the output of this program will not have an entry for the source of variation $A_1 A_2 \cdots A_m$.

2. *Programs assuming no replicates*   In this case the output from a typical program will list the sums of squares, degrees of freedom, and mean squares for the variation due to each factor and each k-way interaction, $k = 2, ..., m$, but there is no entry for a residual component. If in the experiment $N = 1$, then the m-way interaction component is taken to be the residual component, so that

$$SS_R = SS_{A_1 \cdots A_m}, \qquad v_R = v_{A_1 \cdots A_m}, \qquad \text{and} \qquad MS_R = MS_{A_1 \cdots A_m}.$$

We then test for differential effects using Eq. (4.4.2).

On the other hand, if the experiment is such that $N > 1$, we define a new factor A_{m+1} to be the "replicate" factor and use the anova program for an $(m+1)$-way layout. From the output of the program we obtain sources of variation due to each factor and each k-way interaction, $k = 2, ..., m, m+1$. We then pool the sums of squares and degrees of freedom for A_{m+1} and all interactions containing the factor A_{m+1}. These pooled quantities are the residual sum of squares SS_R and degrees of freedom v_R for the m-way layout with N replicates. Then $MS_R = SS_R/v_R$, and we test for differential effects using Eq. (4.4.2).

For example, consider the one-way layout [Eq. (4.2.1)] with one factor A_1 and N replicates for each level of A_1. We define A_2 to be the "replicate" factor and perform the analysis for a two-way layout using the program. The resulting output will list sums of squares $SS_{A_1}, SS_{A_2}, SS_{A_1 A_2}$, and degrees of freedom v_{A_1}, v_{A_2}, and $v_{A_1 A_2}$ for the components A_1, A_2, and $A_1 A_2$, respectively. Then SS_R and v_R for the one-way layout are given by

$$SS_R = SS_{A_2} + SS_{A_1 A_2} \qquad \text{and} \qquad v_R = v_{A_2} + v_{A_1 A_2},$$

so that

$$MS_R = \frac{SS_R}{v_R}. \tag{4.4.3}$$

COMMENT 4.4.1   Pooling becomes a complex question when m is large. If the m-way interaction is not pooled, we recommend that none of the k-way interactions be pooled, $k = 2, ..., m-1$.

EXAMPLE 4.4.1   In an experiment[†] 12 groups of $N = 4$ rats were subjected to the following experimental design: (1) From birth to weaning, groups 1–6 of the nurselings were left with their mother, while groups 7–12 were separated from their mother for a certain period each day; (2) From weaning to maturity, the animals in groups 1–3 and 7–9 were maintained as siblings, while those in groups 4–6 and 10–12 were kept in isolation; (3) From the attainment of full adult life (a) the animals in groups 1 and 7 were maintained as siblings, (b) the animals in groups 4 and 10 were isolated, while (c) the animals in the remaining groups were placed in an interactive box system. At the conclusion of the experiment, the adrenal weight Y (mgm/100gm) was measured.

TABLE A

| Factor A_2:
weaning to
maturity | Factor A_1: birth to weaning | | | | | |
| | With mother
Factor A_3: adult period | | | Isolated from mother
Factor A_3: adult period | | |
	With siblings	Isolated	Box system	With siblings	Isolated	Box system
With siblings	3.2 2.6 2.2 2.6	1.9 2.3 2.2 2.0	4.0 4.6 5.7 5.7	4.6 4.8 4.6 4.4	3.3 4.0 5.0 3.2	6.3 7.2 4.6 7.2
Isolated from siblings	3.2 2.5 3.0 3.3	2.7 2.8 2.4 2.7	4.8 4.8 5.4 3.8	4.5 4.2 4.4 4.3	2.4 3.6 3.0 3.0	3.8 4.4 4.8 5.8

Here, factor A_1 relates to the first time period (from birth to weaning) with $I_1 = 2$ levels—isolated or not isolated from the mother. Factor A_2 relates to the second time period (weaning to maturity) with $I_2 = 2$ levels—isolated or not isolated from siblings. Finally, factor A_3 relates to the adult period with $I_3 = 3$ levels, that is, isolated, not isolated, or placed in interactive box. The data are given in Table A. The 3-way model for this data is therefore:

$$y_{i_1 i_2 i_3 n} = \mu + (\alpha_1)_{i_1} + (\alpha_2)_{i_2} + (\alpha_3)_{i_3} + (\alpha_1 \alpha_2)_{i_1 i_2}$$
$$+ (\alpha_1 \alpha_3)_{i_1 i_3} + (\alpha_2 \alpha_3)_{i_2 i_3} + (\alpha_1 \alpha_2 \alpha_3)_{i_1 i_2 i_3} + e_{i_1 i_2 i_3 n},$$
$$i_1, i_2 = 1, 2, \quad i_3 = 1, 2, 3, \quad n = 1, ..., 4,$$

[†] An experiment of this design was performed by Dr. James Henry, Department of Physiology, USC, Los Angeles, California.

where α_j is the effect due to $A_j, j = 1, 2, 3$. Using a factorial program which permits replicates we obtain the anova table, Table B. The tests of hypotheses are given in Table C. Thus, there are significant effects at the 0.01 level due to factors A_2 and A_3, and also the interaction of A_1 and A_2. The remaining effects have $P > 0.05$.

TABLE B

Source of variation	Sum of squares	Degrees of freedom	Mean square
A	1.505	1	1.505
B	15.075	1	15.075
C	43.006	2	21.503
AB	3.797	1	3.797
AC	2.322	2	1.161
BC	1.922	2	0.961
ABC	0.380	2	0.190
Residual	13.992	36	0.389
Total	81.999	47	

TABLE C

H_0: all $(\alpha_1 \alpha_2 \alpha_3)_{i_1 i_2 i_3} = 0$	H_0: all $(\alpha_2 \alpha_3)_{i_2 i_3} = 0$	H_0: all $(\alpha_1 \alpha_3)_{i_1 i_3} = 0$
$F = 0.49$	$F = 2.5$	$F = 3.0$
$v_1 = 2$	$v_1 = 2$	$v_1 = 2$
$v_2 = 36$	$v_2 = 36$	$v_2 = 36$
NS	NS	NS

H_0: all $(\alpha_1 \alpha_2)_{i_1 i_2} = 0$	H_0: all $(\alpha_3)_{i_3} = 0$	H_0: all $(\alpha_2)_{i_2} = 0$	H_0: all $(\alpha_1)_{i_1} = 0$
$F = 9.8$	$F = 55$	$F = 39$	$F = 3.9$
$v_1 = 1$	$v_1 = 2$	$v_1 = 1$	$v_1 = 1$
$v_2 = 36$	$v_2 = 36$	$v_2 = 36$	$v_2 = 36$
$P < 0.01$	$P < 0.01$	$P < 0.01$	NS

4.4.2 Application of a Factorial Program to Other Models

A factorial program may be used to generate analysis of variance tables for designs other than the m-factorial layout (see Hartley's article in Ralston and Wilf, 1960). The only requirement is that the design have an equal number of observations N for each cell. The trick is to reformulate any given design as a factorial design, to generate the factorial anova table, and then to pool selected sums of squares and degrees of freedom for the original design. Once this has been accomplished, the mean squares are obtained by dividing each pooled sum of squares by its pooled degrees of freedom. The usual tests of hypotheses are then made from the resulting anova table.

This section discusses the application of this procedure to two of the designs introduced so far—the randomized blocks and the nested designs. In addition, two new designs—the *split plot* and the *Latin square* designs—are introduced. When necessary, we distinguish between the two categories of factorial programs already mentioned.

1. *Randomized blocks*   The randomized blocks model given by Eq. (4.3.11) may be treated as a two-way layout with $N = 1$ observation for each pair of levels. The anova table is then obtained from any factorial program and the EMS are given by Table 4.3.8.

The replicated randomized blocks model given by Eq. (4.3.12) may be treated as a two-way layout with $N > 1$ replicates when using a factorial program which permits replicates. If the factorial program does not permit replicates, then we reformulate the model as follows: Let A_1 be the treatment factor, A_2 be the "block" factor, and A_3 be the "replicate" factor. Using the factorial program for a three-way layout, the residual sum of squares for the original model is given by

$$SS_R = SS_{A_3} + SS_{A_1A_3} + SS_{A_2A_3} + SS_{A_1A_2A_3}, \tag{4.4.4}$$

where the quantities on the right are from the anova table for the three-way layout. A similar formula holds for the residual degrees of freedom. The EMS are given by Table 4.3.9.

2. *Two-way nested model*   For the two-way nested model given by Eq. (4.3.13) or (4.3.14), we have two factors A_1 and A_2, where A_2 is nested within A_1. Furthermore, there are N observations for each cell. If a factorial program which allows replicates is used, we may view this model as a two-way layout with N replicates. The sum of squares for factor A_2 within A_1 is given by

$$SS_{A_2(A_1)} = SS_{A_2} + SS_{A_1A_2}, \tag{4.4.5}$$

where the quantities on the right are from the anova table for the two-way layout. A similar formula holds for the degrees of freedom $v_{A_2(A_1)}$. The quantities SS_{A_1} and SS_R and their degrees of freedom are obtained directly from the program.

Using a factorial program not permitting replicates, we let A_3 be the "replicate" factor and view this model as a three-way layout. The residual sum of squares for the nested model is given by

$$SS_R = SS_{A_3} + SS_{A_1A_3} + SS_{A_2A_3} + SS_{A_1A_2A_3}, \tag{4.4.6}$$

with a similar formula for v_R. The quantity $SS_{A_2(A_1)}$ is given by Eq. (4.4.5), and SS_{A_1} and v_{A_1} are obtained directly from the program. The EMS are given in Table 4.3.12.

3. *Split plot design*   In this situation we have I_1 levels of a main treatment (factor A_1), I_2 levels of a subtreatment (factor A_2), and I_3 blocks (factor A_3) (see Brownlee, 1965). Each block is divided into I_1 homogeneous plots, and then each plot is divided into I_2 subplots. The levels of factor A_1 are assigned at random to each plot, and then within each plot, the levels of A_2 are assigned at random to each subplot. For example a block for the case $I_1 = 3$, $I_2 = 2$ might be as shown in the accompanying tabulation, where the subscript i_j denotes the level

Plot 1	Plot 2	Plot 3	
$i_1 = 2$	$i_1 = 3$	$i_1 = 1$	Subplot 1
$i_2 = 2$	$i_2 = 2$	$i_2 = 1$	
$i_1 = 2$	$i_1 = 3$	$i_1 = 1$	Subplot 2
$i_2 = 1$	$i_2 = 1$	$i_2 = 2$	

Block

of the jth factor, $j = 1$ or 2. The purpose of this design is to reduce the number of treatment combinations within a block. The model for this design is

$$y_{i_1 i_2 i_3} = \mu + (\alpha_1)_{i_1} + (\alpha_2)_{i_2} + (\alpha_3)_{i_3} + (\alpha_1 \alpha_2)_{i_1 i_2}$$

$$+ (\alpha_2 \alpha_3)_{i_2 i_3} + e^{(1)}_{i_1 (i_3)} + e_{i_1 i_2 i_3}, \tag{4.4.7}$$

$$i_1 = 1, ..., I_1, \quad i_2 = 1, ..., I_2, \quad i_3 = 1, ..., I_3,$$

where α_1 is the fixed effect due to the treatment, α_2 is the fixed effect due to the subtreatment, α_3 is the random block effect, $(\alpha_1 \alpha_2)$ is the interaction of the treatment and the subtreatment, and $(\alpha_2 \alpha_3)$ is the interaction of the subtreatment with the block. The term $e^{(1)}$ is the random error of plots within blocks, and e is the random error of subplots within plots. Let $e^{(1)}$ be independent $N(0, \sigma_1{}^2)$ and e be $N(0, \sigma^2)$. The side conditions are

$$\sum_{i_1} (\alpha_1)_{i_1} = \sum_{i_2} (\alpha_2)_{i_2} = 0, \qquad \sum_{i_1} (\alpha_1 \alpha_2)_{i_1 i_2} = \sum_{i_2} (\alpha_1 \alpha_2)_{i_1 i_2} = 0$$

for any value of i_1 and i_2. Furthermore, $(\alpha_3)_{i_3}$ is $N(0, \sigma_{\alpha_3}^2)$ and $(\alpha_2 \alpha_3)_{i_2 i_3}$ is $N(0, \sigma_{\alpha_2 \alpha_3}^2)$ for all values of i_2 and i_3.

The analysis of variance table for this design is given by Table 4.4.1, and the tests of hypotheses are given in Table 4.4.2. The degrees of freedom for the F ratio are obtained from the corresponding quantities in the anova table. If $H_0 : \sigma_1{}^2 = 0$ is accepted, then the "poolers" may pool $SS_{R(1)}$ and SS_R to obtain a new residual sum of squares. This quantity would be used for testing H_0: $(\alpha_1)_{i_1} = 0$ and $H_0 : \sigma_{\alpha_3}^2 = 0$.

TABLE 4.4.1. *Split Plot Analysis of Variance Table*

Source of variation	Sum of squares[a]	Degrees of freedom	Mean square	EMS Model I
A_3 (blocks)	$SS_{A_3} = \sum (\bar{y}_{..i_3} - \bar{y}_{...})^2$	$\nu_{A_3} = I_3 - 1$	MS_{A_3}	$\sigma^2 + I_2 \sigma_1^2 + I_1 I_2 \sigma_{z_3}^2$
A_1 (main treatment)	$SS_{A_1} = \sum (\bar{y}_{i_1..} - \bar{y}_{...})^2$	$\nu_{A_1} = I_1 - 1$	MS_{A_1}	$\sigma^2 + I_2 \sigma_1^2 + I_2 I_3 \dfrac{\sum(\alpha_1)_{i_1}^2}{I_1 - 1}$
$R^{(1)}$ (plot error)	$SS_{R^{(1)}} = \sum (\bar{y}_{i_1.i_3} - \bar{y}_{i_1..} - \bar{y}_{..i_3} + \bar{y}_{...})^2$	$\nu_{R^{(1)}} = (I_1-1)(I_3-1)$	$MS_{R^{(1)}}$	$\sigma^2 + I_2 \sigma_1^2$
A_2 (subtreatment)	$SS_{A_2} = \sum (\bar{y}_{.i_2.} - \bar{y}_{...})^2$	$\nu_{A_2} = I_2 - 1$	MS_{A_2}	$\sigma^2 + I_1 \sigma_{z_2 z_3}^2 + I_1 I_3 \dfrac{\sum(\alpha_2)_{i_2}^2}{I_2 - 1}$
$A_1 A_2$ (main × sub)	$SS_{A_1 A_2} = \sum (\bar{y}_{i_1 i_2.} - \bar{y}_{i_1..} - \bar{y}_{.i_2.} + \bar{y}_{...})^2$	$\nu_{A_1 A_2} = (I_1-1)(I_2-1)$	$MS_{A_1 A_2}$	$\sigma^2 + I_3 \dfrac{\sum\sum(\alpha_1 \alpha_2)_{i_1 i_2}^2}{(I_1-1)(I_2-1)}$
$A_2 A_3$ (sub × block)	$SS_{A_2 A_3} = \sum (\bar{y}_{.i_2 i_3} - \bar{y}_{.i_2.} - \bar{y}_{..i_3} + \bar{y}_{...})^2$	$\nu_{A_2 A_3} = (I_2-1)(I_3-1)$	$MS_{A_2 A_3}$	$\sigma^2 + I_1 \sigma_{z_2 z_3}^2$
R (subplot error)	$SS_R = SS_T - $ (sum of above SS)	$\nu_R = (I_1-1)(I_2-1)(I_3-1)$	MS_R	σ^2
Total	$SS_T = \sum (y_{i_1 i_2 i_3} - \bar{y}_{...})^2$	$\nu_T = I_1 I_2 I_3 - 1$		

[a] Summation is over $i_1 i_2$ and i_3.

TABLE 4.4.2. *Tests for the Split Plot Design*

H_0: $\sigma^2_{\alpha_2\alpha_3}=0$	H_0: all $(\alpha_1\alpha_2)_{i_1i_2}=0$	H_0: all $(\alpha_2)_{i_2}=0$
$F=\dfrac{MS_{A_2A_3}}{MS_R}$	$F=\dfrac{MS_{A_1A_2}}{MS_R}$	$F=\dfrac{MS_{A_2}}{MS_{A_2A_3}}$
H_0: $\sigma_1^2=0$	H_0: all $(\alpha_1)_{i_1}=0$	H_0: $\sigma^2_{\alpha_3}=0$
$F=\dfrac{MS_{R(1)}}{MS_R}$	$F=\dfrac{MS_{A_1}}{MS_{R(1)}}$	$F=\dfrac{MS_{A_3}}{MS_{R(1)}}$

To utilize any factorial program to generate Table 4.4.1, we view the split plot design as a three-way layout with $N=1$ observation for each triple of levels. Using the standard notation for the sources of variation of the factorial model, we have the following relationships between the two sets of sources of variation:

Split plot   Factorial

$$SS_{A_j} \quad = \quad SS_{A_j}, \quad j=1,2,3,$$

$$SS_{R(1)} \quad = \quad SS_{A_1A_3},$$

$$SS_{A_1A_2} \quad = \quad SS_{A_1A_2}, \quad\quad\quad (4.4.8)$$

$$SS_{A_2A_3} \quad = \quad SS_{A_2A_3},$$

$$SS_R \quad = \quad \begin{cases} SS_{A_1A_2A_3}, & \text{if program permits no replicates,} \\ SS_R, & \text{if program permits replicates.} \end{cases}$$

4. *Latin square design*   In this situation we have three Model I factors A_1, A_2, and A_3 with the same number of levels for all three factors, that is, $I_1=I_2=I_3=I\geqslant 3$. We also assume that there are no interactions between the factors. The requirement for this design is that each level of A_1 is used exactly once with each level of A_2 and exactly once with each level of A_3. To actually construct a Latin square, we list in order the I levels of A_1 and the I levels of A_2 in a two-dimensional array; the levels of A_1 are the rows, and the levels of A_2 are the columns. In each cell we assign one of the levels of A_3 under the previous requirement, that is, that each level of A_3 appears only once in each row and once in each column of the square. The observation in the i_1i_2th cell, customarily denoted by $y_{i_1i_2}$, is now denoted by $y_{i_1i_2(i_3)}$. Here i_3 is the level of A_3 which is associated with i_1 and i_2 according to the Latin square.

Two examples of Latin squares for $I=3$ are given in Table 4.4.3. The numerical entries refer to the levels of A_3, and the literal entries refer to the observations. Note that for example, $f=y_{23(1)}$ and $f'=y_{23(2)}$, since the 2,3 combination of A_1 and A_2 is associated with level 1 of A_3 in the first square, and with level 2 of A_3 in the second square.

TABLE 4.4.3. *Two Examples of Latin Squares* ($I = 3$)

		A_2					A_2	
	1	2	3		1	2	3
A_1 1	1 *a*	2 *b*	3 *c*		2 *a'*	1 *b'*	3 *c'*
A_1 2	2 *d*	3 *e*	1 *f*	A_1 2	1 *d'*	3 *e'*	2 *f'*
3	3 *g*	1 *h*	2 *i*	3	3 *g'*	2 *h'*	1 *i'*

For any I there are many possible Latin squares. Ideally, the experimenter should make a list of all possible Latin squares and choose one at random for his experiment. To facilitate this operation, we may use the list of Latin squares prepared by Fisher and Yates (1963).

This design has an advantage over the three-way factorial design, since for the Latin square we need I^2 observations as compared to I^3 observations for the factorial design. The disadvantage of this economical design is that it assumes no interactions.

The model for the Latin square design is

$$y_{i_1 i_2 (i_3)} = \mu + (\alpha_1)_{i_1} + (\alpha_2)_{i_2} + (\alpha_3)_{i_3} + e_{i_1 i_2 (i_3)}, \tag{4.4.9}$$

where α_j is the differential effect of A_j, $j = 1, 2, 3$ and $e_{i_1 i_2 (i_3)}$ are independent $N(0, \sigma^2)$. The side conditions are

$$\sum_{i_1} (\alpha_1)_{i_1} = \sum_{i_2} (\alpha_2)_{i_2} = \sum_{i_3} (\alpha_3)_{i_3} = 0.$$

The combinations of subscripts $i_1 i_2 (i_3)$ in Eq. 4.4.9 are those specified by the Latin square used in the experiment. For example, in the first Latin square in Table 4.4.3, we have the set of subscripts $\{(1, 1, 1), (1, 2, 2), (1, 3, 3), (2, 1, 2), (2, 2, 3), (2, 3, 1), (3, 1, 3), (3, 2, 1), (3, 3, 2)\}$ while in the second Latin square we have $\{(1, 1, 2), (1, 2, 1), (1, 3, 3), (2, 1, 1), (2, 2, 3), (2, 3, 2), (3, 1, 3), (3, 2, 2), (3, 3, 1)\}$. Note that there are I^2, rather than I^3, subscripts.

The analysis of variance table for this design and the tests of hypotheses are given in Tables 4.4.4 and 4.4.5, respectively.

To obtain this anova table using a factorial program, we view the Latin square as two different two-way layouts with $N = 1$ and calculate an anova table for each. First we use the program for the factors A_1 and A_2 thus obtaining sums of squares and degrees of freedom for A_1, A_2, and $A_1 A_2$. We obtain the total sums of squares SS_T and degrees of freedom v_T by

$$SS_T = SS_{A_1} + SS_{A_2} + SS_{A_1 A_2}, \quad \text{and} \quad v_T = v_{A_1} + v_{A_2} + v_{A_1 A_2}.$$
$$\tag{4.4.10}$$

TABLE 4.4.4. *Latin Square Analysis of Variance Table*

Source of variation	Sum of squares	Degrees of freedom	Mean square	EMS
A_1	$SS_{A_1} = I \sum_{i_1} (\bar{y}_{i_1 \cdot (\cdot)} - \bar{y}_{\cdot \cdot (\cdot)})^2$	$\nu_{A_1} = I - 1$	MS_{A_1}	$\sigma^2 + \dfrac{I \sum (\alpha_1)^2_{i_1}}{I-1}$
A_2	$SS_{A_2} = I \sum_{i_2} (\bar{y}_{\cdot i_2 (\cdot)} - \bar{y}_{\cdot \cdot (\cdot)})^2$	$\nu_{A_2} = I - 1$	MS_{A_2}	$\sigma^2 + \dfrac{I \sum (\alpha_2)^2_{i_2}}{I-1}$
A_3	$SS_{A_3} = I \sum_{i_3} (\bar{y}_{\cdot \cdot (i_3)} - \bar{y}_{\cdot \cdot (\cdot)})^2$	$\nu_{A_3} = I - 1$	MS_{A_3}	$\sigma^2 + \dfrac{I \sum (\alpha_3)^2_{i_3}}{I-1}$
Residual	$SS_R = SS_T - (SS_{A_1} + SS_{A_2} + SS_{A_3})$	$\nu_R = (I-1)(I-2)$	MS_R	σ^2
Total	$SS_T = \sum_{i_1} \sum_{i_2} \sum_{i_3} (y_{i_1 i_2 (i_3)} - \bar{y}_{\cdot \cdot (\cdot)})^2$	$\nu_T = I^2 - 1$		

TABLE 4.4.5. *Tests for Latin Square Design*

H_0: all $(\alpha_1)_{i_1} = 0$	H_0: all $(\alpha_2)_{i_2} = 0$	H_0: all $(\alpha_3)_{i_3} = 0$
$F = \dfrac{MS_{A_1}}{MS_R}$	$F = \dfrac{MS_{A_2}}{MS_R}$	$F = \dfrac{MS_{A_3}}{MS_R}$

We then rewrite the Latin square so that factor A_3 is the column factor and factor A_2 is the treatment factor with A_1 remaining as the row factor. Thus, we make the transformation

$$y_{i_1 (i_2) i_3} = y_{i_1 i_2 (i_3)},\tag{4.4.11}$$

so that we obtain the same combinations of subscripts of the three factors. For example, the first square in Table 4.4.3 is then as given in the accompanying tabulation, where the numerical entries now refer to the levels of A_2. Thus,

$d = y_{21(2)}$ in the old square now becomes $d = y_{2(1)2}$ in the new square. We once again use the factorial program for the factors A_1 and A_3 obtaining sums of squares and degrees of freedom for A_1, A_3, and $A_1 A_3$.

From the two anova tables thus obtained, we now obtain the anova table for the Latin square. The sums of squares are

$$
\begin{array}{llll}
\text{Latin Square} & \text{Factorial} & & \\
\text{SS}_{A_1} & = & \text{SS}_{A_1}, & \text{from run 1 or run 2,} \\
\text{SS}_{A_2} & = & \text{SS}_{A_2}, & \text{from run 1,} \\
\text{SS}_{A_3} & = & \text{SS}_{A_3}, & \text{from run 2,} \qquad (4.4.12) \\
\text{SS}_{\text{T}} & = & \text{SS}_{A_1} + \text{SS}_{A_2} + \text{SS}_{A_1 A_2} & \text{from run 1,} \\
\text{SS}_{\text{R}} & = & \text{SS}_{\text{T}} - (\text{SS}_{A_1} + \text{SS}_{A_2} + \text{SS}_{A_3}), &
\end{array}
$$

with similar formulas for v_{A_1}, v_{A_2}, v_{A_3}, and v_{R}.

EXAMPLE 4.4.2   In this example we demonstrate the pooling technique of this section for a nested design. In an experiment it was desired to evaluate the difference in the average pH level among $I_1 = 3$ disease states (factor A_1). A nested design was employed in which a random sample of $I_2 = 3$ dogs (factor A_2) was subjected to the i_1th disease, $i_1 = 1, 2, 3$. On each dog $N = 2$ measurements of $Y = p$H were made. The model for this experiment is

$$
y_{i_1 i_2 n} = \mu + (\alpha_1)_{i_1} + (\alpha_2)_{i_2(i_1)} + e_{i_1 i_2 n}, \qquad i_1 = 1, \ldots, 3, \quad i_2 = 1, \ldots, 3, \quad n = 1, 2,
$$

where $(\alpha_1)_{i_1}$ is the effect due to the fixed factor A_1 (diseases), and $(\alpha_2)_{i_2}$ is the effect due to the random factor A_2 (dogs). The data were as shown in Table A.

TABLE A

Disease	Dog	pH
1	1(1)	7.08, 7.02
	2(1)	7.04, 7.07
	3(1)	7.07, 6.98
2	1(2)	7.29, 7.18
	2(2)	7.42, 7.32
	3(2)	7.08, 7.28
3	1(3)	7.74, 7.54
	2(3)	7.53, 7.50
	3(3)	7.51, 7.63

TABLE B

Disease	Dog 1	Dog 2	Dog 3
1	7.08 7.02	7.04 7.07	7.07 6.98
2	7.29 7.18	7.42 7.32	7.08 7.28
3	7.74 7.54	7.53 7.50	7.51 7.63

TABLE C

Source of variation	Sums of squares	Degrees of freedom
A_1 (diseases)	0.8570	2
A_2 (dogs)	0.0111	2
$A_1 A_2$ (diseases × dogs)	0.0439	4
Residual	0.0650	9
Total	0.9770	17

The data were then arranged in a two-way layout with $N = 2$ replicates as shown in Table B. Using a factorial program which permits replicates we obtained the anova table, Table C. To obtain the sum of squares for dogs within diseases, we use Eq. (4.4.5) to get $SS_{A_2(A_1)} = 0.0111 + 0.0439 = 0.0550$, with $v_{A_2(A_1)} = 6$ degrees of freedom. Thus, the anova table for the nested model is shown in Table D. Using the test statistics given in Table 4.3.13, we construct Table E. Therefore, there are significant differences among the three diseases.

TABLE D

Source of variation	Sums of squares	Degrees of freedom	Mean squares
A_1 (diseases)	0.8570	2	0.4285
$A_2(A_1)$ (dogs within diseases)	0.0550	6	0.0092
Residual	0.0650	9	0.0072
Total	0.9770	17	

TABLE E

H_0: $\sigma^2_{\alpha_2(\alpha_1)} = 0$ (no dog within disease variation)	H_0: all $(\alpha_1)_{i_1} = 0$ (no disease differential effects)
$F = \dfrac{0.0092}{0.0072} = 1.3$	$F = \dfrac{0.4285}{0.0092} = 46.5$
$v_1 = 6$	$v_1 = 2$
$v_2 = 9$	$v_2 = 6$
NS	$P < 0.001$

COMMENT 4.4.2   The nesting relationship is *transitive*. This means, for example, that for three factors A, B, and C, if factor B is nested within C, and factor A is nested within B, then A is nested within C also. This last nesting relationship is denoted by $A(BC)$. To obtain the sum of squares $SS_{A(BC)}$ and degrees of freedom $v_{A(BC)}$ for $A(BC)$ from a three-way factorial program, we pool all sums of squares and degrees of freedom for sources of variation containing the letter A. Thus, we have

$$SS_{A(BC)} = SS_A + SS_{AB} + SS_{AC} + SS_{ABC} \qquad \text{and}$$

$$v_{A(BC)} = v_A + v_{AB} + v_{AC} + v_{ABC}.$$

As before, $SS_{B(C)} = SS_B + SS_{BC}$. This procedure applies to any nesting relationship $A_1(A_2 A_3 \cdots A_m)$.

4.5 Anova via Regression

This section discusses the use of a multiple linear regression program for solving any of the anova problems introduced in this chapter. The reasons for this presentation are that some packages include multiple regression and not anova programs, and that multiple regression programs, unlike factorial programs, permit unequal numbers of replicates per cell. The case of unequal replicates per cell frequently arises, either in designed experiments where some observations are missing, or in experiments which were not designed, for example, the attitudinal study mentioned in the introduction to this chapter. In that study, it is unlikely that there are the same number of observations for each combination of levels of the socio-economic and ethnic groups.

Since the technique described in this section utilizes a modified form of the general linear model formulation of the anova model, it is assumed that each factor is Model I. As will be seen in Comment 4.5.3.2, Model II factors may be treated as Model I factors to obtain the anova table. Tests of hypotheses then require the original interpretation of the factors.

We first discuss the technique for a single hypothesis H_0 and then for all of the hypotheses of a full anova table. Briefly, the technique is to rewrite the original anova model in terms of an overall mean μ and a *minimal set* of differential effects where this minimal set is determined by the side conditions. We then write this new anova model in the form of the general linear model. From a multiple regression program we obtain the residual sum of squares SS_R and degrees of freedom v_R for the original anova model. To test H_0, we delete appropriate variables in the general linear model, rerun the regression program, and obtain the sums of squares SS_R' and degrees of freedom v_R' for the anova model under H_0. The test statistic for H_0 is

$$F = \frac{(SS_R' - SS_R)/(v_R' - v_R)}{SS_R/v_R}. \qquad (4.5.1)$$

This statistic, which was introduced earlier in Sections 3.2 and 4.1, has an F distribution with $v_H = v_R' - v_R$ and v_R degrees of freedom. The P value is the area to the right of F under the $F(v_H, v_R)$ frequency curve.

For a given hypothesis H_0 about the differential effects, the steps of the analysis are:

Step 1   Write down the original fixed effects anova model and the side conditions for the differential effects.

Step 2   For each side condition, rewrite one of the effects as a linear combination of the other effects. For example, if the effects of factor A are $\alpha_1, ..., \alpha_I$, and the side condition is $\alpha_1 + \alpha_2 + \cdots + \alpha_I = 0$, then we may write $\alpha_I = -\alpha_1 - \alpha_2 - \cdots - \alpha_{I-1}$. We then substitute all such linear combinations into the anova model. The resulting model is now expressed in terms of μ and a minimal set of p differential effects. The value of p can be determined according to Comments 4.5.1.4 and 4.5.1.5.

Step 3   Rewrite the anova model from Step 2 in a modified form of the general linear model. This entails renumbering the observations in some convenient manner as $y_1, ..., y_n$, where n is the total number of observations. The differential effects in the minimal set are given new names $\theta_1, ..., \theta_p$, say. The model is then written as

$$y_i = \mu + \theta_1 x_{1i} + \cdots + \theta_p x_{pi} + e_i, \qquad i = 1, ..., n, \qquad (4.5.2)$$

where e_i are independent $N(0, \sigma^2)$ error terms. The x_{ji} are determined by the model in Step 2.

Step 4   Regarding $x_1, ..., x_p$ as independent variables and y as the dependent variable, use a packaged multiple linear regression program to obtain SS_R and v_R for the original anova model. As discussed in Section 3.2, these quantities are the residual sum of squares and degrees of freedom from the anova table for the multiple regression.

Step 5   The hypothesis H_0 is expressible as H_0: some θ's $= 0$. Deleting the x's corresponding to these θ's from Eq. (4.5.2), rerun the regression program, obtaining SS_R' and v_R' from the resulting anova table. We then test H_0 using Eq. (4.5.1).

Exception: If H_0 specifies that $\theta_1 = \cdots = \theta_p = 0$, then $SS_R' = \sum_{i=1}^{n}(y_i - \bar{y})^2$ where $\bar{y} = (1/n)\sum_{i=1}^{n} y_i$. This quantity is the total sum of squares SS_T from the anova table for the multiple linear regression of Step 4. The quantity $v_R' = n - 1$.

Alternatively, we can obtain the differences $SS_R' - SS_R$ and $v_R' - v_R$ for Eq. (4.5.1) from this anova table. They are the "due regression" sum of squares and degrees of freedom, respectively.

COMMENTS 4.5.1　　　　1.　Step 5 may be repeated for any other hypothesis about the differential effects expressible as H_0: some other θ's $= 0$.

2.　Some multiple linear regression programs will solve a series of multiple linear regression problems. This permits the user to solve the regression problems of Steps 4 and 5 in one run. The variables for either problem are specified by a *selection* card which "selects" the dependent and independent variables. Thus, in Step 4 all x's are selected as independent variables, while in Step 5, none of the x's corresponding to the θ's under H_0 are selected as independent variables.

3.　Another advantage of this technique is that as Step 4 the user obtains least-squares estimates of the parameters $\mu, \theta_1, ..., \theta_p$ defined in Step 3. By making a correspondence between $\theta_1, ..., \theta_p$ and the differential effects of the model in Step 2, he obtains least-squares estimates of the minimal set of effects. Estimates for the remaining effects are obtained from the linear combinations of Step 2. For example, if $\alpha_I = -\alpha_1 - \cdots - \alpha_{I-1}$, then the least-squares estimate $\hat{\alpha}_I = -\hat{\alpha}_1 - \cdots - \hat{\alpha}_{I-1}$, where $\hat{\alpha}_1, ..., \hat{\alpha}_{I-1}$ are the least-squares estimates from Step 4.

4.　If the anova model is such that its residual degrees of freedom v_R is known a priori, then the number p of differential effects is $p = n - v_R - 1$, where n is the total number of observations.

5.　In general, we know from Comment 4.1.1.2 that $v_R = n - \text{rank } \mathbf{X}'$ where $\mathbf{X}'$ is the design matrix of the general linear model formulation of the anova model. Then $p = n - v_R - 1 = \text{rank } \mathbf{X}' - 1$.

EXAMPLE 4.5.1　　　　For the purposes of clarifying the above technique, we assume that we have one Model I factor with three levels. At each of the first two levels, we make two observations, and at the third level, we make one observation. Paralleling the steps outlined previously, we have:

Step 1　　　The original model is

$$y_{ij} = \mu + \alpha_i + e_{ij}, \quad i = 1, 2, 3, \quad j = 1, ..., J_i,$$

where $J_1 = J_2 = 2$ and $J_3 = 1$. The side condition customarily used is

$$2\alpha_1 + 2\alpha_2 + \alpha_3 = 0.$$

Step 2　　　From the side condition, we have

$$\alpha_3 = -2\alpha_1 - 2\alpha_2.$$

Substituting into the original model, we obtain

$$y_{11} = \mu + \alpha_1 + e_{11}, \qquad y_{12} = \mu + \alpha_1 + e_{12},$$
$$y_{21} = \mu + \alpha_2 + e_{21}, \qquad y_{22} = \mu + \alpha_2 + e_{22},$$
$$y_{31} = \mu + \alpha_3 + e_{31} = \mu - 2\alpha_1 - 2\alpha_2 + e_{31},$$

which is now in terms of μ and the minimal set of differential effects α_1 and α_2. Note that, since it is known that $v_R = 5 - 3 = 2$, then, according to Comment 4.5.1.4, the number of differential effects in the minimal set is $p = n - v_R - 1 = 5 - 2 - 1 = 2$.

Step 3   Renumber

$$y_1 = y_{11}, \qquad y_2 = y_{12}, \qquad y_3 = y_{21}, \qquad y_4 = y_{22}, \qquad \text{and} \qquad y_5 = y_{31}.$$

Define $\theta_1 = \alpha_1$ and $\theta_2 = \alpha_2$. The model is now written as

$$y_1 = \mu + 1\theta_1 + 0\theta_2 + e_1$$
$$y_2 = \mu + 1\theta_1 + 0\theta_2 + e_2$$
$$y_3 = \mu + 0\theta_1 + 1\theta_2 + e_3$$
$$y_4 = \mu + 0\theta_1 + 1\theta_2 + e_4$$
$$y_5 = \mu + (-2)\theta_1 + (-2)\theta_2 + e_5$$

where the coefficients of θ_1 and θ_2 are the x's determined by the model in Step 2.

Step 4 and Comment 4.5.1.3   A packaged multiple linear regression program with input

y	x_1	x_2
y_1	1	0
y_2	1	0
y_3	0	1
y_4	0	1
y_5	-2	-2

will give the desired SS_R and v_R. We also obtain least-squares estimates $\hat{\mu}$, $\hat{\theta}_1$, and $\hat{\theta}_2$. Thus, in terms of the original model $\hat{\alpha}_1 = \hat{\theta}_1, \hat{\alpha}_2 = \hat{\theta}_2$, and $\hat{\alpha}_3 = -2\hat{\alpha}_1 - 2\hat{\alpha}_2$.

Step 5   The hypothesis of interest is $H_0: \alpha_1 = \alpha_2 = \alpha_3 = 0$, that is, that there is no differential effect due to factor A. In terms of θ's, this hypothesis is $H_0: \theta_1 = \theta_2 = 0$. Since this hypothesis specifies that all θ_i are zero, we apply the exception to Step 5 and obtain $SS_R' = $ total sum of squares of the anova table of the multiple regression in Step 4, and $v_R' = n - 1 = 4$. Alternatively, $SS_R' - SS_R$ is the due regression sum of squares and $v_R' - v_R = 3$ is the due regression degrees of freedom.

COMMENTS 4.5.2   1.   This technique may be used to generate the anova table for any anova problem. A given source of variation in the table (not including the residual) corresponds to a hypothesis H_0 about some differential effects of the anova model. This hypothesis, in turn, corresponds to a hypothesis of the form H_0: some θ's $= 0$ for the general linear model. We then obtain SS_R' and v_R' as in Step 5. The sum of squares SS_H and degrees of freedom v_H for the source of variation are then

$$SS_H = SS_R' - SS_R \quad \text{and} \quad v_H = v_R' - v_R.$$

Then, as usual, the mean square is

$$MS_H = \frac{SS_H}{v_H}.$$

We repeat this process for each source of variation and generate the anova table.

2.   Most of the anova problems presented in this chapter assume equal numbers of observations per cell. This assumption determines formulas for both the number of degrees of freedom and the EMS for each source of variation in the anova table. Since the regression technique of this section permits unequal numbers of observations per cell, the formulas for the degrees of freedom and the EMS for each source of variation become more complicated and are not given in this book. However, the terms of the EMS for each source of variation remain the same as in the case of equal replicates, but the coefficient of each term may be different. This implies that the numerator and denominator of the F ratios for tests of hypotheses are the same as in the case of equal replicates.

EXAMPLE 4.5.2   Assume that we have two Model I factors A and B with $I = 2$ and $J = 3$ levels, respectively. Furthermore, assume that we make only one observation per cell. We wish to generate the appropriate anova table (Table 4.3.5). Paralleling the previous steps, we have:

Step 1   The original model is

$$y_{ij} = \mu + \alpha_i + \beta_j + e_{ij}, \quad i = 1, 2, \quad j = 1, 2, 3$$

The side conditions are $\alpha_1 + \alpha_2 = 0$ and $\beta_1 + \beta_2 + \beta_3 = 0$.

Step 2   From the side conditions we have $\alpha_2 = -\alpha_1$ and $\beta_3 = -\beta_1 - \beta_2$. Substituting into the original model, we obtain the model

$$y_{11} = \mu + \alpha_1 + \beta_1 + e_{11},$$

$$y_{12} = \mu + \alpha_1 + \beta_2 + e_{12},$$

$$y_{13} = \mu + \alpha_1 - (\beta_1 + \beta_2) + e_{13},$$

$$y_{21} = \mu - \alpha_1 + \beta_1 + e_{21},$$

$$y_{22} = u - \alpha_1 + \beta_2 + e_{22},$$

$$y_{23} = \mu - \alpha_1 - (\beta_1 + \beta_2) + e_{23},$$

which is now in terms of μ and the minimal set of three differential effects α_1, β_1, and β_2.

Step 3   Renumber

$$y_1 = y_{11}, \qquad y_2 = y_{12}, \qquad y_3 = y_{13}, \qquad y_4 = y_{21},$$

$$y_5 = y_{22}, \qquad \text{and} \qquad y_6 = y_{23}.$$

Define $\theta_1 = \alpha_1$, $\theta_2 = \beta_1$, and $\theta_3 = \beta_2$. The model becomes

$$y_1 = \mu + 1\theta_1 + 1\theta_2 + 0\theta_3 + e_1$$

$$y_2 = \mu + 1\theta_1 + 0\theta_2 + 1\theta_3 + e_2$$

$$y_3 = \mu + 1\theta_1 + (-1)\theta_2 + (-1)\theta_3 + e_3$$

$$y_4 = \mu + (-1)\theta_1 + 1\theta_2 + 0\theta_3 + e_4$$

$$y_5 = \mu + (-1)\theta_1 + 0\theta_2 + 1\theta_3 + e_5$$

$$y_6 = \mu + (-1)\theta_1 + (-1)\theta_2 + (-1)\theta_3 + e_6$$

where the coefficients of θ_1, θ_2, and θ_3 are the x's determined by the model in Step 2.

Step 4 and Comment 4.5.1.3   A packaged multiple linear regression program with input

y	x_1	x_2	x_3
y_1	1	1	0
y_2	1	0	1
y_3	1	-1	-1
y_4	-1	1	0
y_5	-1	0	1
y_6	-1	-1	-1

will give the desired SS_R and v_R. We also obtain least-squares estimates $\hat{\mu}$, $\hat{\theta}_1$, $\hat{\theta}_2$, and $\hat{\theta}_3$. Thus, in terms of the original model

$$\hat{\alpha}_1 = \hat{\theta}_1, \qquad \hat{\alpha}_2 = -\hat{\alpha}_1, \qquad \hat{\beta}_1 = \hat{\theta}_2, \qquad \hat{\beta}_2 = \hat{\theta}_3, \qquad \text{and} \qquad \hat{\beta}_3 = -\hat{\beta}_1 - \hat{\beta}_2.$$

Step 5 and Comment 4.5.2.1   To the source of variation due to factor A corresponds the hypothesis $H_0: \alpha_1 = \alpha_2 = 0$. This, in turn, corresponds to $H_0: \theta_1 = 0$. We thus delete x_1 and use the multiple regression program with input

y	x_2	x_3
y_1	1	0
y_2	0	1
y_3	-1	-1
y_4	1	0
y_5	0	1
y_6	-1	-1

to obtain SS_R' and v_R'. Then $SS_A = SS_R' - SS_R$ and $v_A = v_R' - v_R$ are the entries for this source of variation. For the source of variation due to factor B, we have the hypothesis $H_0: \beta_1 = \beta_2 = \beta_3 = 0$, which corresponds to $H_0: \theta_2 = \theta_3 = 0$. Deleting x_2 and x_3, the input is

y	x_1
y_1	1
y_2	1
y_3	1
y_4	-1
y_5	-1
y_6	-1

and we obtain $SS_B = SS_R' - SS_R$ and $v_B = v_R' - v_R$. From these quantities we compute the mean squares and generate Table 4.3.5.

COMMENTS 4.5.3   1. In some packages there are programs, such as BMD05V, called *general linear hypothesis* programs. They perform the calculations described in this section and permit all hypotheses of interest to be tested in one run.

2. To use the technique of this section for Model II factors, we view them as Model I factors and write down the usual side conditions for these Model I factors. We then generate the anova table for this Model I formulation, using the procedure outlined in Comment 4.5.2.1. To perform the tests of hypotheses, we write down the EMS for the original model, and compute the appropriate F ratios. As mentioned in Comment 4.5.2.2, if the numbers of observations per cell are unequal, then the EMS for this model are different from those for the case of equal replicates per cell. The tests of hypotheses, however, are the same in either case. The user should not use the EMS from the equal replicates case to estimate the components of variance for the Model II factors.

EXAMPLE 4.5.3 $\qquad$ We assume that we have a Model I factor A at three levels and a Model II factor B at two levels. There are an unequal number of observations per cell, which prevents us from using a standard anova program. The

	B	
A	1	2
1	$y_{111} = 17.5$ $y_{112} = 16.2$	$y_{121} = 10.1$ $y_{122} = 8.6$ $y_{123} = 11.3$
2	$y_{211} = 13.2$	$y_{221} = 5.4$ $y_{222} = 3.7$
3	$y_{311} = 12.8$ $y_{312} = 10.4$ $y_{313} = 9.9$	$y_{321} = 10.3$

data for this hypothetical example are given in the accompanying table. Paralleling the steps outlined previously, we have:

Step 1 and Comment 4.5.3.2 $\qquad$ Although we have unequal cell replicates, the anova model is similar to that given by Eq. (4.3.1), namely,

$$y_{ijk} = \mu + \alpha_i + b_j + \gamma_{ij} + e_{ijk}, \qquad i = 1, 2, 3, \quad j = 1, 2, \quad k = 1, \ldots, K_{ij},$$

where $\gamma_{ij} = (\alpha b)_{ij}$ and $K_{11} = 2$, $K_{21} = 1$, $K_{31} = 3$, $K_{12} = 3$, $K_{22} = 2$, and $K_{32} = 1$. The side condition for factor A is $\alpha_1 + \alpha_2 + \alpha_3 = 0$. Viewing factor B as Model I, the side conditions for factor B and AB interactions are

$$b_1 + b_2 = 0 \qquad \text{and}$$

$$\gamma_{11} + \gamma_{21} + \gamma_{31} = \gamma_{12} + \gamma_{22} + \gamma_{32} = \gamma_{11} + \gamma_{12} = \gamma_{21} + \gamma_{22} = \gamma_{31} + \gamma_{32} = 0.$$

Step 2 $\qquad$ From the side conditions, we have

$$\alpha_3 = -\alpha_1 - \alpha_2, \qquad b_2 = -b_1,$$

$$\gamma_{12} = -\gamma_{11}, \qquad \gamma_{22} = -\gamma_{21},$$

$$\gamma_{31} = -\gamma_{11} - \gamma_{21}, \qquad \gamma_{32} = -\gamma_{31} = \gamma_{11} + \gamma_{21}.$$

Substituting into the original model, we obtain the model

$$y_{11k} = \mu + \alpha_1 + b_1 + \gamma_{11} + e_{11k}, \qquad k = 1, 2,$$

$$y_{211} = \mu + \alpha_2 + b_1 + \gamma_{21} + e_{211},$$

$$y_{31k} = \mu - \alpha_1 - \alpha_2 + b_1 - \gamma_{11} - \gamma_{21} + e_{31k}, \qquad k = 1, 2, 3,$$

$$y_{12k} = \mu + \alpha_1 - b_1 - \gamma_{11} + e_{12k}, \qquad k = 1, 2, 3,$$

$$y_{22k} = \mu + \alpha_2 - b_1 - \gamma_{21} + e_{22k}, \qquad k = 1, 2,$$

$$y_{321} = \mu - \alpha_1 - \alpha_2 - b_1 + \gamma_{11} + \gamma_{21} + e_{321},$$

which is now in terms of μ and the minimal set of 5 differential effects α_1, α_2, b_1, γ_{11}, and γ_{21}.

Step 3   Renumber $y_1 = y_{111}$, $y_2 = y_{112}$, ..., $y_{12} = y_{321}$. Define $\theta_1 = \alpha_1$, $\theta_2 = \alpha_2$, $\theta_3 = b_1$, $\theta_4 = \gamma_{11}$, and $\theta_5 = \gamma_{21}$. The model is now written as

$$y_k = \mu + 1\theta_1 + 0\theta_2 + 1\theta_3 + 1\theta_4 + 0\theta_5 + e_k, \qquad k = 1, 2$$

$$y_3 = \mu + 0\theta_1 + 1\theta_2 + 1\theta_3 + 0\theta_4 + 1\theta_5 + e_3,$$

$$y_k = \mu + (-1)\theta_1 + (-1)\theta_2 + 1\theta_3 + (-1)\theta_4 + (-1)\theta_5 + e_k, \qquad k = 4, 5, 6$$

$$y_k = \mu + 1\theta_1 + 0\theta_2 + (-1)\theta_3 + (-1)\theta_4 + 0\theta_5 + e_k, \qquad k = 7, 8, 9$$

$$y_k = \mu + 0\theta_1 + 1\theta_2 + (-1)\theta_3 + 0\theta_4 + (-1)\theta_5 + e_k, \qquad k = 10, 11$$

$$y_{12} = \mu + (-1)\theta_1 + (-1)\theta_2 + (-1)\theta_3 + 1\theta_4 + 1\theta_5 + e_{12}.$$

Step 4 and Comment 4.5.1.3   The data for the packaged multiple regression program is given in the accompanying table. From the multiple regression program, we obtain $SS_R = 10.757$ and $\nu_R = 6$. Furthermore, we have $\hat{\mu} = 10.99$, $\hat{\theta}_1 = \hat{\alpha}_1 = 2.44$, $\hat{\theta}_2 = \hat{\alpha}_2 = -2.11$, so that $\hat{\alpha}_3 = -\hat{\alpha}_1 - \hat{\alpha}_2 = -0.33$. Although estimates are computed for θ_3, θ_4, and θ_5, they correspond to random effects and are not reported here.

k	y	x_1	x_2	x_3	x_4	x_5
1	17.5	1	0	1	1	0
2	16.2	1	0	1	1	0
3	13.2	0	1	1	0	1
4	12.8	-1	-1	1	-1	-1
5	10.4	-1	-1	1	-1	-1
6	9.9	-1	-1	1	-1	-1
7	10.1	1	0	-1	-1	0
8	8.6	1	0	-1	-1	0
9	11.3	1	0	-1	-1	0
10	5.4	0	1	-1	0	-1
11	3.7	0	1	-1	0	-1
12	10.3	-1	-1	-1	1	1

Step 5 and Comments 4.5.2.1 and 4.5.3.2   We wish to generate Table 4.3.1. For each source of variation, we delete the appropriate x's and obtain the results summarized in Table A. Thus, we construct the anova table (Table B) with the

TABLE A

Source of variation	Hypothesis H_0	Delete from input	SS_R'	v_R'	$SS_H = SS_R' - 10.757$	$v_H = v_R' - 6$
Factor A	$\theta_1 = \theta_2 = 0$	x_1, x_2	48.996	8	38.239	2
Factor B	$\theta_3 = 0$	x_3	82.626	7	71.869	1
Interaction AB	$\theta_4 = \theta_5 = 0$	x_4, x_5	36.591	8	25.834	2

TABLE B

Source of variation	Sum of squares	Degrees of freedom	Mean square	EMS
Factor A	38.239	2	19.119	$\sigma^2 + k_1 \Sigma \alpha_i^2$
Factor B	71.869	1	71.869	$\sigma^2 + k_1 \sigma_{ab}^2 + k_2 \sigma_b^2$
Interaction AB	25.834	2	12.917	$\sigma^2 + k_1 \sigma_{ab}^2$
Residual	10.757	6	1.793	σ^2
Total	146.699	11		

appropriate EMS column (that is, assuming factor B and interaction AB are Model II). In the EMS column (see Table 4.3.9), σ_b^2 and σ_{ab}^2 are the variances of the random effects due to B and AB, respectively. The values of the constants k_1 and k_2 are not easily available, since we have unequal replicates per cell. Thus, we cannot easily estimate σ_b^2 and σ_{ab}^2. We can, however, make appropriate tests of hypotheses at level $\alpha = 0.05$. These are

(a) $H_0: \sigma_{ab}^2 = 0$, $F = MS_{AB}/MS_R = 7.2$. Since $F_{0.95}(2, 6) = 5.14$, we reject H_0.
(b) $H_0: \sigma_b^2 = 0$, $F = MS_B/MS_{AB} = 5.6$. Since $F_{0.95}(1, 2) = 18.51$, we accept H_0.
(c) $H_0: \alpha_1 = \alpha_2 = \alpha_3 = 0$, $F = MS_A/MS_R = 10.7$. Since $F_{0.95}(2, 6) = 5.14$, we reject H_0.

In addition to the anova table, the technique of this section is then used to test the single hypothesis $H_0: \sigma_b^2 = 0$ and $\sigma_{ab}^2 = 0$. To obtain SS_R' we delete x_3, x_4, and x_5 from the regression input, and obtain $SS_R' = 117.348$ and $v_R' = 9$. Using Eq. (4.5.1), we have

$$F = \frac{(117.348 - 10.757)/(9 - 6)}{10.757/6} = 19.8.$$

Since $F_{0.95}(3, 6) = 4.76$, we reject the hypothesis.

4.6 Description of a General Anova Program

In Section 4.4 it was shown how a factorial program could be used for solving general analysis of variance problems. The main requirements of this technique were that there be an equal number of replicates per cell and that the layout of the anova design be transformable to that of a factorial design. The sums of squares and degrees of freedom of the resulting factorial anova table were then pooled to generate the anova table for the original design.

In Section 4.5 it was shown how a multiple regression program could be used to solve general anova problems. This technique, which allowed unequal numbers of replicates per cell, consisted of expressing the anova model in terms of a multiple linear regression model. The regression coefficients were elements of a minimal set of differential effects. By solving a series of regression problems, the anova table for the original design was generated.

Both techniques may require considerable effort on the part of the user. In the first case, the user must correctly transform the layout of the anova design to that of a factorial design and carefully pool the resulting degrees of freedom. In the second case, he must accurately determine a minimal set of effects, carefully transform the anova model to the regression model, and patiently express each observation in terms of the regression model. Furthermore, he must solve a series of regression problems, a potentially expensive process to obtain the sums of squares and degrees of freedom for the anova table. In both cases he must also refer to this or another book to find the form of the EMS for each source of variation. For mixed models, or for designs with unequal replicates, the correct EMS may not be easily found. Thus, the testing of hypotheses and the estimation of the components of variance may not be easy tasks.

To facilitate the analysis of any *equal* replicate design and to free the user from these additional efforts, some very general programs have been written. These programs allow Model I and/or Model II factors, and factorial, nested, or *partially nested* designs. (This last design occurs when some factors are crossed while other factors are nested.) In this section we discuss one such program—BMD08V—which has the additional advantage of computing the EMS for each source of variation. For this program, the user need only specify the number of factors, the number of levels for each factor, the identification of each factor (that is, Model I or Model II), and the design describing the relationship between the factors (that is, which factors, if any, are nested). The program then computes and prints the analysis of variance table, the EMS and the F ratio for each source of variation, and the estimates of the variance components.

Since this program introduces some concepts not yet discussed in this chapter, we now explain some of the details of the input and output.

1. *Identification of the factors*　　As discussed earlier in this chapter, identify-

ing the factors as Model I or Model II determines the EMS for each source of variation in the anova table. To compute these quantities this program utilizes the technique of *finite population models* (see Brownlee, 1965, p. 489; Cornfield and Tukey, 1956; and Bennett and Franklin, 1954). For any factor A with I levels, this technique assumes that the main effects $\alpha_1, ..., \alpha_I$ of A represent a random sample of size I from a population of size N_A. If A is a Model I factor, then $N_A = I$, that is, $\alpha_1, \alpha_2, ..., \alpha_I$ represent the whole population. On the other hand, if A is a Model II factor then $N_A = \infty$, that is, $\alpha_1, ..., \alpha_I$ represent a random sample from an infinite population. It is also possible to identify a third type of factor which is defined by $I < N_A < \infty$, that is, $\alpha_1, ..., \alpha_I$ represent a random sample from a finite population. The interpretation of this factor, sometimes called a *Model III factor*, is not discussed in this book. It suffices to say that this program permits all three types of factors and computes the EMS according to the theory described in the above references. It should be noted that this program regards replicates as a factor. Replicates should be identified as a Model II factor.

2. *The design* Recall that two factors A and B are crossed, denoted by $A \times B$, if all possible pairs of levels of the factors are present. Also recall that B is nested within A, denoted by $B(A)$, if each level of B occurs with at most one level of A. Defining these relationships between the factors (the *design*) determines the sources of variation in the anova table. These sources of variation include all the factors and appropriate interactions. A general rule for determining appropriate interactions is that a factor does not interact with a factor which it nests. For example, if A and B are crossed, the resulting sources of variation in the anova table are A, B, and AB. However, if A nests B, then the resulting sources are A and $B(A)$. We do not have the interaction $AB(A)$.

In this program each crossed factor is designated by a single letter. For example, the design A, B, C implies that A is crossed with B, B is crossed with C, and A is crossed with C. A nested factor is designated by a letter followed by the letter(s) of the nesting factor(s) enclosed in parentheses. For example, the design $A, B, C(B)$ implies that A is crossed with B and C is nested within B. The design $A, B, C(AB)$ implies that A and B are crossed and C is nested within AB, that is, each level of C occurs with at most one pair of levels of A and B. It should be noted that the nesting relationships are transitive, that is, if B is nested within A, and C is nested within B, then C is nested within AB. In this case the design is written as $A, B(A)$, and $C(AB)$, and not as $A, B(A), C(B)$. Finally, it should be noted that for replicated factorial designs, the replicate factor is nested within all factors. This is also true for a *completely nested* design, that is, a design of the form $A, B(A), C(AB), D(ABC)$, and so forth.

3. *The analysis of variance table* The analysis of variance table printed by the program lists the sums of squares, degrees of freedom, and mean squares for all of the appropriate sources of variation. These sources of variation include the

factors specified by the design, the appropriate interactions, as well as the variation due to the overall mean μ. The sources of variation are numbered $1, 2,\ldots$ to facilitate interpretation of the EMS as shown below.

   4. *The EMS*   The analysis of variance table also includes the EMS for each source of variation. These quantities are printed in the form

$$k_1(n_1) + k_2(n_2) + \cdots + k_m(n_m). \tag{4.6.1}$$

(The plus signs are assumed and therefore not printed.) The coefficients k_i are constants determined by the design and calculated by the program, $i = 1,\ldots, m$. The notation (n_i) is a symbol for a variance term corresponding to the n_ith source of variation, $i = 1,\ldots, m$. If the n_ith source of variation is Model I, this variance term is a sum of squares of differential effects divided by its degrees of freedom, for example,

$$\sum_{i=1}^{I} \frac{\alpha_i^2}{(I-1)}, \qquad \sum_{i=1}^{I}\sum_{j=1}^{J} \frac{(\alpha\beta)_{ij}^2}{(I-1)(J-1)},$$

and so forth. If the source of variation is the mean μ, then this variance term is μ^2. Finally, if the source of variation is Model II (or Model III), it is the variance of the infinite (or finite) population of differential effects, for example, σ_a^2, σ_{ab}^2, and so forth. Two examples will elucidate these ideas.

EXAMPLE 4.6.1   Consider the one-way layout with a factor A at I levels and a replicate factor R at J levels. This design is denoted by $A, R(A)$. If A is Model I, the anova table would include the quantities shown in the accompanying table.

n_i	Source of variation	EMS
1	Mean μ	$IJ(1) + 1(3)$
2	A	$J(2) + 1(3)$
3	$R(A)$	$1(3)$

   The EMS notation for $R(A)$—"within groups"—implies that a variance term corresponding to the third source of variation—namely, $R(A)$—is multiplied by $k_1 = 1$. Since $R(A)$ is a Model II factor, this variance term is the error variance σ^2 —the variance of the population of effects due to replication. Similarly, the EMS notation for A—"between groups"—implies that σ^2 is added to the product of $k_1 = J$ and a variance term due to A. Since A is Model I this variance term is $\sum_{i=1}^{I} \alpha_i^2/(I-1)$, where $\alpha_1,\ldots, \alpha_I$ are the differential effects due to A. Thus, EMS for A is $J\sum_{i=1}^{I} \alpha_i^2/(I-1) + \sigma^2$. Finally, the EMS for μ is $IJ\mu^2 + \sigma^2$.

n_i	Source of variation	EMS
1	Mean μ	$IJ(1) + J(2) + 1(3)$
2	A	$J(2) + 1(3)$
3	$R(A)$	$1(3)$

Now if A is Model II, the anova table would include the quantities in the accompanying table. In this case the EMS for $R(A)$, A, and μ are σ^2, $J\sigma_a^2 + \sigma^2$, and $IJ\mu^2 + J\sigma_a^2 + \sigma^2$, respectively. The variance term σ_a^2 is the variance of the infinite population of effects due to factor A.

EXAMPLE 4.6.2   Consider the two-way nested layout with a Model II factor A at I levels, a Model II factor B nested within A at J levels, and a replicate factor R at K levels. This design is denoted by A, $B(A)$, $R(AB)$, and the anova table includes the quantities in Table A. Denoting the variance of the population of effects due to A by σ_a^2, and the variance of the population of effects due to B within A by $\sigma_{b(a)}^2$, these EMS are translated into the quantities shown in Table B.

TABLE A

n_i	Source of variation	EMS
1	Mean μ	$IJK(1) + JK(2) + K(3) + 1(4)$
2	A	$JK(2) + K(3) + 1(4)$
3	$B(A)$	$K(3) + 1(4)$
4	$R(AB)$	$1(4)$

TABLE B

Source of variation	EMS
Mean	$IJK\mu^2 + JK\sigma_a^2 + K\sigma_{b(a)}^2 + \sigma^2$
A	$JK\sigma_a^2 + K\sigma_{b(a)}^2 + \sigma^2$
$B(A)$	$K\sigma_{b(a)}^2 + \sigma^2$
$R(AB)$	σ^2

5. *Testing hypotheses*   Let MS_H denote the mean square corresponding to a given source of variation n_1 and let v_H be its degrees of freedom. Assume that

the EMS for this source of variation is $k_1(n_1)+k_2(n_2)+\cdots+k_m(n_m)$. To test the hypothesis that the appropriate effect of this source of variation is zero, an F ratio is used. The numerator of the F ratio is MS_H and its denominator is the mean square for a source of variation whose EMS is $k_2(n_2)+\cdots+k_m(n_m)$, that is, the same as the above EMS except for the term $k_1(n_1)$. In the program, this source of variation is called the *error term*. Denoting its mean square and degrees of freedom by MS_E and ν_E, respectively, the F ratio is then

$$F = \frac{MS_H}{MS_E}. \tag{4.6.2}$$

We reject the hypothesis if $F > F_{1-\alpha}(\nu_H, \nu_E)$.

This program computes and prints in the anova table the name of the error term and the F statistic for each source of variation. For the mean μ this statistic tests H_0: $\mu = 0$. For each Model I factor (or interaction) it tests whether the differential effects due to the factor (or interaction) are all zero. For each Model II factor (or interaction) it tests whether the variance of the population effects due to the factor (or interaction) is zero. (A k-way interaction is Model I if all k factors in the interaction are Model I; otherwise, it is Model II or mixed.)

For some sources of variation (such as the residual term), the EMS may consist of a single variance term. In such a case there is no F test. For some other sources of variation, neither an error term nor an F statistic is printed. In this case the error term is not a single source of variation, but rather a linear combination of more than one source of variation (see Comment 4.6.1).

EXAMPLE 4.6.1 (*continued*)   In the one-way layout with factor A assumed to be Model I, the anova table would include the quantities in Table A. For both of the sources of variation due to μ and A, the source of variation due to $R(A)$ was chosen as the error term. Thus, to test H_0: $\mu = 0$, we compare $F = MS_1/MS_3$ with $F_{1-\alpha}(1, IJ-1)$, and to test H_0: all $\alpha_i = 0$, we compare $F = MS_2/MS_3$ with $F_{1-\alpha}(I-1, IJ-1)$.

TABLE A

n_i	Source of variation	Error term	F	Degrees of freedom	Mean square
1	Mean μ	$R(A)$	$\dfrac{MS_1}{MS_3}$	1	MS_1
2	A	$R(A)$	$\dfrac{MS_2}{MS_3}$	$I-1$	MS_2
3	$R(A)$			$IJ-1$	MS_3

TABLE B

n_i	Source of variation	Error term	F	Degrees of freedom	Mean square
1	Mean μ	A	$\dfrac{MS_1}{MS_2}$	1	MS_1
2	A	$R(A)$	$\dfrac{MS_2}{MS_3}$	$I-1$	MS_2
3	$R(A)$			$IJ-1$	MS_3

When factor A is assumed to be Model II, the anova table would include the quantities in Table B. In this case, the error term for μ is now the source of variation due to A. Thus, to test $H_0: \mu = 0$, we compare $F = MS_1/MS_2$ with $F_{1-\alpha}(1, I-1)$. To test $H_0: \sigma_a^2 = 0$, we compare $F = MS_2/MS_3$ with $F_{1-\alpha}(I-1, IJ-1)$.

6. *Estimates of variance components*   Also included in the output are estimates of the variance components for each source of variation. If the source of variation is due to a Model I factor (or interaction) the estimate is for a sum of squares of differential effects divided by its degrees of freedom. If the source of variation is μ, then the estimate is for μ^2. Finally, if the source of variation is due to a Model II factor (or interaction), the estimate is for the variance of the population of differential effects. Since these quantities are obtained by computing linear combinations of mean squares, they may be negative. In that case, the estimate should be taken as zero.

EXAMPLE 4.6.1 (*continued*)   Corresponding to the sources of variation in the one-way layout, we have the quantities in the accompanying table. The quantity $\hat{\mu}^2$ estimates μ^2, $\hat{\sigma}_a^2 = \sum_{i=1}^{I} \hat{\alpha}_i^2/(I-1)$ estimates $\sum_{i=1}^{I} \alpha_i^2/(I-1)$ if A is Model I, and $\hat{\sigma}_a^2$ estimates σ_a^2 if A is Model II. Finally, $\hat{\sigma}^2$ estimates σ^2.

n_i	Estimate of variance component
1	$\hat{\mu}^2$
2	$\hat{\sigma}_a^2$
3	$\hat{\sigma}^2$

COMMENT 4.6.1   If in the output of the program there is no error term for a given source of variation (and hence, no F statistic for the corresponding hypothesis H_0), the user must find a linear combination of mean squares according to the following recipe (Welch, 1937):

As before, let MS_H, v_H, and $k_1(n_1) + k_2(n_2) + \cdots + k_m(n_m)$ be the mean square, degrees of freedom, and EMS of the n_ith source of variation, respectively. The user must find a linear combination of the EMS which is equal to $k_2(n_2) + \cdots + k_m(n_m)$. Let this linear combination be

$$c_1 \, EMS_1 + \cdots + c_k \, EMS_k.$$

Then the denominator mean square MS_E and degrees of freedom v_E are

$$MS_E = c_1 \, MS_1 + \cdots + c_k \, MS_k \qquad \text{and} \qquad v_E = \frac{(MS_E)^2}{\sum_{i=1}^{k} c_i^2 ((MS_i)^2 / v_i)},$$

where v_i is the number of degrees of freedom of MS_i, $i = 1, \ldots, k$. The F statistic for H_0 is

$$F = \frac{MS_H}{MS_E},$$

which has an *approximate* F distribution with v_H and v_E degrees of freedom. We reject H_0 if $F > F(v_H, v_E)$. Note that v_E is not necessarily an integer, and, therefore, some interpolation in the F tables may be necessary to obtain $F_{1-\alpha}(v_H, v_E)$.

EXAMPLE 4.6.3   In an experiment to investigate the side effects of two drugs to treat asthma, one of the variables measured was blood pressure (mm Hg). For a given patient, the following procedure was applied: His blood pressure was measured, one of the two drugs was administered, and his pressure was measured after 15, 30, and 60 minutes. After an appropriate rest period, the same procedure was repeated for the other drug. Then the two drugs were administered in random order for a second time. Thus, each patient received the same drug twice. This process was replicated on five randomly selected patients.

We identify "patients" as a Model II factor P at 5 levels, "drugs" as a Model I factor D at 2 levels, "time" as a Model I factor T at 4 levels, and the "replicates" as a Model II factor R at 2 levels. Furthermore, P, D, and T are crossed, and R is nested within DT. Thus, the design is $P, D, T, R(DT)$, a partially nested design.

The data for this experiment are given in Table 4.6.1. Using BMD08V, the resulting output is given in Table 4.6.2. Tests of hypotheses were made for P, PD, PT, $R(DT)$, and PDT. Thus, for example, for the hypothesis $H_0: \sigma_p^2 = 0$, where σ_p^2 is the variance due to the patient effect, we have $F = 49.4$. When compared to $F_{0.95}(4, 32)$ we see that this result is highly significant. The P value for the patient–time interaction is almost exactly 0.05, while the remaining sources of variation for which F is printed are nonsignificant.

TABLE 4.6.1

Time	Patient 1 Drug 1	Patient 1 Drug 2	Patient 2 Drug 1	Patient 2 Drug 2	Patient 3 Drug 1	Patient 3 Drug 2
0	120, 140	160, 130	70, 85	90, 75	90, 85	90, 100
15	145, 140	155, 140	80, 70	70, 75	90, 90	100, 100
30	120, 140	150, 110	80, 70	70, 70	100, 90	120, 100
60	125, 120	110, 110	100, 90	75, 70	100, 90	90, 90

Time	Patient 4 Drug 1	Patient 4 Drug 2	Patient 5 Drug 1	Patient 5 Drug 2
0	110, 150	110, 110	90, 110	80, 110
15	100, 120	110, 120	120, 140	110, 95
30	100, 130	110, 120	105, 110	90, 104
60	110, 140	120, 120	110, 120	120, 110

TABLE 4.6.2. *Analysis of Variance for Dependent Variable I*

Source	Error term	F	Sum of squares	Degrees of freedom	Mean square
1. Mean			899728.2	1	899728.2
2. P	$PR(DT)$	49.3961	27960.91	4	6990.227
3. D			140.4500	1	140.4500
4. T			184.1000	3	61.36665
5. PD	$PR(DT)$	1.3799	781.0967	4	195.2742
6. PT	$PR(DT)$	2.0924	3653.173	12	296.0977
7. DT			285.8496	3	95.28320
8. $R(DT)$	$PR(DT)$	1.0552	1194.600	8	149.3250
9. PDT	$PR(DT)$	0.8886	1508.911	12	125.7426
10. $PR(DT)$			4528.434	32	141.5185

Expected mean square				Estimates of variance components	
80.000 (1)	16.000 (2)	5.000 (8)	1.000 (10)	(1)	11159.13
16.000 (2)	1.000 (10)			(2)	428.0444
40.000 (3)	8.000 (5)	5.000 (8)	1.000 (10)	(3)	−1.565890
20.000 (4)	4.000 (6)	5.000 (8)	1.000 (10)	(4)	−12.12712
8.000 (5)	1.000 (10)			(5)	6.720078
4.000 (6)	1.000 (10)			(6)	38.64603
10.000 (7)	5.000 (8)	2.000 (9)	1.000 (10)	(7)	−3.827081
5.000 (8)	1.000 (10)			(8)	1.562292
2.000 (9)	1.000 (10)			(9)	−7.885498
1.000 (10)				(10)	141.5135

Note that the mean μ, D, T, and DT do not have error terms computed. For the sake of example, let us test H_0: $\delta_1 = \delta_2 = 0$, that is, there is no effect δ_i to the ith drug, $i = 1$ or 2. We apply Comment 4.6.i to the EMS of the source of variation due to the drug D. This quantity is $40(3) + 8(5) + 5(8) + 1(10)$. Thus, we need a linear combination of the other EMS which is equal to $8(5) + 5(8) + 1(10)$. Note that

$$1\,\mathrm{EMS}_5 + 1\,\mathrm{EMS}_8 - 1\,\mathrm{EMS}_{10}$$

yields this quantity. Thus, the denominator mean square MS_E is

$$\mathrm{MS}_E = 1\,\mathrm{MS}_5 + 1\,\mathrm{MS}_8 - 1\,\mathrm{MS}_{10}$$

$$= 195.3 + 149.3 - 141.5 = 203.1,$$

with

$$\nu_E = \frac{(203.1)^2}{\dfrac{(195.3)^2}{4} + \dfrac{(149.3)^2}{8} + \dfrac{(141.5)^2}{32}} = 3.19 \simeq 3$$

degrees of freedom. The approximate F statistic for H_0 is $F = 140.5/203.1 = 0.7$ which is not significant when compared to $F_{0.95}(1, 3)$.

Finally, note that the estimates of the components of variance for D, T, DT, and PDT are negative. They should be regarded as zero. The estimate of σ_p^2 is $\hat{\sigma}_p^2 = 428.0$.

4.7 The Analysis of Covariance

In this section we discuss a technique, called the *one-way analysis of covariance*, which utilizes both the concepts of the one-way analysis of variance and simple linear regression analysis. Suppose that we have a factor A (usually called a treatment factor) which has I levels. Denote by y_{ij} the observation made on the jth experimental unit subjected to the ith level of A, $j = 1, ..., J_i$, $i = 1, ..., I$. If y_{ij} is assumed to be distributed as $N(\mu_i, \sigma^2)$, then we have the familiar one-way anova model

$$y_{ij} = \mu + \alpha_i + e_{ij}, \qquad j = 1, ..., J_i, \quad i = 1, ..., I, \tag{4.7.1}$$

where μ is an overall mean, α_i is the differential effect of the ith level of factor A, where $\mu_i = \mu + \alpha_i$, and e_{ij} are independent $N(0, \sigma^2)$ error variables. As is customarily done, we impose the side condition

$$\sum_{i=1}^{I} J_i \alpha_i = 0, \tag{4.7.2}$$

in order to obtain unique least-squares estimators for μ and $\alpha_1, ..., \alpha_I$.

Now suppose that before we subjected the jth experimental unit to the ith level of the factor, we made an observation x_{ij} on another variable, called a *covariate* (or *concomitant variate*), which is assumed to be linearly related to y_{ij}. Then we may write the model

$$y_{ij} = \mu + \alpha_i + \beta(x_{ij} - \bar{x}_{..}) + e_{ij}, \qquad j = 1, ..., J_i, \quad i = 1, ..., I, \quad (4.7.3)$$

where

$$\bar{x}_{..} = \frac{1}{n} \sum_{i=1}^{I} \sum_{j=1}^{J_i} x_{ij} \qquad \text{and} \qquad n = \sum_{i=1}^{I} J_i.$$

This model, called the *one-way analysis of covariance model*, expresses the ijth observation as a sum of the overall mean μ, a fixed differential effect α_i due to the ith level of the factor, a term $\beta(x_{ij} - \bar{x}_{..})$ attributable to the linear association with x_{ij}, and an error term e_{ij}. Note that we may write this equation as the one-way anova model

$$y_{ij}^* = \mu + \alpha_i + e_{ij}, \qquad j = 1, ..., J_i, \quad i = 1, ..., I, \qquad (4.7.4)$$

where

$$y_{ij}^* = y_{ij} - \beta(x_{ij} - \bar{x}_{..}) \qquad (4.7.5)$$

is the value of y_{ij} after adjustment for the linear regression on x_{ij}. Thus, α_i may be regarded as the true differential effect of the ith level of factor A *after* adjustment for the linear regression on the covariate x_{ij}.

To see the advantages of this model we consider the following example.

EXAMPLE 4.7.1　　　In an experiment,[†] $n = 40$ subjects were given a behavioral approach test to determine how close they could walk towards a phobic object (live snake) before they felt uncomfortable or anxious. Each subject was then put into one of $I = 4$ equal size treatment groups, each group representing a different degree and type of placebo potentiality. One of the four groups was a control. Subsequent to the treatment, each subject underwent a posttreatment behavioral approach test determining how close he could now walk to the phobic object without feeling anxious or uncomfortable.

In this example the factor A is the treatment factor, the levels are the four treatments, each administered to a group of size $J_i = 10$, $i = 1, ..., 4$. The response being analyzed is $y_{ij} = $ distance to snake after the ith treatment, and the covariate is $x_{ij} = $ distance to snake before this treatment. Assuming a linear relationship between y_{ij} and x_{ij}, we may perform a one-way covariance analysis, measuring and comparing the differential effects of the 4 treatments after adjusting for pretreatment differences in the subjects.

† An experiment of this nature was done by Dr. Stephen Zahm, University of Portland, Portland, Oregon.

Thus, it is seen from this example that one purpose of the analysis of covariance is to increase the precision of the measurement of interest y_{ij} by taking out the effect of a related pretreatment variable x_{ij}. The gain in precision depends largely on the size of the correlation between the two variables.

Since the one-way analysis of covariance model can be put into the form of the general linear model of Section 4.1, the least-squares procedure produces unbiased estimators of the parameters which have minimum variance among all unbiased linear estimators. Using this technique we obtain

$$\hat{\mu} = \bar{y}_{..} \quad \text{and} \quad \hat{\beta} = \frac{E_{xy}}{E_{xx}}, \tag{4.7.6}$$

the estimated *within group regression coefficient*, where

$$E_{xx} = \sum_{i=1}^{I} \sum_{j=1}^{J_i} (x_{ij} - \bar{x}_{i.})^2, \quad \text{and} \quad E_{xy} = \sum_{i=1}^{I} \sum_{j=1}^{J_i} (x_{ij} - \bar{x}_{i.})(y_{ij} - \bar{y}_{i.}).$$

Also, we have the *estimated adjusted differential effect* of the ith level

$$\hat{\alpha}_i = (\bar{y}_{i.} - \bar{y}_{..}) - \hat{\beta}(\bar{x}_{i.} - \bar{x}_{..}) \quad \text{for} \quad i = 1, ..., I. \tag{4.7.7}$$

Note that the *adjusted mean* $\mu + \alpha_i$ corresponding to the ith level is estimated by

$$\hat{\mu} + \hat{\alpha}_i = \bar{y}_{i.} - \hat{\beta}(\bar{x}_{i.} - \bar{x}_{..}), \quad i = 1, ..., I. \tag{4.7.8}$$

A typical packaged analysis of covariance program computes and prints an analysis of covariance table similar to that of Table 4.7.1. The entries of this table are the sums of squares and cross products for the sources of variation *between levels* (*groups* or *means*), *within levels* (*groups* or *means*), also called the *residual* (or *error*), and the *total*. Note that the entries for the columns labeled XX and YY are the usual sums of squares for a one-way analysis of variance on the variables x_{ij} and y_{ij}, respectively. The entries for the column XY are analagous expressions for the product of the two variables. From the residual source of variation, we obtain the quantities E_{xx} and E_{xy} for the estimate of the within group regression coefficient β [using Eq. (4.7.6)]. From this or a descriptive program, we obtain the means $\bar{x}_{i.}$, $\bar{x}_{..}$, $\bar{y}_{i.}$, and $\bar{y}_{..}$ in order to estimate $\hat{\mu}$ and $\hat{\alpha}_i$, $i = 1, ..., I$. Finally, for each i, $i = 1, ..., I$, we can also estimate a regression line, called the *within group regression line*, given by

$$\hat{y} = \bar{y}_{i.} + \hat{\beta}(x - \bar{x}_{i.}). \tag{4.7.9}$$

This is the least-squares line for the subpopulation corresponding to the ith level of the factor. Since the slopes of all I estimated lines are equal to β, the lines are parallel.

TABLE 4.7.1. *One-Way Analysis of Covariance*

Source of variation	Degrees of freedom	Sums of squares and cross products		
		XX	XY	YY
Between levels (means)	$v_M = I - 1$	$M_{xx} = \sum_{i=1}^{I} J_i(\bar{x}_{i.} - \bar{x}_{..})^2$	$M_{xy} = \sum_{i=1}^{I} J_i(\bar{x}_{i.} - \bar{x}_{..})(\bar{y}_{i.} - \bar{y}_{..})$	$M_{yy} = \sum_{i=1}^{I} J_i(\bar{y}_{i.} - \bar{y}_{..})^2$
Within levels (means) (Residual)	$v_E = n - I$	$E_{xx} = \sum_{i=1}^{I} \sum_{j=1}^{J_i}(x_{ij} - \bar{x}_{i.})^2$	$E_{xy} = \sum_{i=1}^{I} \sum_{j=1}^{J_i}(x_{ij} - \bar{x}_{i.})(y_{ij} - \bar{y}_{i.})$	$E_{yy} = \sum_{i=1}^{I} \sum_{j=1}^{J_i}(y_{ij} - \bar{y}_{i.})^2$
Total	$v_T = n - 1$	$T_{xx} = \sum_{i=1}^{I} \sum_{j=1}^{J_i}(x_{ij} - \bar{x}_{..})^2$	$T_{xy} = \sum_{i=1}^{I} \sum_{j=1}^{J_i}(x_{ij} - \bar{x}_{..})(y_{ij} - \bar{y}_{..})$	$T_{yy} = \sum_{i=1}^{I} \sum_{j=1}^{J_i}(y_{ij} - \bar{y}_{..})^2$

From the analysis of covariance table we may obtain two additional regression lines. The first line, called the *means regression line*, is the least-squares line through the sample means $(\bar{x}_{1.}, \bar{y}_{1.}), \ldots, (\bar{x}_{I.}, \bar{y}_{I.})$ corresponding to the I levels of the factor. The least-squares equation is

$$\hat{y} = \bar{y}_{..} + b_M(x - \bar{x}_{..}),$$ (4.7.10)

where

$$b_M = \frac{M_{xy}}{M_{xx}},$$

the ratio of the between means sum of squares for XY and XX, is called the *means regression coefficient*. The second line, called the *total regression line*, is the regression of y on x when the samples for each of the I levels are combined into one sample of size n. The least-squares equation is

$$\hat{y} = \bar{y}_{..} + b_T(x - \bar{x}_{..}),$$ (4.7.11)

where

$$b_T = \frac{T_{xy}}{T_{xx}},$$

the ratio of total sum of squares for XY and XX, is called the *total regression coefficient*.

Finally, we obtain an unbiased estimator of the error variance σ^2 from the expression

$$MS_E = \frac{1}{n - I - 1}\left[E_{yy} - \frac{E_{xy}^2}{E_{xx}}\right].$$ (4.7.12)

EXAMPLE 4.7.1 (*continued*)   The data collected for this experiment are given in Table A.

TABLE A

| (x, y) for subject | \multicolumn{4}{c}{Treatment group} |
	1	2	3	4 (control)
1	25, 25	17, 11	32, 24	10, 8
2	13, 25	9, 9	30, 18	29, 17
3	10, 12	19, 16	12, 2	7, 8
4	25, 30	25, 17	30, 24	17, 12
5	10, 37	6, 1	10, 2	8, 7
6	17, 25	23, 12	8, 0	30, 26
7	9, 31	7, 4	5, 0	5, 8
8	18, 26	5, 3	11, 1	29, 29
9	27, 28	30, 26	5, 1	5, 29
10	17, 29	19, 20	25, 10	13, 9

From a descriptive program we obtained the estimates

$$\bar{x}_{1.} = 17.1, \qquad \bar{x}_{2.} = 16.0, \qquad \bar{x}_{3.} = 16.8, \qquad \bar{x}_{4.} = 15.3,$$

$$\bar{y}_{1.} = 26.8, \qquad \bar{y}_{2.} = 11.9, \qquad \bar{y}_{3.} = 8.2, \qquad \bar{y}_{4.} = 15.3,$$

$$\bar{x}_{..} = 16.3, \qquad \hat{\mu} = \bar{y}_{..} = 15.55.$$

From a packaged one-way analysis of covariance table, we obtained Table B.

TABLE B

Source of variation	Degrees of freedom	Sums of squares and cross products		
		XX	XY	YY
Between means	3	19.80	66.69	1939.69
Within means (residual)	36	3170.60	2037.70	2630.21
Total	39	3190.40	2104.39	4569.90

Thus, we have the estimates

$$\hat{\beta} = \frac{E_{xy}}{E_{xx}} = \frac{2037.7}{3170.6} = 0.643,$$

$$b_M = \frac{M_{xy}}{M_{xx}} = \frac{66.69}{19.80} = 3.37, \qquad b_T = \frac{T_{xy}}{T_{xx}} = \frac{2104.4}{3190.4} = 0.660,$$

$$MS_E = \frac{1}{35}\left[2630.21 - \frac{(2037.70)^2}{3170.60}\right] = 37.73.$$

From these estimates we calculated the estimated adjusted differential effects of each level. Thus,

$$\hat{\alpha}_1 = (\bar{y}_{1.} - \bar{y}_{..}) - \hat{\beta}(\bar{x}_{1.} - \bar{x}_{..}) = (26.8 - 15.5) - 0.643(17.1 - 16.3) = 10.74$$

is the differential effect of the first treatment adjusted by the regression of y on x. Similarly, $\hat{\alpha}_2 = -3.46$, $\hat{\alpha}_3 = -7.67$, $\hat{\alpha}_4 = 0.39$ are the other differential effects, all adjusted by the regression of y on x.

The 4 estimated within group regression equations are:

$$\hat{y} = 26.8 + 0.643(x - 17.1) \qquad \text{for treatment group 1,}$$

$$\hat{y} = 11.9 + 0.643(x - 16.0) \qquad \text{for treatment group 2,}$$

$$\hat{y} = 8.2 + 0.643(x - 16.8) \qquad \text{for treatment group 3,}$$

$$\hat{y} = 15.3 + 0.643(x - 15.3) \qquad \text{for control group 4.}$$

The means regression line is

$$\hat{y} = 15.55 + 3.37(x - 16.3)$$

and the total regression line is

$$\hat{y} = 15.55 + 0.660(x - 16.3).$$

These lines are plotted in Fig. 4.7.1.

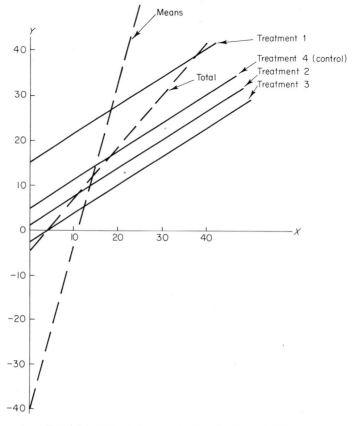

FIG. 4.7.1. *Estimated regression lines for Example 4.7.1.*

We now discuss the testing of hypotheses. From the analysis of covariance table, we can test the hypothesis that the I population means of the covariate are all equal. Let μ_{xi} denote the covariate mean for the ith level, $i = 1, ..., I$. Then the hypothesis is $H_0: \mu_{x1} = \cdots = \mu_{xI}$. The test statistic is

$$F = \frac{M_{xx}/\nu_M}{E_{xx}/\nu_E}, \tag{4.7.13}$$

which has an F distribution with ν_M and ν_E degrees of freedom. The quantities in Eq. (4.7.13) are the between and within sums of squares and degrees of freedom in the XX column of Table 4.7.1. The P value is the area to the right of F under the $F(\nu_M, \nu_E)$ frequency curve. This hypothesis tests the randomness of the assignment of experimental units to the I levels of the factor.

We can also test the hypothesis that the I population means of the observation variable are all equal. Let μ_{yi} denote the mean of the observation variable for the ith level, $i = 1, ..., I$. Then the hypothesis $H_0: \mu_{y1} = \cdots = \mu_{yI}$ is tested by

$$F = \frac{M_{yy}/\nu_M}{E_{yy}/\nu_E},\qquad(4.7.14)$$

where the quantities in this equation are read from the YY column in the table. The P value is the area to the right of F under the $F(\nu_M, \nu_E)$ frequency curve.

The more interesting hypothesis is whether the I population means of the adjusted variable

$$y^* = y - \hat{\beta}(x - \bar{x}_{..})\qquad(4.7.15)$$

are all equal. This hypothesis may be written as $H_0: \alpha_1 = \cdots = \alpha_I = 0$, where, as before, α_i is the adjusted differential effect of the ith level, $i = 1, ..., I$. To test this hypothesis a new table is formed from Table 4.7.1 as follows: The residual sum of squares E_{yy} for y is divided into two parts—the sum of squares E_{xy}^2/E_{xx} *due regression* and the sum of squares $E_{yy} - E_{xy}^2/E_{xx}$ *about regression*. The residual degrees of freedom ν_E is correspondingly divided into 1 and $\nu_E - 1$ degrees of freedom, respectively. A similar partitioning is made for the total sum of squares T_{yy} and its degrees of freedom ν_T. These quantities are the sums of squares and degrees of freedom for the first two sources of variation shown in Table 4.7.2,

TABLE 4.7.2. *Partitioning of the Residual and Total Sum of Squares*

Source of variation	Sum of squares (Table 4.7.1) df	Sum of squares (Table 4.7.1) SS	Due regression df	Due regression SS	Deviation about regression df	Deviation about regression SS	Deviation about regression MS
Within levels (residual)	ν_E	E_{yy}	1	$\dfrac{E_{xy}^2}{E_{xx}}$	$\nu_E - 1$	$E_{yy} - \dfrac{E_{xy}^2}{E_{xx}}$	MS_E
Total	ν_T	T_{yy}	1	$\dfrac{T_{xy}^2}{T_{xx}}$	$\nu_T - 1$	$T_{yy} - \dfrac{T_{xy}^2}{T_{xx}}$	
Difference for testing among adjusted means					$\nu_M = \nu_T - \nu_E$	$M_{yy} - \dfrac{T_{xy}^2}{T_{xx}} + \dfrac{E_{xy}^2}{E_{xx}}$	MS_M

that is, the within levels and the total. The sum of squares and the degrees of freedom for the remaining source of variation, the *difference for testing among adjusted means*, is obtained by subtracting the within sum of squares and degrees of freedom from the total sum of squares and degrees of freedom. (Recall that $v_M + v_E = v_T$ and $SS_M + SS_E = SS_T$.) The mean squares in the last column are obtained by dividing the deviations about regression sums of squares by its degrees of freedom. Thus, we obtain the within levels mean square MS_E, which is the unbiased estimate for σ^2 given in Eq. (4.7.12), and the mean square MS_M. To test for differences among adjusted means, or equivalently, $H_0: \alpha_1 = \cdots = \alpha_I = 0$, we use the F ratio

$$F = \frac{MS_M}{MS_E}, \qquad (4.7.16)$$

which has an F distribution with v_M and $v_E - 1$ degrees of freedom. The P value is the area to the right of F under the $F(v_M, v_E - 1)$ frequency curve.

Finally, we can test the hypothesis $H_0: \beta = 0$, that is, the within group regression coefficient is zero, by comparing the residual due regression mean square to the residual about regression mean square. Thus, we use the F ratio

$$F = \frac{E_{xy}^2/E_{xx}}{MS_E}, \qquad (4.7.17)$$

which has an F distribution with 1 and $v_E - 1$ degrees of freedom. The P value is the area to the right of F under the $F(1, v_E - 1)$ frequency curve.

A typical packaged analysis of covariance program will print all or part of Table 4.7.2. and calculate and print the four F ratios.

EXAMPLE 4.7.1 (*continued*)   To test that the pretest means are the same for the four groups, that is, $H_0: \mu_{x1} = \cdots = \mu_{x4}$, we calculate

$$F = \frac{M_{xx}/v_M}{E_{xx}/v_E} = \frac{19.80/3}{3171/36} = 0.0749.$$

Since $F_{0.95}(3, 36) = 2.9$, we accept H_0 at $\alpha = 0.05$ and conclude that the assignment to treatment groups seems random, as desired. To test that the posttest means are the same for the four groups, that is, $H_0: \mu_{y1} = \cdots = \mu_{y4}$, we calculate

$$F = \frac{M_{yy}/v_M}{E_{yy}/v_E} = \frac{1939/3}{2630/36} = 8.84.$$

This test is significant at $P < 0.001$.

The more interesting test is to test that the adjusted means are equal (or equivalently, $H_0: \alpha_1 = \cdots = \alpha_4 = 0$). The partitioned table is given in the accompanying table. The F ratio is $F = MS_M/MS_E = 16.44$. Note that P is much less than the above P value of 0.001.

Source of variation	Sum of squares df	Sum of squares SS	Due regression df	Due regression SS	Deviations about regression df	Deviations about regression SS	Deviations about regression Mean square
Within means (residual)	36	2630.21	1	1309.61	35	1320.60	37.731
Total	39	4569.90	1	1388.07	38	3181.83	
Difference for testing among adjusted means					3	1861.23	620.410

Finally, to test $H_0: \beta = 0$, we calculate

$$F = \frac{E_{xy}^2/E_{xx}}{MS_E} = \frac{1309.61}{37.73} = 34.71.$$

This result is also significant at $P < 0.001$.

COMMENTS 4.7.1   1.   The one-way analysis of covariance model extends to a one-way model with multiple covariates, that is, a model of the form

$$y_{ij} = \mu + \alpha_i + \beta(x_{ij} - \bar{x}_{..}) + \gamma(z_{ij} - \bar{z}_{..}) + \cdots + e_{ij}, \qquad j = 1, ..., J_i, \quad i = 1, ..., I.$$

Here, $x_{ij}, z_{ij}, ...,$ are covariates, each linearly related to y_{ij}. The analysis would estimate μ, each of the coefficients $\beta, \gamma, ...,$ and the adjusted differential effects α_i. We then could test whether these effects are equal.

2.   The one-way analysis of covariance model also extends to an m-way model with multiple covariates. For example, the model

$$y_{ij} = \mu + \alpha_i + \beta_j + \gamma(x_{ij} - \bar{x}_{..}) + e_{ij},$$

has two factors and one covariate linearly related to y_{ij}. For detailed description of an m-way model with multiple covariates see Scheffé (1958) and various papers in *Biometrics* **13**, 1957.

3.   The one-way model presented in this section assumes that the within group coefficient β is equal for all I levels of the factor. A generalization of this model is to assume that these coefficients are not the same for the I groups. For this more general model, it is possible to test equality of these coefficients and then to proceed with the analysis as presented here (see Brownlee, 1965, Ch. 11).

Problems

Section 4.2

4.2.1.   Three strains of rats were selectively bred for differences in arterial blood pressure in order to determine the possible effect of heredity on the blood pressure variable. Ten rats from each strain were observed and their pressures were measured in mm Hg. The sample means are $\bar{x}_A = 84.5$, $\bar{x}_B = 88.0$, and $\bar{x}_C = 91.1$ for the strains A, B, and C, respectively. The "within groups" sum of squares is 270.

(a) Construct the one-way anova table.

(b) Test the hypothesis that there is no significant differences among the strains.

(c) Test the hypothesis that there is no significant difference between strain B and strain C, assuming that this is the only comparison you are making.

(d) Comment about making further tests about differences between the strains.

4.2.2.   In a large factory, many workers are employed in the assembly line. Four workers were selected randomly, and the time required to assemble a particular piece of equipment was recorded several times for each of the four workers. The data are given in the accompanying table.

Time (Min) to Assemble Product

| | Worker | | |
1	2	3	4
24.2	19.4	19.0	19.9
22.2	21.1	23.1	15.7
24.5	16.2	23.8	15.2
21.1	21.2	22.7	19.8
22.0	21.6		18.9
	17.8		16.1
	19.6		16.2
			18.5

(a) Estimate the components of variance, within and between groups, and the grand mean.

(b) Compute the analysis of variance table, including the expected mean squares.

(c) Is there a significant between-worker variation?

Section 4.3

4.3.1   Four female pregnant mice delivered three baby mice each. Three diet plans are to be compared. Each of the three diets was randomly fed to one of the baby mice for three weeks and the gain in weight (gm) was recorded. The data are given in the accompanying table.

	Diet		
Mother	1	2	3
1	5.2	7.4	9.1
2	11.4	13.0	13.8
3	4.2	9.5	8.8
4	10.7	11.9	13.0

(a) State the appropriate anova model and the necessary assumptions and estimate the parameters of this model.

(b) Compute the anova table and test the relevant hypotheses relating to the diets and to the mothers.

4.3.2. Again we have four mother mice with three baby mice each. Now we would like to compare two diets. We randomly assign two mothers to each diet and feed that diet to all six baby mice. The data are given in the accompanying table.

Diet	Mother	Gain in weight (gm)	Diet	Mother	Gain in weight (gm)
1	1	11.8	2	1	7.4
		10.5			9.7
		12.5			8.2
	2	12.3		2	7.2
		15.5			8.6
		11.4			7.1

(a) State the appropriate model and assumptions and estimate the model parameters.

(b) Compute the anova table and test the relevant hypotheses.

4.3.3. We are still considering mice and diets. We now would like to compare three diets and two different drug injections. We use the twelve baby mice, regardless of their mother, and assign two mice randomly to each diet–drug combination. The measurement is gain in weight (gm) after three weeks. The data are given in the accompanying table.

		Diet	
Drug injection	1	2	3
1	8.2	13.1	10.5
	8.0	12.3	10.1
2	8.4	12.4	9.7
	7.3	13.0	9.4

(a) Write the anova model and perform the standard analysis.

(b) Is pooling the interaction terms advisable? If so, retest for the main effects.

(c) If you accept the hypothesis of no injection effect, it might be advisable to pool the "injection" terms. If you do so, what would be the appropriate analysis?

4.3.4.   Compare the designs used in Problems 4.3.1, 4.3.2, and 4.3.3 and discuss the advantages and disadvantages of each design.

Section 4.4

4.4.1.   Do Problem 4.3.1 using a factorial program.

4.4.2.   Do Problem 4.3.2 using a factorial program.

4.4.3.   Do Problem 4.3.3 using a factorial program.

4.4.4.   The research department of an automobile factory would like to compare the wear of four brands of tires on its four models of automobiles. The Latin square design is quite appropriate since there are four tire positions on each auto. The design used (with tire brand indicated in parentheses) and the amount of wear (mm) after 10,000 miles of identical driving are given in the accompanying table. Compute the anova table and perform the standard analyses.

Tire position	Automobile 1	2	3	4
1	(b) 2.12	(a) 1.73	(d) 1.65	(c) 1.89
2	(c) 1.83	(b) 2.28	(a) 1.67	(d) 2.01
3	(d) 1.83	(c) 2.27	(b) 2.18	(a) 2.03
4	(a) 1.85	(d) 1.93	(c) 2.24	(b) 2.52

Section 4.5

4.5.1. Do Problem 4.2.2 using a regression program.

4.5.2. Do Problem 4.3.1 using a regression program.

4.5.3. Do Problem 4.3.3 using a regression program.

4.5.4. Do Problem 4.4.4 using a regression program.

Section 4.6

4.6.1. Do Problem 4.3.1 using a general anova program.

4.6.2. Do Problem 4.3.2 using a general anova program.

4.6.3. Do Problem 4.3.3 using a general anova program.

4.6.4. We would like to compare two diets as in Problem 4.3.2 in conjunction with three different strains of mice. Thus we perform the same design used in Problem 4.3.2 on mice from the first strain and then repeat the same experiment on mice from the second and third strains. The resulting data are given in the accompanying table.

				Strain				
		1			2			3
Diet	Mother	Gain in weight (gm)	Mother	Gain in weight (gm)		Mother	Gain in weight (gm)	
1	1	10.7	1	11.1		1	9.4	
		11.0		10.4			8.3	
		11.9		9.9			11.6	
	2	13.2	2	12.2		2	10.9	
		12.2		12.5			14.0	
		11.8		12.0			12.0	
2	1	8.1	1	9.0		1	8.2	
		9.7		7.7			9.9	
		9.3		8.1			7.8	
	2	8.2	2	5.2		2	5.3	
		7.7		6.0			5.5	
		7.7		6.7			4.4	

(a) Write down the anova model, including the appropriate nesting and crossing relationships and the necessary assumptions.

(b) Use a general anova program to obtain the anova table and test all the appropriate hypotheses. Note that you may have to use a linear combination of mean squares for testing some hypotheses.

Section 4.7

4.7.1.   Four drugs to reduce arterial pressure are to be compared using analysis of covariance. The response variable is systolic pressure (SP) measured on an experimental animal after treatment, and the covariate is SP before treatment. The data are given in the accompanying table.

Drug	Systolic pressure Before treatment	After treatment	Drug	Systolic pressure Before treatment	After treatment
1	194	157	3	172	136
	162	136		196	182
	183	145		158	134
	180	153			
2	154	124	4	158	124
	184	123		165	124
	173	143		186	132
	170	136		182	133

(a) Obtain estimates of the within group, total and means regression lines.

(b) Test the randomness of assignment of animals to the four drugs.

(c) Perform an analysis of variance on the response variable, ignoring the covariate.

(d) Perform an analysis of covariance on the response variable. Can we conclude that the four drugs are significantly different?

(e) Compare the results of (c) and (d).

4.7.2.   Do Problem 4.7.1 (d) using a regression program.

5 MULTIVARIATE STATISTICAL METHODS

In Chapters 2 and 4 we were mostly concerned with statistical techniques appropriate for analyzing observations on a single random variable. As we have seen in Chapter 3, observations on more than one random variable may be made for each individual in the sample. In particular, this situation occurs in multiple regression problems when the independent variables are considered as random variables along with the dependent variable. This is the first example of a multivariate statistical method. In regression problems emphasis is placed upon the relationship between the dependent variable on one hand and the set of independent variables on the other hand. In other multivariate methods, however, all of the random variables are analyzed simultaneously as a random vector having a multivariate distribution. As will be seen, some multivariate techniques (for example, testing hypotheses about mean vectors) are generalizations of univariate techniques, while other techniques (for example, classification of a random vector into one of k populations) are unique to multivariate analysis.

Historically, statistical analysis of more than one variable considered each variable separately. This procedure was limited, since overall inference statements could not be easily made from the individual inference statements. Multivariate techniques make such overall inference statements possible. It should be noted that the calculation of most multivariate statistics is more difficult than their univariate analogs. In fact, in some analyses it is imperative to use a digital computer.

Most of the continuous multivariate techniques assumes that the underlying distribution of the random vector is multivariate normal. The justifications of this assumption, similar to those in the univariate case, are: (a) many observable

phenomena follow an approximate multivariate normal distribution; (b) transformations of some or all of the components of the random vector sometimes
induce a multivariate normal distribution; and (c) the central limit theorem for
one random variable extends to the multivariate case, that is, summations of
many independent and identically distributed random vectors approach multivariate normality (see Anderson, 1958, Theorem 4.2.3).

In this chapter, Section 5.1 discusses methods for identifying outliers; Section
5.2 describes the Hotelling T^2 statistic for testing hypotheses about one or two
mean vectors; Section 5.3 introduces the problem of classifying an observation
vector into one of two multivariate populations; Section 5.4 extends the classification problem to k populations, $k \geqslant 2$; and Section 5.5 describes the stepwise
classification procedure. The last two sections are concerned with examining the
structure of multivariate observations. Section 5.6 deals with principal component
analysis, while Section 5.7 discusses factor analysis. The reader is urged to review
Section I.6 before proceeding with this chapter.

5.1 The Analysis of Outliers

In the univariate case, if the random variable Y is $N(\mu, \sigma^2)$, then the quantity
$(Y - \mu)^2 / \sigma^2$ is distributed as $\chi^2(1)$. In the multivariate case, it can be shown that
if the $p \times 1$ random vector $\mathbf{X}$ is multivariate normal with mean vector $\boldsymbol{\mu}^{p \times 1}$ and
covariance matrix $\boldsymbol{\Sigma}^{p \times p}$, then the quantity

$$\chi^2 = (\mathbf{x} - \boldsymbol{\mu})' \boldsymbol{\Sigma}^{-1} (\mathbf{x} - \boldsymbol{\mu}) \tag{5.1.1}$$

is distributed as $\chi^2(p)$. If $\boldsymbol{\mu}$ and $\boldsymbol{\Sigma}$ are assumed known, then this statistic may be
used to test whether or not an observation vector $\mathbf{x}$ is an outlier. The P value is
the area to the right of the computed χ^2 under the $\chi^2(p)$ frequency curve. If P is
less than a preselected significance level α, we regard $\mathbf{x}$ as an outlier, and the
components of this observation vector should be checked for errors. Indeed, in
this way we may screen each observation vector in a random sample.

EXAMPLE 5.1.1   In a computerized patient monitoring system, measurements of systolic, diastolic, and mean arterial pressures, mean venous pressure,
respiratory rate, heart rate, and rectal temperatures are made every minute on a
critically ill patient. The χ^2 statistic of Eq. (5.1.1) and the corresponding P value
is calculated on each vector of measurements. The parameters $\boldsymbol{\mu}$ and $\boldsymbol{\Sigma}$ are known
from previous experience with healthy patients. For each observation vector, if
$P > 0.2$, the patient is considered within normal range, if $0.05 < P \leqslant 0.2$, the
patient is slightly abnormal, if $0.01 < P \leqslant 0.05$, an alert light is turned on, and if

$P \leqslant 0.01$, then an alarm is sounded. This assists the attending physician and staff in detecting changes in the status of a patient or alerts them to possible malfunctioning of the monitoring equipment. For extensions of this procedure see Afifi *et al.* (1971).

In most applications of detecting outliers, the parameters $\boldsymbol{\mu}$ and $\boldsymbol{\Sigma}$ are not known, so that the χ^2 of Eq. (5.1.1) is not a valid test statistic. An alternative test procedure utilizes a statistic which is the sample analog of Eq. (5.1.1). Let $\mathbf{x}_1, \ldots, \mathbf{x}_k$ be a random sample from $N(\boldsymbol{\mu}, \boldsymbol{\Sigma})$. Then the sample mean vector and covariance matrix are

$$\bar{\mathbf{x}} = \frac{1}{k} \sum_{i=1}^{k} \mathbf{x}_i \tag{5.1.2}$$

and

$$\mathbf{S} = \frac{1}{k-1} \sum_{i=1}^{k} (\mathbf{x}_i - \bar{\mathbf{x}})(\mathbf{x}_i - \bar{\mathbf{x}})', \tag{5.1.3}$$

respectively. If $\mathbf{x}$ is another observation vector from $N(\boldsymbol{\mu}, \boldsymbol{\Sigma})$, then the sample analog of Eq. (5.1.1), called the *sample Mahalanobis distance*, is

$$D^2 = (\mathbf{x} - \bar{\mathbf{x}})' \mathbf{S}^{-1} (\mathbf{x} - \bar{\mathbf{x}}). \tag{5.1.4}$$

It can be shown that

$$F = \frac{(k-p)k}{(k^2-1)p} D^2 \tag{5.1.5}$$

has an F distribution with p and $k-p$ degrees of freedom.

The procedure for testing for outliers uses the statistic of Eq. (5.1.5), where $\bar{\mathbf{x}}$ and $\mathbf{S}$ are calculated from a subset of the same sample being screened. This procedure, applied to a random sample $\mathbf{x}_1, \ldots, \mathbf{x}_n$ of size n, is as follows:

1. For each observation vector $\mathbf{x}_i$, $i = 1, \ldots, n$, calculate the sample mean vector $\bar{\mathbf{x}}_i$ and covariance matrix $\mathbf{S}_i$ using the remaining $k = n-1$ observation vectors. Calculate from Eq. (5.1.4) the sample Mahalanobis distance D_i^2 between $\mathbf{x}_i$ and $\bar{\mathbf{x}}_i$ using $\mathbf{S}_i$. Calculate F_i from Eq. (5.1.5) with $k = n-1$ and compute the corresponding P value $P_i = \Pr(F(p, k-p) > F_i)$.

2. Examine $P_1, P_2, \ldots, P_n$. If all $P_i > \alpha$ for some preselected α, then no outliers are identified and we end the process. If some $P_i < \alpha$, then the observation vector corresponding to the smallest P value is identified as an outlier and is excluded from the sample. The procedure is then repeated on the reduced sample with n redefined as $n-1$.

EXAMPLE 5.1.2   The data in Table 5.1.1 are fifteen systolic (X_1) and diastolic (X_2) pressures in mm Hg. The observations were analyzed for outliers using the program BMDX74 with $\alpha = 0.05$. The program first found the sample means and standard deviations to be $\bar{x}_1 = 120.6$, $s_1 = 20.9$, $\bar{x}_2 = 81.0$, and $s_2 = 21.7$. Using the above procedure with $n = 15$, the program detected and removed the observation vector $\mathbf{x}_9 = (132, 125)'$, since it yielded the smallest P value, $P_9 = 0.0003$. Using the procedure on the reduced sample of $n = 14$ observations, the program identified and removed $\mathbf{x}_7 = (93, 43)'$, since $P_7 = 0.0264$ was the smallest P value less than α. No other outliers were detected on the sample of $n = 13$ observations. The sample means and standard deviations were recalculated as $\bar{x}_1 = 121.9$, $s_1 = 20.7$, $\bar{x}_2 = 80.5$, and $s_2 = 16.3$. On examining the removed data it was apparent that the difference between systolic and diastolic pressure was unusually small for $\mathbf{x}_9$ and suspiciously large for $\mathbf{x}_7$. Examination of the records showed that the correct values are $\mathbf{x}_7 = (93, 54)'$ and $\mathbf{x}_9 = (132, 94)'$.

TABLE 5.1.1. *15 Hypothetical Systolic and Diastolic Pressures (mm Hg)*

Observation	1	2	3	4	5	6	7	8	9	10	11	12	13	14	15
X_1: systolic pressure	154	136	91	125	133	125	93	80	132	107	142	115	114	120	141
X_2: diastolic pressure	108	90	54	89	93	77	43	50	125	76	96	74	79	71	90

5.2 Tests of Hypotheses on Mean Vectors

This section discusses the multivariate analogs of the univariate tests on means presented in Chapter 2. Section 5.2.1 discusses testing $H_0: \boldsymbol{\mu} = \boldsymbol{\mu}_0$ when $\boldsymbol{\Sigma}$ is known. Section 5.2.2 discusses this test when $\boldsymbol{\Sigma}$ is not known. Finally, Section 5.2.3 presents the two-sample test $H_0: \boldsymbol{\mu}_1 = \boldsymbol{\mu}_2$ with $\boldsymbol{\Sigma}$ unknown.

5.2.1 Tests of Hypotheses about Mean Vectors (Known Covariance Matrix)

In the univariate case, if the random variable Y is $N(\mu, \sigma^2)$, σ^2 assumed known, and if $y_1, \ldots, y_n$ is a random sample from this distribution, then to test the hypothesis $H_0: \mu = \mu_0$ against $H_1: \mu \neq \mu_0$, we use the test statistic $z = \sqrt{n}(\bar{y} - \mu_0)/\sigma$, where $\bar{y}$ is the sample mean. We reject H_0 if $|z| > z_{1-(\alpha/2)}$ for some preassigned value of α. In the multivariate case, we assume that $\mathbf{X}$ is

$N(\boldsymbol{\mu}, \boldsymbol{\Sigma})$, and that $\mathbf{x}_1, \ldots, \mathbf{x}_n$ is a random sample from this distribution. Furthermore, if $\boldsymbol{\Sigma}$ is known, then to test the hypothesis that the mean vector is equal to a given vector, that is, $H_0: \boldsymbol{\mu} = \boldsymbol{\mu}_0$ against $H_1: \boldsymbol{\mu} \neq \boldsymbol{\mu}_0$, we use the test statistic

$$\chi^2 = n(\overline{\mathbf{x}} - \boldsymbol{\mu}_0)' \boldsymbol{\Sigma}^{-1} (\overline{\mathbf{x}} - \boldsymbol{\mu}_0), \tag{5.2.1}$$

where $\overline{\mathbf{x}}$ is the sample mean vector. Under H_0, χ^2 has a χ^2 distribution with p degrees of freedom. The P value is the area to the right of the computed χ^2 under the $\chi^2(p)$ frequency curve. Equation (5.2.1) follows from the fact that the sampling distribution of $\overline{\mathbf{x}}$ under H_0 is $N(\boldsymbol{\mu}_0, (1/n)\boldsymbol{\Sigma})$. Note that one-sided tests are meaningless in the multivariate case.

COMMENT 5.2.1   The χ^2 statistic of Eq. (5.2.1) is easily programmed by using one of the many packaged matrix multiplication and inversion routines. Letting $A = \mathbf{x} - \boldsymbol{\mu}_0$, $B = \boldsymbol{\Sigma}$ written in column form, and $IP = p$, then Eq. (5.2.1) can be programmed in three FØRTRAN statements using IBM subprograms from the Scientific Subroutine Package. For example, CALL MINV (B, IP, D, L, M) replaces the matrix B by its inverse B^{-1}. (The quantity D is the determinant of B, and L and M are working vectors of size IP.) Then, CALL GMPRD$(A, B, C, 1, IP, IP)$ calculates $C = AB$, and CALL GMPRD $(C, A, E, 1, IP, 1)$ calculates the value $E = CA$; then $\chi^2 = nE$.

EXAMPLE 5.1.2 (*continued*)   The two errors of Example 5.1.2 were corrected so that $\mathbf{x}_7 = (93, 54)'$ and $\mathbf{x}_9 = (132, 94)'$. Assuming that the standard deviations are $\sigma_1 = 20$ and $\sigma_2 = 15$, and that the correlation coefficient is $\rho = 0.8$, then

$$\boldsymbol{\Sigma} = \begin{bmatrix} 400 & 240 \\ 240 & 225 \end{bmatrix}.$$

We wish to test the hypothesis that these 15 patients come from the population of healthy patients with mean systolic and diastolic pressures equal to 120 and 80, respectively. Thus, $H_0: \boldsymbol{\mu} = (120, 80)'$ and $H_1: \boldsymbol{\mu} \neq (120, 80)'$. From the data, it was found that $\overline{\mathbf{x}} = (120.6, 79.7)'$. Using the notation of Comment 5.2.1, we have

$$A = \overline{\mathbf{x}} - \boldsymbol{\mu}_0 = \begin{bmatrix} 0.6 \\ -0.3 \end{bmatrix} \quad \text{and} \quad B = \begin{bmatrix} 400 \\ 240 \\ 240 \\ 225 \end{bmatrix}.$$

Hence, $\chi^2 = 0.006$ and the test is nonsignificant.

5.2.2 Tests of Hypotheses about Mean Vectors (Unknown Covariance Matrix)

In most practical problems the variances and covariances are not known and must be estimated from the sample. In the univariate case, using the notation defined in Section 5.2.1, we test $H_0: \mu = \mu_0$ against $H_1: \mu \neq \mu_0$ by using the test statistic $t = \sqrt{n}(\bar{y} - \mu_0)/s$, where s is the sample standard deviation. We reject H_0 if $|t| > t_{1-(\alpha/2)}(n-1)$. In the multivariate case, we calculate the unbiased estimate $\mathbf{S}$ of Σ as

$$\mathbf{S} = \frac{1}{n-1} \sum_{i=1}^{n} (\mathbf{x}_i - \bar{\mathbf{x}})(\mathbf{x}_i - \bar{\mathbf{x}})' = \begin{bmatrix} s_1^2 & \cdots & s_{1p} \\ & \vdots & \\ s_{p1} & \cdots & s_p^2 \end{bmatrix}. \tag{5.2.2}$$

Note that s_i^2 may also be denoted by s_{ii}. Then, *Hotelling's T^2* (Hotelling, 1931), is given by

$$T^2 = n(\bar{\mathbf{x}} - \mathbf{\mu}_0)' \mathbf{S}^{-1} (\bar{\mathbf{x}} - \mathbf{\mu}_0). \tag{5.2.3}$$

Under $H_0: \mathbf{\mu} = \mathbf{\mu}_0$, the quantity

$$F = \frac{n-p}{p(n-1)} T^2 \tag{5.2.4}$$

has an F distribution with p and $n-p$ degrees of freedom. The P value is the area to the right of F under the $F(p, n-p)$ frequency curve.

COMMENTS 5.2.2   1.   The suggestions for programming χ^2 of Eq. (5.2.1) are applicable for programming T^2 of Eq. (5.2.3). Simply define B of Comment 5.2.1 to be $B = \mathbf{S}$. To obtain F we multiply the resulting T^2 by the constant $(n-p)/p(n-1)$.

2.   In addition to testing hypotheses about $\mathbf{\mu}$, the researcher may form multiple confidence intervals for linear combinations of the components of $\mathbf{\mu}$. For a given choice of constants $a_1, \ldots, a_p$, the multiple confidence interval for $\sum_{i=1}^{p} a_i \mu_i$ is

$$\sum_{i=1}^{p} a_i \bar{x}_i \pm \left[\frac{(n-1)p}{(n-p)n} F_{1-\alpha}(p, n-p) \sum_{i=1}^{p} \sum_{j=1}^{p} a_i s_{ij} a_j \right]^{1/2}.$$

The overall confidence level is $1 - \alpha$ for all choices of $a_1, \ldots, a_p$.

Thus, for example, the multiple confidence interval for μ_i, the ith component of $\mathbf{\mu}$, is

$$\bar{x}_i \pm \left[\frac{(n-1)p}{(n-p)n} F_{1-\alpha}(p, n-p) s_i^2 \right]^{1/2}, \qquad i = 1, \ldots, p.$$

Hence, we may obtain confidence intervals similar to those obtained using the Student's t distribution, but utilizing the multivariate structure of the data. As in the case of analysis of variance, shortness of the intervals is sacrificed for the sake of obtaining an overall confidence level of $1 - \alpha$.

EXAMPLE 5.1.2 (*continued*)   The sample covariance matrix is

$$S = \begin{bmatrix} 438.26 & 343.02 \\ 343.02 & 291.38 \end{bmatrix},$$

so that the value of T^2 for testing $H_0: \mu = (120, 80)'$ is [from Eq. (5.2.3)]

$$T^2 = 15[0.6, -0.3] \begin{bmatrix} 0.02902 & -0.03417 \\ -0.03417 & 0.04365 \end{bmatrix} \begin{bmatrix} 0.6 \\ -0.3 \end{bmatrix} = 0.400.$$

Thus, from Eq. (5.2.4)

$$F = \frac{15-2}{2(15-1)}(0.400) = 0.186,$$

which is also nonsignificant.

Now, suppose we wish to test the hypothesis that these 15 patients come from the population of hypersensitive patients with mean systolic and diastolic pressures equal to 90 and 60, respectively. Thus, $H_0: \mu = (90, 60)'$ and $H_1: \mu \neq (90, 60)'$. In this case, $T^2 = 43.76$ and $F = 20.32$. Since $P < 0.005$, H_0 is rejected.

Using Comment 5.2.2.2 we form 95% multiple confidence intervals for μ_1, μ_2, and $\mu_1 - 2\mu_2$. Thus, with overall 95% confidence, the interval

$$120.6 \pm \left[\frac{14(2)}{13(15)}(3.81)(438.26) \right]^{\frac{1}{2}} = (105.1, 136.1)$$

includes μ_1; the interval

$$79.7 \pm \left[\frac{14(2)}{13(15)}(3.81)(291.38) \right]^{\frac{1}{2}} = (67.1, 92.3)$$

includes μ_2; and the interval

$$120.6 - 2(79.7) \pm \left[\frac{14(2)}{13(15)}(3.81)\{438.26 - 2(2)(343.02) + 4(291.38)\} \right]^{\frac{1}{2}}$$

$$= (-50.1, -27.5)$$

includes $\mu_1 - 2\mu_2$.

5.2.3 Tests of Hypotheses about Two Mean Vectors (Unknown Covariance Matrix)

In the univariate case with $i = 1$ or 2, let the random variable Y_i be $N(\mu_i, \sigma^2)$ and let $y_{i1}, ..., y_{in_i}$ be a random sample from this distribution. To test the hypothesis $H_0: \mu_1 = \mu_2$ against $H_1: \mu_1 \neq \mu_2$ with σ^2 unknown, we use the test statistic

$$t = \frac{(\bar{y}_1 - \bar{y}_2)}{\sqrt{s_p^2(n_1^{-1} + n_2^{-1})}},$$

where $\bar{y}_i$ is the ith sample mean, $i = 1$ or 2, and s_p^2 is the pooled variance. We reject H_0 if $|t| > t_{1-(\alpha/2)}(n_1 + n_2 - 2)$ for some preassigned α. The multivariate analog of this two-sample Student's t statistic is the *two-sample Hotelling T^2 statistic*. Assume for $i = 1$ or 2 that the random vector $\mathbf{X}_i$ is $N(\boldsymbol{\mu}_i, \boldsymbol{\Sigma})$. Let $\mathbf{x}_{i1}, ..., \mathbf{x}_{in_i}$ be a random sample from each distribution. We estimate $\boldsymbol{\Sigma}$ by the *pooled sample covariance matrix* $\mathbf{S}$ defined by

$$\mathbf{S} = \frac{1}{n_1 + n_2 - 2}[(n_1 - 1)\mathbf{S}_1 + (n_2 - 1)\mathbf{S}_2], \tag{5.2.5}$$

where $\mathbf{S}_i$ is the ith sample covariance matrix. Then the two-sample T^2 statistic is

$$T^2 = \frac{n_1 n_2}{n_1 + n_2}(\bar{\mathbf{x}}_1 - \bar{\mathbf{x}}_2)' \mathbf{S}^{-1}(\bar{\mathbf{x}}_1 - \bar{\mathbf{x}}_2), \tag{5.2.6}$$

where

$$\bar{\mathbf{x}}_i = \frac{1}{n_i}\sum_{j=1}^{n_i} \mathbf{x}_{ij}, \quad i = 1 \quad \text{or} \quad 2 \tag{5.2.7}$$

is the estimate of $\boldsymbol{\mu}_i$. Under $H_0: \boldsymbol{\mu}_1 = \boldsymbol{\mu}_2$, the quantity

$$F = \frac{n_1 + n_2 - p - 1}{(n_1 + n_2 - 2)p} T^2 \tag{5.2.8}$$

has an F distribution with p and $n_1 + n_2 - p - 1$ degrees of freedom.

COMMENTS 5.2.3   1. The sample mean vectors and covariance matrices may be obtained from a descriptive program for each sample. The pooled covariance matrix may be obtained from Eq. (5.2.5). The program of Comment 5.2.1 may then be modified to calculate Eq. (5.2.6) by defining $A = \bar{\mathbf{x}}_1 - \bar{\mathbf{x}}_2$ and $B = \mathbf{S}$.

2. We may also obtain the F statistic directly from the output of a two-sample discriminant analysis program (see Comment 5.3.2.3). For this reason, examples will be given in that section.

5.3 Classification of an Individual into One of Two Populations

The *classification problem* consists in classifying an unknown individual w into one of k populations $W_1, W_2, ..., W_k$ on the basis of measurements $x_1, ..., x_p$ on p characteristics. The following two examples illustrate the nature of this problem.

EXAMPLE 5.3.1.   The admissions committee of a college wishes to classify an applicant into $W_1 =$ the population of students who successfully complete college, or $W_2 =$ the population of students who do not complete college. Its decision is based on the applicant's scores on p entrance examinations.

EXAMPLE 5.3.2   A physician wishes to diagnose a patient as having one of k diseases on the basis of the existence or absence of p symptoms. This example will be discussed in Section 5.4.3.

In this section we consider the special case of classifying an individual into one of two populations, that is, $k = 2$. Section 5.3.1 studies this problem assuming multivariate normal populations with known parameters, while Section 5.3.2 presents the same problem, assuming that the parameters are not known. Section 5.3.3 discusses the techniques of estimating the probabilities of misclassification, and Section 5.3.4 discusses the estimation of the posterior probabilities.

5.3.1 Classification into One of Two Multivariate Normal Populations with Known Parameters

The standard classification procedure for p continuous variables assumes that the observations come from one of two multivariate normal populations. We write the observations $x_1, x_2, ..., x_p$ as a vector $\mathbf{x} = (x_1, ..., x_p)'$, and assume that W_1 is $N(\boldsymbol{\mu}_1^{p \times 1}, \boldsymbol{\Sigma}_1^{p \times p})$ and W_2 is $N(\boldsymbol{\mu}_2^{p \times 1}, \boldsymbol{\Sigma}_2^{p \times p})$, where $\boldsymbol{\mu}_i = (\mu_{i1}, ..., \mu_{ip})'$, $i = 1$ or 2. Another simplifying assumption is that

$$\boldsymbol{\Sigma}_1 = \boldsymbol{\Sigma}_2 = \boldsymbol{\Sigma} = (\sigma_{ij}), \qquad i = 1, ..., p, \quad j = 1, ..., p.$$

Hence, $\mathbf{x}$ is $N(\boldsymbol{\mu}_1, \boldsymbol{\Sigma})$ or $N(\boldsymbol{\mu}_2, \boldsymbol{\Sigma})$.

In order to develop the classification theory, we first assume that the parameters $\boldsymbol{\mu}_1$, $\boldsymbol{\mu}_2$, and $\boldsymbol{\Sigma}$ are known. Intuitively, it seems reasonable to find a linear combination of the observations, called a *discriminant function*, given by

$$z = \alpha_1 x_1 + \alpha_2 x_2 + \cdots + \alpha_p x_p, \tag{5.3.1}$$

where $\alpha_1, ..., \alpha_p$ are some constants, and to classify $\mathbf{x}$ into W_1 if

$$z \geqslant c \tag{5.3.2}$$

and into W_2 if

$$z < c, \tag{5.3.3}$$

where c is another constant. The problem then reduces to determining the values of $\alpha_1, ..., \alpha_p$, and c which minimize the probabilities of making an incorrect classification. We first choose these constants on an intuitive basis, and then we show that the resulting classification procedure is theoretically optimal.

If $\mathbf{x}$ is from W_1, then z is normal with mean

$$\zeta_1 = \sum_{j=1}^{p} \alpha_j \mu_{1j} \tag{5.3.4}$$

and variance

$$\sigma_z^2 = \sum_{i=1}^{p} \sum_{j=1}^{p} \alpha_i \sigma_{ij} \alpha_j. \tag{5.3.5}$$

Similarly, if $\mathbf{x}$ is from W_2, then z is normal with mean

$$\zeta_2 = \sum_{j=1}^{p} \alpha_j \mu_{2j} \tag{5.3.6}$$

and the same variance σ_z^2. An intuitive criterion for choosing $\alpha_1, ..., \alpha_p$ is that they separate ζ_1 as far from ζ_2 as possible, relative to σ_z^2. To achieve this, we define the *Mahalanobis distance*

$$\Delta^2 = \frac{(\zeta_1 - \zeta_2)^2}{\sigma_z^2}. \tag{5.3.7}$$

This quantity was first proposed by Mahalanobis (1936) as a measure of the "distance" between the two populations. Thus, we wish to find coefficients $\alpha_1, ..., \alpha_p$ which maximize Δ^2. Following the arguments of Fisher (1936), we find that the α_i are solutions to the set of linear equations

$$\begin{aligned}
\alpha_1 \sigma_{11} + \alpha_2 \sigma_{12} + \cdots + \alpha_p \sigma_{1p} &= \mu_{11} - \mu_{21}, \\
\alpha_1 \sigma_{21} + \alpha_2 \sigma_{22} + \cdots + \alpha_p \sigma_{2p} &= \mu_{12} - \mu_{22}, \\
&\vdots \\
\alpha_1 \sigma_{p1} + \alpha_2 \sigma_{p2} + \cdots + \alpha_p \sigma_{pp} &= \mu_{1p} - \mu_{2p}.
\end{aligned} \tag{5.3.8}$$

Once the α_i's have been found, evaluation of Eq. (5.3.1) for an individual whose measurements are $x_1, ..., x_p$ yields the *discriminant score* z for that individual.

To determine the constant c, we examine Fig. 5.3.1, which shows the two distributions of z along with an arbitrary constant c. If $\mathbf{x}$ is from W_2 but $z = \sum_{i=1}^{p} \alpha_i x_i \geqslant c$, then we would classify $\mathbf{x}$ into W_1, thus committing an error.

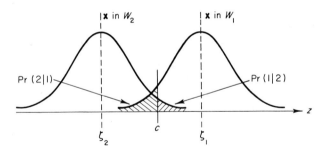

FIG. 5.3.1. *Distribution of z when* **x** *is from* W_1 *and when* **x** *is from* W_2. W_i *is* $N(\mu_i, \Sigma)$, $i = 1$ *or* 2.

The probability $\Pr(1|2)$ of making this error is shown in the figure. Also shown is the probability $\Pr(2|1)$ of falsely classifying an individual from W_1 into W_2. Intuitively, we would like to find c such that the sum of these probabilities $\Pr(1|2) + \Pr(2|1)$ is minimized. This is achieved by choosing c half way between the two means, that is,

$$c = \frac{\zeta_1 + \zeta_2}{2}. \tag{5.3.9}$$

In summary, the *intuitive classification procedure* is to calculate $\alpha_1, \ldots, \alpha_p$ from Eq. (5.3.8), to evaluate ζ_1 and ζ_2 from Eqs. (5.3.4) and (5.3.6), and to compute c from Eq. (5.3.9). For any observation vector **x**, we calculate the discriminant score z from Eq. (5.3.1) and classify **x** into W_1 if Eq. (5.3.2) holds; otherwise we classify **x** into W_2.

A more theoretical solution of the classification problem, based on the Bayes theorem, is now presented. We first define the *a priori* (or *prior*) *probability* q_i as the probability that an individual comes from W_i, $i = 1$ or 2. We assume that the sum of the prior probabilities $q_1 + q_2$ is equal to 1. For example, in Example 5.3.1 we may know from previous records that $\frac{1}{3}$ of all students who apply to the college have graduated. Hence, $q_1 = \frac{1}{3}$ and $q_2 = \frac{2}{3}$. In Example 5.3.2 with $k = 2$, we may know a priori in the population of patients with disease A or B, that 20% have disease A while 80% have disease B. Hence, $q_1 = 0.2$ and $q_2 = 0.8$.

Next, we define the conditional probability $\Pr(\mathbf{x}|W_i)$ of obtaining a particular observation vector **x** given that the individual comes from W_i, $i = 1$ or 2. We also denote by $\Pr(W_i|\mathbf{x})$ the conditional probability that the individual comes from W_i given his observation vector **x**, $i = 1$ or 2. The quantities $\Pr(W_1|\mathbf{x})$ and $\Pr(W_2|\mathbf{x})$ are called the *a posteriori* (or *posterior*) *probabilities*. The difference in meaning between the prior and posterior probabilities is that before we examine the observation vector of a given individual, we know he belongs to W_i with probability q_i. After we examine his observation vector, we know he belongs to W_i with probability $\Pr(W_i|\mathbf{x})$.

We now state the form of the Bayes theorem appropriate for the classification problem.

Bayes theorem   Using the above definitions with $i = 1$ or 2,

$$\Pr(W_i|\mathbf{x}) = \frac{q_i \Pr(\mathbf{x}|W_i)}{q_1 \Pr(\mathbf{x}|W_1) + q_2 \Pr(\mathbf{x}|W_2)}. \tag{5.3.10}$$

This equation is true regardless of the distribution of $\mathbf{X}$. When $\mathbf{X}$ is multivariate normal $N(\boldsymbol{\mu}_1, \boldsymbol{\Sigma})$ or $N(\boldsymbol{\mu}_2, \boldsymbol{\Sigma})$, then the densities $f_1(\mathbf{x})$ and $f_2(\mathbf{x})$ are substituted for $\Pr(\mathbf{x}|W_1)$ and $\Pr(\mathbf{x}|W_2)$, respectively. Hence, we have for $i = 1$ or 2,

$$\Pr(W_i|\mathbf{x}) = \frac{q_i f_i(\mathbf{x})}{q_1 f_1(\mathbf{x}) + q_2 f_2(\mathbf{x})}. \tag{5.3.11}$$

The *Bayes classification procedure* consists of classifying $\mathbf{x}$ into W_1 if

$$\Pr(W_1|\mathbf{x}) \geqslant \Pr(W_2|\mathbf{x})$$

and classifying $\mathbf{x}$ into W_2 if

$$\Pr(W_1|\mathbf{x}) < \Pr(W_2|\mathbf{x}).$$

Substituting Eq. (5.3.11) into these equations results in classifying $\mathbf{x}$ into W_1 if

$$\frac{q_1 f_1(\mathbf{x})}{q_2 f_2(\mathbf{x})} \geqslant 1 \tag{5.3.12}$$

and into W_2 if

$$\frac{q_1 f_1(\mathbf{x})}{q_2 f_2(\mathbf{x})} < 1. \tag{5.3.13}$$

It may be shown (for example, Rao, 1965; and Anderson, 1958) that this classification procedure minimizes the *expected probability of misclassification*

$$q_1 \Pr(2|1) + q_2 \Pr(1|2). \tag{5.3.14}$$

Note that this quantity is the probability that an individual comes from W_1 and is classified into W_2, or that he comes from W_2 and is classified into W_1.

Some algebraic manipulations of Eq. (5.3.12) show that this procedure is equivalent to classifying $\mathbf{x}$ into W_1 if

$$\sum_{i=1}^{p} \alpha_i x_i \geqslant \frac{\zeta_1 + \zeta_2}{2} + \ln\left(\frac{q_2}{q_1}\right) \tag{5.3.15}$$

and into W_2 if

$$\sum_{i=1}^{p} \alpha_i x_i < \frac{\zeta_1 + \zeta_2}{2} + \ln\left(\frac{q_2}{q_1}\right). \tag{5.3.16}$$

The constants α_i are solutions to Eqs. (5.3.8) and ζ_1 and ζ_2 are from Eqs. (5.3.4) and (5.3.6), respectively. Note that if $q_1 = q_2 = \frac{1}{2}$, then the Bayes classification procedure is the same as the intuitive classification procedure [Eqs. (5.3.2) and (5.3.3)].

A further refinement of the Bayes procedure is to incorporate the concepts of the *costs of misclassification*. This involves introducing a quantity $C(2|1)$ which is the cost or loss due to classifying $\mathbf{x}$ from W_1 into W_2. Similarly, the quantity $C(1|2)$ is the cost of classifying an individual from W_2 into W_1. In Example 5.3.1, $C(1|2)$ is the cost of educating a student who drops out, and $C(2|1)$ is the cost of losing a potentially successful student.

The *generalized Bayes classification procedure* consists in classifying $\mathbf{x}$ into W_1 if

$$\sum_{i=1}^{p} \alpha_i x_i \geq \frac{\zeta_1 + \zeta_2}{2} + \ln \frac{q_2 C(1|2)}{q_1 C(2|1)} \tag{5.3.17}$$

and into W_2 if

$$\sum_{i=1}^{p} \alpha_i x_i < \frac{\zeta_1 + \zeta_2}{2} + \ln \frac{q_2 C(1|2)}{q_1 C(2|1)}. \tag{5.3.18}$$

This solution minimizes the *expected cost of misclassification*

$$q_1 C(2|1) \Pr(2|1) + q_2 C(1|2) \Pr(1|2). \tag{5.3.19}$$

Note that this procedure reduces to the Bayes procedure when the costs are equal, and reduces to the intuitive procedure when, in addition, $q_1 = q_2 = \frac{1}{2}$.

For the generalized Bayes procedure, the probabilities of misclassification are

$$\Pr(2|1) = \Phi\left(\frac{K - \frac{1}{2}\Delta^2}{\Delta}\right) \tag{5.3.20}$$

and

$$\Pr(1|2) = \Phi\left(\frac{-K - \frac{1}{2}\Delta^2}{\Delta}\right), \tag{5.3.21}$$

where

$$K = \ln \frac{q_2 C(1|2)}{q_1 C(2|1)} \tag{5.3.22}$$

and Δ^2 is given in Eq. (5.3.7). Note that when $C(1|2) = C(2|1)$ and $q_1 = q_2 = \frac{1}{2}$, then

$$\Pr(2|1) = \Pr(1|2) = \Phi\left(-\frac{\Delta}{2}\right). \tag{5.3.23}$$

COMMENTS 5.3.1   1.   A solution of Eq. (5.3.8) may be obtained by using any of the subroutines for solving linear equations, for example, subroutine SIMQ of the IBM Scientific Subroutine Package.

2.   It may be shown that any positive multiple of the coefficients $\alpha_1, ..., \alpha_p$ obtained by solving Eq. (5.3.8) also maximizes Δ^2. If for any reason $\alpha_1, ..., \alpha_p$ are multiplied by a positive constant, then the quantity

$$K = \ln \frac{q_2 \, C(1|2)}{q_1 \, C(2|1)}$$

appearing in the classification procedure is also multiplied by the same constant.

3.   Note that $\Pr(2|1)$ and $\Pr(1|2)$ are decreasing functions of Δ^2, that is, the probabilities of misclassification decrease as the "distance" between the two populations increases.

*4.   The solution of Eq. (5.3.8) may be expressed in matrix notation as

$$\alpha = \Sigma^{-1}(\mu_1 - \mu_2), \qquad \text{where} \quad \alpha = (\alpha_1, ..., \alpha_p)'.$$

Substitution into Eq. (5.3.7) yields

$$\Delta^2 = (\mu_1 - \mu_2)' \Sigma^{-1} (\mu_1 - \mu_2)$$

as the Mahalanobis distance.*

5.   If $\mathbf{x}$ comes from one of two known populations with arbitrary densities $f_1(\mathbf{x})$ and $f_2(\mathbf{x})$, respectively, then the generalized Bayes solution is to classify $\mathbf{x}$ into W_1 if

$$\frac{q_1 \, C(2|1) f_1(\mathbf{x})}{q_2 \, C(1|2) f_2(\mathbf{x})} \geqslant 1,$$

otherwise, classify $\mathbf{x}$ into W_2.

EXAMPLE 5.3.1 (*continued*)   Assume that the admissions committee considers scores of applicants on $p = 2$ tests. Let $\mathbf{x} = (x_1, x_2)'$ be the vector of the applicant's scores. From previous experience it is known that

$$\mu_1 = (60, 57)', \qquad \mu_2 = (42, 39)', \qquad \text{and} \qquad \Sigma = \begin{bmatrix} 100 & 70 \\ 70 & 100 \end{bmatrix}.$$

Assume that $q_1 = \frac{1}{3}$, $q_2 = \frac{2}{3}$, and, for the sake of illustration, let $C(1|2) = \$2000$ and $C(2|1) = \$3000$. Solving Eq. (5.3.8),

$$100\alpha_1 + 70\alpha_2 = 18, \qquad 70\alpha_1 + 100\alpha_2 = 18,$$

we obtain $\alpha_1 = \alpha_2 = 54/510$. Hence, the discriminant function is $z = (54/510)(x_1 + x_2)$. From Eq. (5.3.4), we calculate $\zeta_1 = (54/510)(60 + 57) = 12.39$, and from

Eq. (5.3.6), we calculate $\zeta_2 = 8.58$. Hence, from Eq. (5.3.9), $c = (12.39 + 8.58)/2 = 10.49$, and from Eq. (5.3.22), $K = \ln \frac{4}{3} = 0.288$. Thus, the generalized Bayes classification procedure is to classify $\mathbf{x}$ into W_1

$$\frac{54}{510}(x_1 + x_2) \geqslant 10.49 + 0.288,$$

or, equivalently, if

$$x_1 + x_2 \geqslant 101.79.$$

Similarly, we classify $\mathbf{x}$ into W_2 if

$$x_1 + x_2 < 101.79.$$

The quantity $\sigma_z{}^2$ [Eq. (5.3.5)] equals 3.81, and the Mahalanobis distance Δ^2 [Eq. (5.3.7)] is also 3.81. Thus, from Eqs. (5.3.20)–(5.3.21), we have the probabilities of misclassification

$$\Pr(2|1) = \Phi(-0.83) = 0.203, \quad \text{and} \quad \Pr(1|2) = \Phi(-1.12) = 0.131.$$

In summary, the applicant is accepted if his combined score is greater than or equal to 101.79 and is rejected otherwise. Following this procedure, a potentially successful student is denied admission 20.3% of the time, and a potentially unsuccessful student is admitted only 13.1% of the time.

5.3.2 Classification into One of Two Multivariate Normal Populations with Unknown Parameters

Suppose that we have an individual with an observation vector $\mathbf{x} = (x_1, x_2, \ldots, x_p)'$, and we wish to classify him into $W_1 : N(\mathbf{\mu}_1^{p \times 1}, \mathbf{\Sigma}^{p \times p})$ or into $W_2 : N(\mathbf{\mu}_2^{p \times 1}, \mathbf{\Sigma}^{p \times p})$ on the basis of $\mathbf{x}$. Assume that we know the prior probabilities and the costs of misclassification, but that the mean vectors $\mathbf{\mu}_1$ and $\mathbf{\mu}_2$ and the covariance matrix $\mathbf{\Sigma}$ are unknown. If $\mathbf{x}_{11}, \ldots, \mathbf{x}_{1n_1}$ and $\mathbf{x}_{21}, \ldots, \mathbf{x}_{2n_2}$ are independent random samples from W_1 and W_2, respectively, then we can estimate $\mathbf{\mu}_i$ by the sample mean vector $\bar{\mathbf{x}}_i = (\bar{x}_{i1}, \ldots, \bar{x}_{ip})'$, $i = 1$ or 2 [Eq. (5.2.7)], and $\mathbf{\Sigma}$ by the pooled covariance matrix $\mathbf{S} = (s_{jk})$, $j = 1, \ldots, p$, $k = 1, \ldots, p$ [Eq. (5.2.5)]. In such a situation, it is not possible to derive a classification procedure which is optimal [in the sense of minimizing the expected cost of misclassification, Eq. (5.3.19)]. However, it may be shown (Anderson, 1958, Theorem 6.5.1) that if consistent estimators are substituted for the parameters in the generalized Bayes procedure [Eqs. (5.3.17)–(5.3.18)], then the resulting expected cost of misclassification becomes minimized as n_1 and $n_2 \to \infty$ with constant ratio. Since the above estimates are consistent, the *estimated generalized Bayes classification procedure* is constructed as follows: First solve Eq. (5.3.8) with μ_{ij} replaced by $\bar{x}_{ij}$, $i = 1$ or 2, $j = 1, \ldots, p$, and σ_{jm} replaced by s_{jm}, $m = 1, \ldots, p$. Use the resulting estimates of $\alpha_1, \ldots, \alpha_p$, denoted by $a_1, \ldots, a_p$, to calculate the estimated discriminant score z_{il} [Eq. (5.3.1)] for each observation vector $\mathbf{x}_{il}$, $l = 1, \ldots, n_i$. Next, estimate ζ_i [Eqs. (5.3.4)–(5.3.6)] by

$$\bar{z}_i = \frac{1}{n_i} \sum_{l=1}^{n_i} z_{il} \tag{5.3.24}$$

and σ_z^2 [Eq. (5.3.5)] by

$$s_z^2 = \sum_{j=1}^p \sum_{m=1}^p a_j s_{jm} a_m. \tag{5.3.25}$$

Thus, the estimated generalized Bayes procedure is to classify $\mathbf{x} = (x_1, \ldots, x_p)'$ into W_1 if

$$z = \sum_{i=1}^p a_i x_i \geqslant \frac{\bar{z}_1 + \bar{z}_2}{2} + \ln \frac{q_2 \, C(1|2)}{q_1 \, C(2|1)}, \tag{5.3.26}$$

otherwise, classify $\mathbf{x}$ into W_2. The *sample Mahalanobis distance* is

$$D^2 = \frac{(\bar{z}_1 - \bar{z}_2)^2}{s_z^2} \tag{5.3.27}$$

and is an estimate of Δ^2 [Eq. (5.3.7)].

Fortunately for Dr. Gesund's eyesight, there are programs which perform these calculations. The output of such programs typically consists of (a) the estimated discriminant function coefficients $a_1, \ldots, a_p$; (b) the value of the discriminant score z_{il} for each observation vector $\mathbf{x}_{il}$, $i = 1$ or 2, $l = 1, \ldots, n_i$; (c) the sample means $\bar{z}_1$ and $\bar{z}_2$; and (d) the sample Mahalanobis distance D^2. This information is sufficient to write down the classification procedure [Eq. (5.3.26)].

COMMENTS 5.3.2　　1. As in the case of known parameters, a positive multiple of the coefficients $a_1, \ldots, a_p$ may be used in defining the discriminant function. In this case,

$$K = \ln \frac{q_2 \, C(1|2)}{q_1 \, C(2|1)}$$

must also be multiplied by the same multiple. Some programs, for example, BMD04M, substitute the sums of squares and cross products of deviations for the pooled variances and covariances in Eq. (5.3.8). This has the effect of dividing the discriminant function coefficients $a_1, \ldots, a_p$ by $(n_1 + n_2 - 2)$. Hence, K must also be divided by $(n_1 + n_2 - 2)$.

2. Often it is difficult to obtain values of the prior probabilities q_1 and q_2. If the samples of sizes n_1 and n_2, respectively, are obtained randomly from the combined population of W_1 and W_2, then

$$\hat{q}_1 = \frac{n_1}{n_1 + n_2} \quad \text{and} \quad \hat{q}_2 = \frac{n_2}{n_1 + n_2}$$

may be used as estimates of q_1 and q_2, respectively.

3. Another quantity usually printed by the program is an F statistic,

$$F = \frac{n_1 + n_2 - p - 1}{(n_1 + n_2 - 2)p} \frac{n_1 n_2}{n_1 + n_2} D^2,$$

which is used for testing $H_0: \Delta^2 = 0$. The degrees of freedom are p and $n_1 + n_2 - p - 1$. Since this hypothesis is equivalent to $H_0: \mu_1 = \mu_2$, this statistic is the same as F given by Eq. (5.2.8). This is the computational aid promised in Comment 5.2.3.2.

EXAMPLE 5.3.3   When a critically ill patient is admitted to an intensive care unit, it is desired to classify him as "severely ill" or "less severely ill." Since death is an imminent possibility for such patients, "severely ill" is defined as "non-survival," and "less severely ill" is defined as "survival." Thus, W_1 = patients who survive and W_2 = patients who do not survive. Furthermore, data are collected on these patients at various stages during their treatment. In order to obtain maximum separation between the two populations, the samples used in this example are those observations collected right before discharge or death of the patient.

Data on $n_1 = 70$ survivors and $n_2 = 43$ nonsurvivors were gathered on $p = 13$ physiological variables. These variables included arterial and venous pressures, blood flow measurements, volumes of blood constituents, and urinary output. Using Comment 5.3.2.2 we estimated q_1 by $\hat{q}_1 = 70/113$, and q_2 by $\hat{q}_2 = 43/113$. Furthermore, $C(2|1) = C(1|2) = 1$, since we had no objective basis for deriving other values for them.

Using BMD04M, we obtained the discriminant function coefficients:

a_1	a_2	a_3	a_4	a_5
-0.00013	0.00183	-0.00006	-0.00043	-0.00167

a_6	a_7	a_8	a_9	a_{10}
0.06193	-0.01437	0.02534	0.00007	0.10206

a_{11}	a_{12}	a_{13}
0.02684	-0.00176	0.00070

Also obtained were $\bar{z}_1 = 0.37487$, $\bar{z}_2 = 0.28851$, and $D^2 = 9.58588$. Referring to Comment 5.3.2.1, we divided $K = \ln(43/70) = -0.49$ by $n_1 + n_2 - 2 = 111$. Thus, the estimated Bayes procedure is to classify $\mathbf{x} = (x_1, ..., x_{13})'$ into W_1 if

$$\sum_{i=1}^{13} a_i x_i \geqslant 0.33169 - 0.00442 = 0.32727,$$

otherwise, we classify $\mathbf{x}$ into W_2.

Finally, to test $H_0: \Delta^2 = 0$, or equivalently, $H_0: \boldsymbol{\mu}_1 = \boldsymbol{\mu}_2$, the program printed $F(13, 99) = 17.52$ which is significant for $P < 0.001$.

A method of classification utilizing the repeated measurements on such patients is given in general form by Azen (1969) and Afifi et al. (1971).

5.3.3 Estimation of the Probabilities of Misclassification

In the situation when the parameters are known, Eqs. (5.3.20)–(5.3.21) give values for the probabilities of misclassification $\Pr(2|1)$ and $\Pr(1|2)$. In the situation where the parameters are estimated, however, there are more than one way of estimating these probabilities. These estimators, along with their advantages and disadvantages, are discussed below. For a more detailed discussion, see Hills (1966), and Lachenbruch and Mickey (1968).

Method 1   Since D^2 is an estimator of Δ^2, D^2 is substituted for Δ^2 in Eqs. (5.3.20) and (5.3.21). It should be noted, however, that these estimators are biased, that is, on the average, the actual probability of misclassification is greater than the estimated one. An advantage is its simplicity since it may be easily calculated from the output of a program.

Method 2   This method consists in classifying each member of the sample of size n_1 from W_1 and the sample of size n_2 from W_2 according to Eq. (5.3.26). If m_1 is the number of observations from W_1 which are classified into W_2, and m_2 is the number of observations from W_2 which are classified into W_1, then $\hat{\Pr}(2|1) = m_1/n_1$ and $\hat{\Pr}(1|2) = m_2/n_2$. This method has a larger bias than that of the first method, and unless the program prints the discriminant score for each individual in the sample, this method is difficult to execute.

Method 3   This method consists of dividing the sample of n_1 observations from W_1 into two subsamples. Similarly, the sample of n_2 observations from W_2 are divided into two subsamples. The members of the first subsample of each pair are used to calculate the discriminant function and the classification procedure, while the members of the second subsample of each pair are classified according to that procedure. The proportions of misclassified individuals are the desired estimates for the probabilities of misclassification. This method has the advantage of producing unbiased estimators, but these estimators have larger variances than those obtained by the other two methods. Another disadvantage is that there is no standard method of dividing the samples.

Method 4   Lachenbruch (1967) suggests this alternative. From the first sample, exclude the first observation and compute the discriminant function on the basis of the remaining observations. Then classify the excluded observation. Do this for each member of the first sample. The proportion of misclassified individuals estimates $\Pr(2|1)$. A similar process applied to the second sample estimates $\Pr(1|2)$. On the basis of Monte Carlo studies, Lachenbruch and Mickey (1968) have shown that the bias of these estimators is negligible. The disadvantage of this method is the computational effort involved.

EXAMPLE 5.3.3 (*continued*)   According to Method 1 we substitute $D^2 = 9.58588$ and $K = -0.49$. We obtain $\hat{\Pr}(2|1) = \Phi(-1.71) = 0.044$, and $\hat{\Pr}(1|2) = \Phi(-1.39) = 0.082$. Applying Method 2 to the 113 discriminant scores, we obtain $m_1 = 5$ and $m_2 = 4$. Hence, $\hat{\Pr}(2|1) = 5/70 = 0.071$ and $\hat{\Pr}(1|2) = 4/43 = 0.093$. Methods 3 and 4 require special computer programs.

5.3.4 Estimation of Posterior Probabilities

In many applications, it is not sufficient to classify the individual into one of the populations and to report the probability of misclassification. A possibly more meaningful quantity to report is the posterior probability of the individual belonging to W_1 or W_2 [Eq. (5.3.10)]. In the case of known multivariate normal populations, the posterior probability that the individual belongs to W_1 may be expressed as

$$\Pr(W_1|\mathbf{x}) = \cfrac{1}{1 + \cfrac{q_2}{q_1} \exp\left\{-z + \cfrac{\zeta_1 + \zeta_2}{2}\right\}}, \tag{5.3.28}$$

where z is given by Eq. (5.3.1), and ζ_1 and ζ_2 are given by Eqs. (5.3.4) and (5.3.6), respectively. The posterior probability $\Pr(W_2|\mathbf{x}) = 1 - \Pr(W_1|\mathbf{x})$. When the parameters are estimated, then we substitute $\bar{z}_i$ [Eq. (5.3.24)] for ζ_i, $i = 1, 2$. Computation of these probabilities becomes particularly easy when a packaged program is used.

EXAMPLE 5.3.3 (*continued*)   As the critically ill patient is monitored, the physician may repetitively compute the posterior probability of survival. This gives him an estimate of the prognosis of the patient's status and also measures the improvement or deterioration in the patient's status. An application of this to a group of patients with myocardial infarction is given by Shubin *et al.* (1968).

5.4 Classification of an Individual into One of k Populations

In this section we consider the case of classifying an unknown observation vector $\mathbf{x}^{p \times 1} = (x_1, \ldots, x_p)'$ into one of k populations $W_i, i = 1, \ldots, k, k \geqslant 2$. Section 5.4.1 discusses the general case where the distributions of W_i are arbitrary with known parameters; Section 5.4.2 studies the case where W_i are assumed multivariate normal; and Section 5.4.3 presents the case where W_i are assumed binomial.

5.4.1 Classification into Arbitrary Known Populations

Let $f_i(\mathbf{x})$ be the density of $\mathbf{x}$ in W_i, and let q_i be the a priori probability that an observation $\mathbf{x}$ comes from $W_i, i = 1, \ldots, k$. Also let $C(i|j)$ be the cost of classifying an observation from W_j into W_i, and let $\Pr(i|j)$ be the probability of classifying an observation from W_j into $W_i, i, j = 1, \ldots, k, i \neq j$. Assuming that all parameters are known, it may be shown that the *generalized Bayes classification procedure* classifies $\mathbf{x}$ into W_i if

$$- \sum_{\substack{j=1 \\ j \neq i}}^{k} q_j f_j(\mathbf{x}) C(i|j) \tag{5.4.1}$$

is maximum, $i = 1, \ldots, k$. (If i_1 and i_2 both yield the same maximum, then we classify $\mathbf{x}$ into W_{i_1} or W_{i_2}.) Equation (5.4.1) is called the *ith discriminant score* and the Bayes procedure minimizes the *expected cost of misclassification* given by

$$\sum_{j=1}^{k} q_j \left\{ \sum_{\substack{i=1 \\ i \neq j}}^{k} C(i|j) \Pr(i|j) \right\}. \tag{5.4.2}$$

When we disregard the costs of misclassification, that is, we assume $C(i|j)$ are all equal, then the Bayes procedure is to classify $\mathbf{x}$ into W_i if

$$q_i f_i(\mathbf{x}) \tag{5.4.3}$$

is maximum, $i = 1, \ldots, k$. This rule minimizes the *expected probability of misclassification* given by

$$\sum_{j=1}^{k} q_j \left\{ \sum_{\substack{i=1 \\ i \neq j}}^{k} \Pr(i|j) \right\}. \tag{5.4.4}$$

Note that this rule is equivalent to classifying $\mathbf{x}$ into W_i if the posterior probability

$$\Pr(W_i|\mathbf{x}) = \frac{q_i f_i(\mathbf{x})}{\sum_{j=1}^{k} q_j f_j(\mathbf{x})} \tag{5.4.5}$$

is maximum.

5.4.2 Classification into Multivariate Normal Populations

Assume that W_i is $N(\mu_i^{p \times 1}, \Sigma^{p \times p})$ and let $f_i(\mathbf{x})$ be the ith density, $i = 1, ..., k$. Also assume that the parameters are known and the costs are all equal. Substituting $f_i(\mathbf{x})$ into Eq. (5.4.3), taking logarithms, and omitting factors which are common, we obtain the ith *linear discriminant score*

$$\delta_i = \alpha_{i1} x_1 + \cdots + \alpha_{ip} x_p + \gamma_i + \ln q_i \qquad \text{for} \quad i = 1, ..., k. \qquad (5.4.6)$$

The matrix form for the coefficients $\alpha_{i1}, ..., \alpha_{ip}$ and the constant γ_i are given in Comment 5.4.1.6. These quantities are functions of μ_i and Σ. We then classify $\mathbf{x}$ into W_i when δ_i is maximum, $i = 1, ..., k$. The posterior probability [Eq. (5.4.5)] is now given by

$$\Pr(W_i|\mathbf{x}) = \frac{q_i e^{\delta_i}}{\sum_{j=1}^{k} q_j e^{\delta_j}} \qquad \text{for} \quad i = 1, ..., k. \qquad (5.4.7)$$

As noted previously, the assumption of known parameters is only a theoretical simplification. In most applications, we have independent random samples from the k populations from which estimates of the parameters may be obtained. In this case, there is no optimal classification procedure, but substituting consistent estimators into Eq. (5.4.6) produces a procedure which is asymptotically optimal. Let n_i, $\bar{\mathbf{x}}_i$, and $\mathbf{S}_i$ be the ith sample size, sample mean vector, and sample covariance matrix, respectively, $i = 1, ..., k$. Then in Eq. (5.4.6) we substitute $\bar{\mathbf{x}}_i$ for μ_i, and the *pooled covariance matrix*

$$\mathbf{S} = \frac{\sum_{i=1}^{k} (n_i - 1) \mathbf{S}_i}{\sum_{i=1}^{k} n_i - k} \qquad (5.4.8)$$

for Σ. Thus, the *i-th estimated linear discriminant score* is

$$d_i = a_{i1} x_1 + \cdots + a_{ip} x_p + c_i + \ln q_i \qquad \text{for} \quad i = 1, ..., k. \qquad (5.4.9)$$

The matrix form for the coefficients $a_{i1}, ..., a_{ip}$ and the constant c_i are given below. We classify $\mathbf{x}$ into W_i if d_i is maximum. Then the estimated posterior probability is given by

$$\hat{\Pr}(W_i|\mathbf{x}) = \frac{q_i e^{d_i}}{\sum_{j=1}^{k} q_j e^{d_j}}. \qquad (5.4.10)$$

Fortunately, Dr. Gesund can utilize a packaged program to perform the computations for a_{ij} and c_i, $i = 1, ..., k, j = 1, ..., p$. Such programs may assume equal a priori probabilities and costs of misclassification. Typical output from a program may include:

(a) the pooled covariance matrix $\mathbf{S}$ and sometimes, the covariance matrix $\mathbf{S}_i$ for population W_i, $i = 1, ..., k$;

(b) the estimated linear discriminant score coefficients $a_{i1}, ..., a_{ip}$, and the constant c_i for population W_i, $i = 1, ..., k$ [the linear discriminant score is sometimes referred to as the (linear) "discriminant function" or simply, the "function"];

(c) the estimated linear discriminant score for each member $\mathbf{x}_{im}$ of the sample from W_i, $m = 1, ..., n_i$, $i = 1, ..., k$;

(d) the estimated posterior probabilities for each population W_j given each member $\mathbf{x}_{im}$ of the sample from W_i, $m = 1, ..., n_i$, $i, j = 1, ..., k$;

(e) the population in which each member $\mathbf{x}_{im}$ of the sample from W_i is classified, $m = 1, ..., n_i$, $i = 1, ..., k$ (this is the population for which the estimated posterior probability, or equivalently, the estimated discriminant score, is largest).

COMMENTS 5.4.1   1. In addition to the above output, some programs print a *classification table* indicating the number n_{ij} of members $\mathbf{x}_{jm}$ of the sample from W_j which are classified into W_i, $m = 1, ..., n_i$, $i, j = 1, ..., k$. Note that $\sum_{i=1}^{k} n_{ij} = n_j$, the jth sample size, $j = 1, ..., k$. From this table we can estimate the probability of misclassification by

$$\hat{\mathrm{Pr}}(i|j) = \frac{n_{ij}}{n_{j.}}, \qquad i, j = 1, ..., k, \quad i \neq j.$$

Although these estimates are biased, other estimates of these probabilities are difficult to calculate in the case of k populations.

2. Some programs compute a quantity called the *generalized Mahalanobis distance V* which generalizes the Mahalanobis D^2. This quantity may be used for testing the hypothesis $H_0: \boldsymbol{\mu}_1 = \cdots = \boldsymbol{\mu}_k$. Under H_0, as the sample sizes $n_i \to \infty$, the distribution of V approaches that of χ^2 with $p(k-1)$ degrees of freedom. Thus, an approximate test of H_0 consists in rejecting H_0 if $\chi^2 > \chi^2_{1-\alpha}(p(k-1))$.

3. Note that testing $H_0: \boldsymbol{\mu}_1 = \cdots = \boldsymbol{\mu}_k$ is the multivariate version of the one-way analysis of variance. The theory for developing an exact test of this and more general hypotheses belongs to the area of multivariate analysis of variance (see Anderson, 1958, or Rao, 1965). Many programs print a statistic, called the *U statistic*, which is an exact test statistic of H_0. Due to the complexity of the distribution of U, an *F approximation to U* is printed along with its degrees of freedom. This F test is exact whenever $p = 1$ or 2 for all k, or $k = 2$ for all p.

4. If the q_i are not equal for $i = 1, ..., k$, we can still use the results of a packaged program which assumes equal q_i. We modify output item (b), the estimated linear discriminant score coefficients for population W_i, by adding the quantity $\ln q_i$ to the constant c_i. We modify (c) by adding $\ln q_i$ to the discriminant score for member $\mathbf{x}_{im}$. We then reclassify him into the population for which these modified scores are highest [item (e)]. Recalculation of (d), the posterior probabilities, is more difficult.

5. The form of the classification problem given in this section applies for $k \geq 2$ with equal costs of misclassification. When $k = 2$, the procedure reduces to comparing

$$d_1 = \sum_{j=1}^{p} a_{1j} x_j + c_1 + \ln q_1 \quad \text{with} \quad d_2 = \sum_{j=1}^{p} a_{2j} x_j + c_2 + \ln q_2.$$

We classify $\mathbf{x}$ into W_1 if $d_1 \geq d_2$ or, equivalently, if

$$\sum_{j=1}^{p} (a_{1j} - a_{2j}) x_j \geq (c_2 - c_1) + \ln \frac{q_2}{q_1}.$$

Defining $a_j = a_{1j} - a_{2j}$ for $j = 1, \ldots, p$, and $(\bar{z}_1 + \bar{z}_2)/2 = c_2 - c_1$, we obtain the linear discriminant function in Eq. (5.3.26) with $C(1|2) = C(2|1)$.

*6. The formulas for the linear discriminant score [Eq. (5.4.6)] are given in matrix notation as

$$\delta_i = (\boldsymbol{\mu}_i' \boldsymbol{\Sigma}^{-1}) \mathbf{x} - \tfrac{1}{2} \boldsymbol{\mu}_i' \boldsymbol{\Sigma}^{-1} \boldsymbol{\mu}_i + \ln q_i,$$

so that

$$\begin{bmatrix} \alpha_{i1} \\ \vdots \\ \alpha_{ip} \end{bmatrix} = \boldsymbol{\mu}_i \boldsymbol{\Sigma}^{-1} \quad \text{and} \quad \gamma_i = -\tfrac{1}{2} \boldsymbol{\mu}_i' \boldsymbol{\Sigma}^{-1} \boldsymbol{\mu}_i \quad \text{for} \quad i = 1, \ldots, k.$$

Similarly, the estimated linear discriminant score [Eq. (5.4.9)] is

$$d_i = (\bar{\mathbf{x}}_i' \mathbf{S}^{-1}) \mathbf{x} - \tfrac{1}{2} \bar{\mathbf{x}}_i' \mathbf{S}^{-1} \bar{\mathbf{x}}_i + \ln q_i$$

with

$$\begin{bmatrix} a_{i1} \\ \vdots \\ a_{ip} \end{bmatrix} = \bar{\mathbf{x}}_i' \mathbf{S}^{-1} \quad \text{and} \quad c_i = -\tfrac{1}{2} \bar{\mathbf{x}}_i' \mathbf{S}^{-1} \bar{\mathbf{x}}_i \quad \text{for} \quad i = 1, \ldots, k.$$

Finally, the matrix expression for V (Comment 5.4.1.2) is

$$V = \sum_{i=1}^{k} n_i (\bar{\mathbf{x}}_i - \bar{\mathbf{x}})' \mathbf{S}^{-1} (\bar{\mathbf{x}}_i - \bar{\mathbf{x}}),$$

where

$$\bar{\mathbf{x}} = \frac{\sum_{i=1}^{k} n_i \bar{\mathbf{x}}_i}{\sum_{i=1}^{k} n_i}.$$

This gives an intuitive interpretation of V as a weighted sum of the "distances" of group means $\bar{\mathbf{x}}_i$ from the overall mean $\bar{\mathbf{x}}$.*

7. When the assumption of equal covariance matrices is not satisfied, that is, W_i is $N(\boldsymbol{\mu}_i, \boldsymbol{\Sigma}_i)$ for $i = 1, \ldots, k$, we obtain a quadratic discriminant score (see Rao, 1965, p. 488). If we do not know the form of $f_i(\mathbf{x})$ at all, we may use a nonparametric technique to classify $\mathbf{x}$ (see Fix and Hodges, 1951, 1952; and Palmersheim, 1970).

EXAMPLE 5.4.1   Measurements on $p = 4$ variables—$X_1 =$ age, $X_2 =$ systolic pressure (mm Hg), $X_3 =$ log cardiac index (liters/min/m^2), and $X_4 =$ urine output (10 cc/cm)—were made on $n = 113$ critically ill patients at the time of admission. These patients came from one of $k = 6$ populations defined by $W_1 =$ nonshock, $W_2 =$ hypovolemic shock, $W_3 =$ cardiogenic shock, $W_4 =$ bacterial shock, $W_5 =$ neurogenic shock, and $W_6 =$ other type of shock. The sample sizes from each population were $n_1 = 34$, $n_2 = 17$, $n_3 = 20$, $n_4 = 16$, $n_5 = 16$, and $n_6 = 10$, respectively. It was desired to derive a classification procedure for identifying the type of illness of any newly admitted patient given that he belongs to one of these six populations. Assuming equal a priori probabilities and equal costs of misclassification, a packaged computer program was used to calculate d_i, $i = 1, ..., 6$. The coefficients are presented in Table 5.4.1. Thus, for example, $d_1 = 0.339x_1 + \cdots + 0.196x_4 - 26.827$ is the estimated discriminant score for W_1. The classification table is given in Table 5.4.2. Thus, for example, $n_{13} = 3$ cases of the $n_3 = 20$ members of the sample from W_3 are classified into W_1. Hence, $\hat{\Pr}(1|3) = 3/20$. The approximate F statistic for testing $H_0: \mathbf{\mu}_1^{4 \times 1} = \cdots = \mathbf{\mu}_6^{4 \times 1}$ is given as $F = 4.21$ with 20 and 345.9 degrees of freedom (Note that the fractional degrees of freedom resulted from the F approximation to the U statistic.) From the F table, we find that $F_{0.999}(20, 345.9) \simeq 2.5$. Hence H_0 is rejected at $P < 0.001$.

TABLE 5.4.1. *Linear Discriminant Score Coefficients for Example 5.4.1*

Coefficients	W_1	W_2	W_3	W_4	W_5	W_6
a_{i1}	0.339	0.331	0.339	0.333	0.250	0.337
a_{i2}	0.197	0.150	0.167	0.151	0.147	0.166
a_{i3}	1.575	1.453	0.916	1.915	1.049	0.999
a_{i4}	0.196	0.135	0.129	0.132	0.142	0.123
c_i	-26.827	-20.491	-21.455	-22.141	-15.362	-21.373

TABLE 5.4.2. *Classification Table for Example 5.4.1*

W_j	$W_i =$ W_1	W_2	W_3	W_4	W_5	W_6	Totals n_j
			Number of cases n_{ij} from W_j classified into population W_i				
W_1	20	1	5	4	4	0	34
W_2	1	2	5	5	3	1	17
W_3	3	0	8	2	5	2	20
W_4	2	2	0	6	4	2	16
W_5	2	1	0	1	11	1	16
W_6	2	1	4	1	1	1	10

5.4.3 Classification into Binomial Populations

An example of classifying an individual into one of k binomial populations was given in Example 5.3.2. In this case an individual was to be diagnosed into one of k disease categories on the basis of the presence or absence of p symptoms. In general, we wish to classify an individual on the basis of the existence or absence of p phenomena. For the jth phenomenon we define the random variable

$$X_j = \begin{cases} 1, & \text{if phenomenon } j \text{ is present,} \\ 0, & \text{if phenomenon } j \text{ is absent,} \end{cases} \quad \text{for } j = 1, \ldots, p. \quad (5.4.11)$$

We assume that $\Pr(X_j = 1 | W_i) = p_{ij}$ and $\Pr(X_j = 0 | W_i) = 1 - p_{ij}$ for $i = 1, \ldots, k$, $j = 1, \ldots, p$. Thus, the probability function of X_j for the population W_i is

$$f_i(x_j) = p_{ij}^{x_j}(1 - p_{ij})^{1 - x_j} \quad \text{for } i = 1, \ldots, k, \quad j = 1, \ldots, p. \quad (5.4.12)$$

If we assume $X_1, \ldots, X_p$ to be independent, then we may write the joint probability function $f_i(\mathbf{x})$ for W_i as

$$f_i(\mathbf{x}) = \prod_{j=1}^{p} f_i(x_j) \quad \text{for } i = 1, \ldots, k, \quad \mathbf{x} = (x_1, \ldots, x_p)'. \quad (5.4.13)$$

Assuming equal costs and a priori probabilities $q_1, \ldots, q_k$ we calculate the posterior probabilities from Eq. (5.4.5) and get

$$\Pr(W_i | \mathbf{x}) = \frac{q_i \prod_{j=1}^{p} p_{ij}^{x_j}(1 - p_{ij})^{1 - x_j}}{\sum_{m=1}^{k} q_m [\prod_{j=1}^{p} p_{mj}^{x_j}(1 - p_{mj})^{1 - x_j}]} \quad \text{for } i = 1, \ldots, k. \quad (5.4.14)$$

We classify $\mathbf{x}$ into W_i if $\Pr(W_i | \mathbf{x})$ is maximum.

EXAMPLE 5.3.2 (*continued*)   This example is one of *computer diagnosis*. Since the p_{ij} are not known, we estimate them from a random sample of n patients taken from the combined population of the k diseases. Let n_i be the number of patients with the ith disease, $i = 1, \ldots, k$. Then $\sum_{i=1}^{k} n_i = n$. Also let n_{ij} be the number of patients with the ith disease who have the jth symptom. Then an estimate of p_{ij} is

$$\hat{p}_{ij} = \frac{n_{ij}}{n_i}, \quad i = 1, \ldots, k, \quad j = 1, \ldots, p.$$

If the a priori probabilities are unknown, then an estimate of q_i is

$$\hat{q}_i = \frac{n_i}{n}, \quad i = 1, \ldots, k.$$

A program may be written to compute the estimated posterior probability for a new individual $\mathbf{x} = (x_1, \ldots, x_p)'$ given by

$$\hat{\Pr}(W_i|\mathbf{x}) = \frac{\dfrac{n_i}{n} \prod_{j=1}^{p} \left(\dfrac{n_{ij}}{n_i}\right)^{x_j} \left(1 - \dfrac{n_{ij}}{n_i}\right)^{1-x_j}}{\sum_{m=1}^{k} \dfrac{n_m}{n} \left[\prod_{j=1}^{p} \left(\dfrac{n_{mj}}{n_m}\right)^{x_j} \left(1 - \dfrac{n_{mj}}{n_m}\right)^{1-x_j}\right]}.$$

Since the assumption of independence of symptoms may be unrealistic in most applications, an alternative procedure for classification when all q_i are equal is to calculate the proportion of individuals in each sample with any possible combination of symptoms. A new individual with a given combination of symptoms is classified into the disease category having the highest proportion associated with his symptoms. For example, let $k = 2$ and $p = 3$, and suppose from two samples of sizes n_1 and n_2, we obtain the accompanying table. Then, a new individual with a combination 110 would be classified into W_1, since $0.20 > 0.10$.

Symptoms	W_1	W_2
000	0.10	0.20
100	0.10	0.15
010	0.20	0.10
001	0.10	0.15
110	0.20	0.10
101	0.10	0.20
011	0.10	0.05
111	0.10	0.05

The difficulty with this method is that there are 2^p parameters to be estimated for each population. For this reason, the assumption of independence is convenient.

5.5 Stepwise Discriminant Analysis

In Sections 5.3 and 5.4 we considered the case of classifying a p-dimensional observation vector $\mathbf{x} = (x_1, \ldots, x_p)'$ into one of k multivariate normal populations W_i: $N(\boldsymbol{\mu}_i^{p \times 1}, \boldsymbol{\Sigma}^{p \times p})$, where $\boldsymbol{\mu}_i = (\mu_{i1}, \ldots, \mu_{ip})'$, $i = 1, \ldots, k$. Since $\mathbf{x}$ is a realization of a random vector $\mathbf{X} = (X_1, \ldots, X_p)'$, the results presented so far used all p variables $X_1, \ldots, X_p$ to discriminate between the k populations. In many applications, however, it is desired to identify a subset of these variables which "best" discriminates between the k populations. This problem is analogous to that of stepwise regression (Section 3.3) in which it was desired to identify a subset of independent variables which "best" predicts a dependent variable Y. In that case an F statistic based on the partial correlation was used to choose variables; in this case an F statistic based on a one-way analysis of variance test is used to choose variables. In both cases the F statistic is called the F to enter for variables not chosen and the F to remove for chosen variables.

In brief, the logic behind the stepwise discriminant procedure is as follows: We first identify the variable for which the mean values in the k populations are "most different." For each variable this difference is measured by a one-way analysis of variance F statistic, and the variable with the largest F is chosen (or entered). On successive steps, we consider the conditional distribution of each variable not entered given the variables entered. Of the variables not entered, we identify the variable for which the mean values of the conditional distributions in the k populations are "most different." This difference is also measured by a one-way analysis of variance F statistic. The stepwise process is stopped when no additional variables significantly contribute to the discrimination between the k populations.

In detail, a program for the stepwise procedure is as follows. Let $\mathbf{x}_{i1}^{p \times 1}, \ldots, \mathbf{x}_{in_i}^{p \times 1}$ be a random sample from W_i, $i = 1, \ldots, k$. Then, using the notation and definitions of Section 5.4, we have the following steps.

Step 0   The F to enter along with its degrees of freedom is computed for each X_j, $j = 1, \ldots, p$. This F to enter is the one-way analysis of variance F statistic for testing H_0: $\mu_{1j} = \cdots = \mu_{kj}$, for $j = 1, \ldots, p$. If all F to enter are less than a prescribed inclusion level, called the F to include, the process is terminated, and we conclude that no variable significantly discriminates between the populations.

Step 1   The variable X_{j_1} having the largest F to enter is chosen as the first variable. The estimated linear discriminant coefficient and constant are calculated for each population W_i, $i = 1, \ldots, k$. The classification table, U statistic, and F approximation to U are also calculated. Also, the F to remove, which is equal to the F to enter, and its degrees of freedom for X_{j_1} is calculated. Then, the F to enter and its degrees of freedom for each variable not entered are calculated. This tests the hypothesis H_0: $\mu_{1j.j_1} = \cdots = \mu_{kj.j_1}$, where $\mu_{ij.j_1}$ is the mean of the conditional distribution in W_i of X_j given X_{j_1}, $i = 1, \ldots, k$, $j = 1, \ldots, p$, $j \neq j_1$. If all the F to enter are less than F to include, then Step S is executed; otherwise, Step 2 is executed.

Step 2   The variable X_{j_2} is chosen for which the F to enter is maximum. The two estimated linear discriminant coefficients and constant are calculated for each population W_i, $i = 1, \ldots, k$. The classification table, U statistic, and F approximation to U are also calculated. Also, the two F to remove and their degrees of freedom for X_{j_1} and X_{j_2} are calculated. These test $H_0: \mu_{1j_1.j_2} = \cdots = \mu_{kj_1.j_2}$ and $H_0: \mu_{1j_2.j_1} = \cdots = \mu_{kj_2.j_1}$, respectively. Finally, the F to enter and its degrees of freedom for each variable not entered are calculated. This tests the hypothesis $H_0: \mu_{1j.j_1j_2} = \cdots = \mu_{kj.j_1j_2}$, where $\mu_{ij.j_1j_2}$ is the mean of the conditional distribution in W_i of X_j given X_{j_1} and X_{j_2}, $i = 1, \ldots, k, j = 1, \ldots, p, j \neq j_1$ or j_2.

Step 3   (a) Let L denote the set of l variables which have been entered. If any of the F to remove for the variables in L are less than a prescribed deletion level, called the *F to exclude*, then the variable with the smallest F to remove is deleted from L and (b) is executed with l replaced by $l-1$. If all the F to enter for the variables not in L are less than F to include, then Step S is executed. Otherwise, the variable with the largest F to enter is chosen and is added to L so that l is replaced by $l+1$.

(b) The l estimated linear discriminant coefficients and constant are calculated for each population $W_i, i = 1, \ldots, k$. The classification table, U statistic, and F approximation to U are also calculated. Also, the F to remove and its degrees of freedom for each variable in L is calculated. This tests $H_0: \mu_{1s.(l-1)} = \cdots = \mu_{ks.(l-1)}$ for each X_s in L given all of the $l-1$ remaining variables in L. The notation $\mu_{is.(l-1)}$ is the mean of the conditional distribution in W_i of X_s given all the variables in L (excluding X_s). Finally, the F to enter and its degrees of freedom for each variable not in L are calculated. This tests the hypothesis $H_0: \mu_{1j.(l)} = \cdots = \mu_{kj.(l)}$, where $\mu_{ij.(l)}$ is the mean of the conditional distribution in W_i of X_j given all the variables in $L, i = 1, \ldots, k, j = 1, \ldots, p, X_j$ not in L.

Steps 4, 5, ...   Repeat Step 3 recursively. Step S is executed when the F to enter is less than the F to include for all variables not in L, or when all the variables have been entered and no F to remove is less than the F to exclude. Some programs also terminate the process when $l = \min(n_i), i = 1, \ldots, k$.

Step S   The posterior probability of belonging to population $W_1, \ldots, W_k$ is calculated for each $\mathbf{x}_{im}, m = 1, \ldots, n_i, i = 1, \ldots, k$. Based on these probabilities, each of the individuals is classified into one of the populations and a classification table is prepared. A summary table is usually available upon request by the user. For each step the quantities printed are the step number, the variable entered or removed, the F to enter or F to remove, the U statistic, and the F approximation to U. The user should amend this table by copying the degrees of freedom v_1 and v_2 for the F to enter or F to remove from the detailed output of each step.

COMMENTS 5.5.1   1. The summary table is extremely useful for selecting which variables should be used in the "best" classification procedure. For a prescribed significance level α, we select a set H of variables as follows:

(a) At Step 1, a variable X_{j_1} was entered. If its F to enter is less than $F_{1-\alpha}(v_1, v_2)$, then we are not able to discriminate significantly between the k populations. Otherwise, $H = \{X_{j_1}\}$.

(b) At Step 2, a variable X_{j_2} was entered. If its F to enter is less than $F_{1-\alpha}(v_1, v_2)$, then H consists only of X_{j_1}. Otherwise, $H = \{X_{j_1}, X_{j_2}\}$.

(c) For each Step q, $q = 3, 4, \ldots$, if a variable is removed, delete it from H and go to Step $q+1$. If a variable is entered, compare its F to enter with $F_{1-\alpha}(v_1, v_2)$. If its F to enter is less than $F_{1-\alpha}(v_1, v_2)$, stop amending H. Otherwise, augment H and go to Step $q+1$.

2. This comment applies only to the case of $k = 2$.

(a) The linear discriminant function of Eq. (5.3.26) is obtained by taking the difference of the two estimated linear discriminant scores [Eq. (5.4.9)]. That is, $a_i = a_{1i} - a_{2i}$ and $(\bar{z}_1 + \bar{z}_2)/2 = c_2 - c_1$, where these symbols are defined by these two equations.

(b) As noted above, the F approximation to the U statistic based on q variables is exact. Furthermore, we may obtain from this F the value of the estimated Mahalanobis distance D_q^2 based on q variables by using the relationship

$$D_q^2 = \frac{q(n_1 + n_2)(n_1 + n_2 - 2)}{n_1 n_2 (n_1 + n_2 - q - 1)} F$$

for $q = 1, \ldots, p$ and sample sizes n_1 and n_2 from W_1 and W_2, respectively.

(c) Suppose the variables $X_1, \ldots, X_q$ have been entered, $q = 1, \ldots, p-1$. To test the hypothesis that the remaining variables $X_{q+1}, \ldots, X_p$ do not contribute to the discrimination achieved by $X_1, \ldots, X_q$, we test the hypothesis $H_0: \Delta_q^2 = \Delta_p^2$, where Δ_m^2 is the population Mahalanobis distance based on m variables, $m = p$ or q, $q = 1, \ldots, p-1$. To test H_0, let D_p^2 and D_q^2 be the sample estimates of the Mahalanobis distances Δ_p^2 and Δ_q^2, respectively. Then,

$$F = \frac{n_1 + n_2 - p - 1}{p - q} \frac{n_1 n_2 (D_p^2 - D_q^2)}{(n_1 + n_2)(n_1 + n_2 - 2) + n_1 n_2 D_q^2}$$

has an $F(p-q, n_1+n_2-p-1)$ distribution under H_0. We reject H_0 if $F > F_{1-\alpha}(p-q, n_1+n_2-p-1)$. This test may be used to develop another selection procedure similar to that of Comment 3.3.1.2 in the case of stepwise regression.

3. As in stepwise regression, the user is given the option of *forcing* certain variables to be included in the classification procedure. The stepwise procedure is then applied to the remaining variables.

EXAMPLE 5.5.1   A stepwise discriminant analysis was performed on a subset of $p = 4$ variables of Example 5.3.3. Data was collected on $n_1 = 70$ individuals from W_1, the population of survivors, and $n_2 = 43$ individuals from W_2, the population of nonsurvivors. The variables of interest were $X_1 = $ mean arterial pressure (mm Hg), $X_2 =$ mean venous pressure (mm Hg), $X_3 =$ urinary output (cc/hr), and $X_4 =$ log plasma volume index (ml/kg). We used BMD07M with an F to include $= 0.01$ and an F to exclude $= 0.005$. Following the steps outlined above, the details are as follows.

Step 0   The F to enter with 1 and 111 degrees of freedom are

Variable	F to enter
X_1	131.41
X_2	14.06
X_3	22.01
X_4	2.49

Since all F to enter are greater than F to include, we proceed to Step 1.

Step 1   Variable X_1 is chosen as the first variable since it has the largest F to enter. The estimated linear discriminant coefficient and constant for each population are

W_i	a_{i1}	c_i
W_1	0.262	-11.627
W_2	0.141	-3.383

The classification table is

	W_1	W_2
W_1	60	10
W_2	8	35

and for the first step, the F approximation to $U = F$ to remove for $X_1 = F$ to enter for $X_1 = 131.41$ with 1 and 111 degrees of freedom. Finally, the F to enter with 1 and 110 degrees of freedom are calculated for X_2, X_3, and X_4. These are

Variable	F to enter
X_2	10.55
X_3	9.52
X_4	17.35

Since all F to enter are greater than F to include, we proceed to Step 2.

Step 2   Since X_4 has the largest F to enter, it is chosen as the next variable entered. The estimated linear discriminant coefficients and constant for each population are

W_i	a_{i1}	a_{i4}	c_i
W_1	0.690	211.621	-213.956
W_2	0.544	199.016	-182.326

The classification table is

	W_1	W_2
W_1	63	7
W_2	6	37

The F approximation to U is 84.06 with 2 and 111 degrees of freedom. The F to remove for X_1 and X_4 with 1 and 110 degrees of freedom are

Variable	F to remove
X_1	162.01
X_4	17.35

Finally, the F to enter with 1 and 109 degrees of freedom are calculated for X_2 and X_3. These are

Variable	F to enter
X_2	9.55
X_3	8.97

Step 3   (a) Let $L = \{X_1, X_4\}$ and $l = 2$. Since both of the F to remove for X_1 and X_4 are greater than F to exclude, and all F to enter are greater than F to include, we augment L by the variable with the largest F to enter. Hence, $L = \{X_1, X_2, X_4\}$ and $l = 3$.

(b) The estimated linear discriminant coefficients and constant for each population are

W_i	a_{i1}	a_{i2}	a_{i4}	c_i
W_1	0.686	0.126	211.481	-214.141
W_2	0.535	0.324	198.654	-183.557

The classification table is

	W_1	W_2
W_1	67	3
W_2	7	36

The F approximation to U is 63.58 with 3 and 109 degrees of freedom. The F to remove for each variable in L with 1 and 109 degrees of freedom are

Variable	F to remove
X_1	152.16
X_2	9.55
X_4	16.24

Finally, the F to enter with 1 and 108 degrees of freedom for X_3 is 7.49.

Step 4   (a) Since no F to remove is less than F to exclude, and the F to enter is greater than F to include, $L = \{X_1, X_2, X_3, X_4\}$, and $l = 4$.
   (b) The estimated discriminant coefficients and constant for each population are

W_i	a_{i1}	a_{i2}	a_{i3}	a_{i4}	c_i
W_1	0.686	0.140	0.013	212.086	-215.524
W_2	0.535	0.331	0.006	198.934	-183.856

The classification table is

	W_1	W_2
W_1	67	3
W_2	6	37

The F approximation to U is 52.39 with 4 and 108 degrees of freedom. The F to remove for each variable in L with 1 and 108 degrees of freedom are

Variable	F to remove
X_1	129.31
X_2	8.06
X_3	7.49
X_4	15.76

Since there are no further variables and all four F to remove are greater than F to exclude, Step S is executed.

Step S   The summary table is shown. Letting $\alpha = 0.05$, we apply Comment 5.5.1.1 to the summary table. At Step 1, the F to enter for X_1 is greater than

Step number	Variable Entered	Variable Removed	F to Enter	F to Remove	v_1	v_2	U
1	X_1		131.41		1	111	0.458
2	X_4		17.35		1	110	0.396
3	X_2		9.55		1	109	0.364
4	X_3		7.49		1	108	0.340

$F_{0.95}(1, 111) \simeq 3.96$. Hence, $H = \{X_1\}$. At Step 2, the F to enter for X_4 is greater than $F_{0.95}(1, 110) \simeq 3.96$. Indeed, at each step, the F to enter is greater than $F_{0.95}(1, 108) \simeq 3.96$, so that $H = \{X_1, X_2, X_3, X_4\}$. If α had been chosen as 0.001, then $F_{0.999}(1, 108) \simeq 11.5$, so that $H = \{X_1, X_4\}$ only.

Since $k = 2$ for this example, we can apply Comment 5.5.1.2. The estimated linear discriminant function of Eq. (5.3.26) is obtained from the coefficients and constants at Step 4. Thus, we classify $\mathbf{x}$ into W_1 if

$$z = 0.151x_1 - 0.191x_2 + 0.007x_3 + 13.152x_4 \geqslant 31.668,$$

$$\text{where} \quad q_1 = q_2 = \tfrac{1}{2}.$$

We can also calculate the estimated Mahalanobis distance D_q^2 for each step q. These are given in the accompanying table where the second column is the F approximation to U (exact) and where the fourth column is the F test statistic for testing $H_0: \Delta_q^2 = \Delta_4^2$ (Comment 5.5.1.2c). The last column is the 95th percentile of the F distribution with $4 - q$ and 108 degrees of freedom. Since the F test statistic (column 4) is greater than $F_{0.95}(4-q, 108)$ for $q = 1$, 2, and 3, we reject the hypotheses $H_0: \Delta_1^2 = \Delta_4^2$, $H_0: \Delta_2^2 = \Delta_4^2$, and $H_0: \Delta_3^2 = \Delta_4^2$. Hence, all four variables significantly contribute to the discrimination of the two populations at the $\alpha = 0.05$ significance level.

Step q	F	D_q^2	F	$F_{0.95}(4-q, 108)$
1	131.41	4.95	12.25	2.7
2	84.06	6.35	8.72	3.1
3	63.58	7.30	7.05	3.9
4	52.39	8.05	—	—

5.6 Principal Component Analysis

Assume that we are given p random variables $X_1, \ldots, X_p$ having a multivariate distribution (not necessarily normal) with mean $\boldsymbol{\mu}^{p \times 1} = (\mu_1, \ldots, \mu_p)'$ and covariance matrix $\boldsymbol{\Sigma}^{p \times p} = (\sigma_{ij})$. In many applications, it is desirable to examine the interrelationship among the variables $X_1, \ldots, X_p$. This interrelationship, called the *dependence structure*, may be measured by the covariances, or equivalently, the variances and correlations between $X_1, \ldots, X_p$. For some applications, it is possible to find a few linear combinations $Y_1, \ldots, Y_q$ of $X_1, \ldots, X_p$ $(q < p)$ which approximately generate the dependence structure between $X_1, \ldots, X_p$. Hence, it is possible to find a parsimonious description of the dependence structure which conveys approximately the same amount of information expressed by the original variables.

This section discusses one method, called *principal component analysis,* for analyzing this dependence structure. The method finds p linear combinations

$$Y_1 = \sum_{j=1}^{p} \alpha_{1j} X_j, \qquad \cdots, \qquad Y_p = \sum_{j=1}^{p} \alpha_{pj} X_j,$$

so that

$$\text{cov}(Y_i, Y_j) = 0 \qquad \text{for} \quad i, j = 1, ..., p, \quad i \neq j, \tag{5.6.1}$$

$$V(Y_1) \geqslant V(Y_2) \geqslant \cdots \geqslant V(Y_p), \tag{5.6.2}$$

and

$$\sum_{i=1}^{p} V(Y_i) = \sum_{i=1}^{p} \sigma_{ii}. \tag{5.6.3}$$

From these equations it is seen that the new variables $Y_1, ..., Y_p$ are uncorrelated and are ordered by their variances. Furthermore, the *total variance* $V = \sum_{i=1}^{p} \sigma_{ii}$ remains the same after the transformation. In this way, a subset of the first q Y_i's may explain most of the total variance and therefore produce a parsimonious description of the dependence structure among the original variables. The method of principal components is then to determine the coefficients α_{ij}, $i, j = 1, ..., p$. We first discuss the details of the method in terms of the population parameters and then in terms of estimating the parameters from samples. Note that it is not necessary to assume a multivariate normal distribution. However, this distribution is convenient since linear combinations of normally distributed variables are also normal and since it is completely determined by μ and Σ. Hence, we may assume $\mu = (0, ..., 0)'$, and, by explaining the dependence structure given by Σ, we in turn completely explain the distribution of $X_1, ..., X_p$.

Assuming Σ known, we let $Y_1 = \alpha_{11} X_1 + \cdots + \alpha_{1p} X_p$. We wish to find $\alpha_{11}, ..., \alpha_{1p}$, so that

$$V(Y_1) = \sum_{i=1}^{p} \sum_{j=1}^{p} \alpha_{1i} \alpha_{1j} \sigma_{ij} \tag{5.6.4}$$

is maximized subject to the condition that $\sum_{j=1}^{p} \alpha_{1j}^2 = 1$. (This condition ensures uniqueness of the solution.) The solution $\alpha_1 = (\alpha_{11}, ..., \alpha_{1p})'$ is called an *eigenvector* and is associated with the largest *eigenvalue* of Σ. This eigenvalue is equal to the variance $V(Y_1)$. The linear combination $Y_1 = \alpha_{11} X_1 + \cdots + \alpha_{1p} X_p$ is called the first *principal component* of $X_1, ..., X_p$, and it explains $100V(Y_1)/V$ percent of the total variance.

Next, we let $Y_2 = \alpha_{21} X_1 + \cdots + \alpha_{2p} X_p$. We wish to find $\alpha_{21}, ..., \alpha_{2p}$ so that

$$V(Y_2) = \sum_{i=1}^{p} \sum_{j=1}^{p} \alpha_{2i} \alpha_{2j} \sigma_{ij} \tag{5.6.5}$$

is maximized subject to the conditions that $\sum_{j=1}^{p} \alpha_{2j}^2 = 1$ and $\text{cov}(Y_1, Y_2) = \sum_{i=1}^{p} \sum_{j=1}^{p} \alpha_{1i} \alpha_{2j} \sigma_{ij} = 0$. The first condition ensures uniqueness of the solution and the second condition ensures that Y_1 and Y_2 are uncorrelated. The solution $\boldsymbol{\alpha}_2 = (\alpha_{21}, \ldots, \alpha_{2p})'$ is the eigenvector associated with the second largest eigenvalue of $\boldsymbol{\Sigma}$. This eigenvalue is equal to $V(Y_2)$, and the linear combination Y_2 is the second principal component of $X_1, \ldots, X_p$. The first two principal components explain $100[V(Y_1) + V(Y_2)]/V$ percent of the total variance.

After $Y_1, \ldots, Y_{q-1}, q = 2, \ldots, p$, have been obtained we find $Y_q = \sum_{j=1}^{p} \alpha_{qj} X_j$ such that

$$V(Y_q) = \sum_{i=1}^{p} \sum_{j=1}^{p} \alpha_{qi} \alpha_{qj} \sigma_{ij}$$

is maximized subject to $\sum_{j=1}^{p} \alpha_{qj}^2 = 1$ and

$$\text{cov}(Y_m, Y_q) = \sum_{i=1}^{p} \sum_{j=1}^{p} \alpha_{qi} \alpha_{mj} \sigma_{ij} = 0 \qquad \text{for} \quad m = 1, \ldots, q-1.$$

This results in $\boldsymbol{\alpha}_q = (\alpha_{q1}, \ldots, \alpha_{qp})'$ as the eigenvector corresponding to the qth largest eigenvalue $V(Y_q)$. Hence, Y_q is the qth principal component, and the variables $Y_1, \ldots, Y_q$ explain $100 \sum_{i=1}^{q} V(Y_i)/V$ percent of the total variance.

A geometric interpretation of the principal components is as follows (see Fig. 5.6.1 for $p = 2$): Each variable $X_1, \ldots, X_p$ is represented by a coordinate axis from the origin $\boldsymbol{\mu} = (\mu_1, \ldots, \mu_p)'$. These p axes form a *p-dimensional space* with

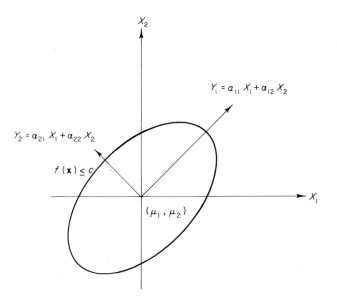

FIG. 5.6.1. *The two principal components for the case $p = 2$.*

each realization $\mathbf{x} = (x_1, \ldots, x_p)'$ represented by a point whose coordinates are $X_1 = x_1, \ldots, X_p = x_p$. In principal component analysis, we search for a *rotation* of these axes so that the variable Y_1 represented by the first of the new axes has maximum variance. The variable Y_2 represented by the second of the new axes is uncorrelated with Y_1 and has maximum variance under this restriction. Similarly, the variable Y_q represented by the qth new axis is uncorrelated with $Y_1, Y_2, \ldots, Y_{q-1}$ and has maximum variance under these restrictions, $q = 3, \ldots, p$. Letting $f(\mathbf{x})$ be the normal density of the random vector $\mathbf{X} = (X_1, \ldots, X_p)'$, then the inequality $f(\mathbf{x}) \leqslant c$, where c is a constant, defines a region in the p-dimensional space called an *ellipsoid of concentration*. It can be shown that these principal components are in the direction of the principal axes of the ellipsoid of concentration.

In Fig. 5.6.1 we have a two-dimensional space defined by X_1 and X_2 with the origin at (μ_1, μ_2). The ellipsoid of concentration is simply an ellipse. The first principal component $Y_1 = \alpha_{11} X_1 + \alpha_{12} X_2$ is in the direction of the major axis of the ellipse, and the second principal component $Y_2 = \alpha_{21} X_1 + \alpha_{22} X_2$ is in the direction of the minor axis of the ellipse.

When Σ is unknown, we assume that we have a random sample $\mathbf{x}_1^{p \times 1}, \ldots, \mathbf{x}_n^{p \times 1}$ and estimate Σ by the sample covariance matrix $\mathbf{S}$. To obtain the *estimated principal components* we apply the above procedure to $\mathbf{S}$ obtaining estimates a_{ij} of α_{ij}, $i, j = 1, \ldots, p$. Thus, the qth estimated principal component is $Y_q = \sum_{j=1}^{p} a_{qj} X_j$ which corresponds to the qth largest eigenvalue of $\mathbf{S}$ and the qth eigenvector $\mathbf{a}_q = (a_{q1}, \ldots, a_{qp})'$, $q = 1, \ldots, p$. The geometric interpretation holds with sample means $\bar{x}_1, \ldots, \bar{x}_p$ replacing $\mu_1, \ldots, \mu_p$.

COMMENTS 5.6.1   1.   It should be noted that if $X_1, \ldots, X_p$ have a multivariate normal distribution, then the principal components $Y_1, \ldots, Y_p$ are mutually independent.

2.   For any two eigenvalues $V(Y_i)$ and $V(Y_j)$ which are unequal, the corresponding eigenvectors are at right angles to each other. This property is called *orthogonality* and is expressed as $\sum_{m=1}^{p} \alpha_{im} \alpha_{jm} = 0$, $i, j = 1, \ldots, p$, $i \neq j$. If two eigenvalues are equal, then their eigenvectors may be chosen so that they are orthogonal. Thus, the total set of p principal components are mutually orthogonal.

3.   The correlation matrix may be used instead of the covariance matrix for deriving principal components. If the p variables are in widely different units, linear combinations of them may have little meaning. Hence, standardizing each variable by

$$Z_i = \frac{X_i - \mu}{\sigma_i} \quad \text{or} \quad Z_i = \frac{X_i - \bar{x}}{s_i} \quad \text{for} \quad i = 1, \ldots, p$$

may be more appealing, since the Z_i are dimensionless. We then analyze the dependence structure of $Z_1, \ldots, Z_p$, which is given by the correlation matrix of

$X_1, ..., X_p$. Note that the total variance $V = p$, the number of variables. In general, the components obtained from the correlation matrix are not the same as those obtained from the covariance matrix. Indeed, any linear transformation of the original variables will produce different principal components.

4. The correlation between X_i and Y_j is given by $\alpha_{ji}[V(Y_j)]^{1/2}/\sigma_i$, where σ_i is the standard deviation of X_i. Therefore, to compare the contribution of $X_1, ..., X_p$ to Y_j, we examine the quantities α_{ji}/σ_i, $i = 1, ..., p$. Furthermore, when the correlation matrix is used, then comparison of the coefficients α_{ji} is all that is necessary. Hence, the larger the coefficient, the larger the contribution of the variable to the principal component.

5. Principal component analysis originally arose from the work of Pearson (1901). Given a set of vector observations, Pearson was interested in finding the straight line which minimized the sum of squares of perpendicular distances of those points from the line. The solution is the straight line through the mean $(\bar{x}_1, ..., \bar{x}_p)$ and the point whose coordinates are $(a_{11}, ..., a_{1p})$—the coefficients of the first estimated principal component.

6. Note that the parsimony afforded by the principal components is only a parsimony of presentation, since all p variables must be measured in order to compute the value of the principal components for a realization $\mathbf{x} = (x_1, ..., x_p)'$. This is in contrast to stepwise procedures which actually do reduce the number of variables necessary to measure.

EXAMPLE 5.6.1   Data on $n = 113$ critically ill patients was collected on admission to an intensive care unit. The variables were $p = 14$ measurements of

TABLE 5.6.1. *Coefficients of the First Five Principal Components for Example 5.6.1*

Variable	1	2	3	4	5
Age	−0.0206	−0.2806 ④	0.4211 ①	0.0267	0.1255
Systolic pressure	−0.1013	0.4757 ②	0.3127	0.1050	0.0750
Mean arterial pressure	−0.1670	0.4988 ①	0.2168	0.1045	0.0835
Heart rate	−0.0244	−0.0352	−0.4000 ②	0.5175 ②	−0.2238
Diastolic pressure	−0.2229 ⑦	0.4679 ③	0.1269	0.1092	0.0764
Mean venous pressure	0.0400	−0.1023	0.2286	0.5637 ①	−0.1177
Log cardiac index	0.3701 ④	0.1630	0.1378	−0.0681	−0.2945
Log appearance time	−0.3101 ⑥	−0.2686 ⑥	0.3416 ④	−0.1659	0.1506
Log mean circulation time	−0.3708 ③	−0.2745 ⑤	0.1939	0.0061	0.1499
Urine output	0.0767	0.1986	−0.2661	−0.5002 ③	0.2111
Log plasma volume index	0.3125 ⑤	−0.0333	0.3559 ③	−0.1521	−0.3662
Log red cell index	−0.2006	0.0205	0.1425	−0.2600	−0.6958 ①
Hemoglobin	−0.4467 ①	0.0166	−0.1760	−0.0587	−0.2305
Hematocrit	−0.4423 ②	0.0122	−0.1662	−0.0574	−0.2326

age, arterial and venous pressures, blood flow, heart rate and volumes (see Table 5.6.1). BMD01M was used with the sample correlation matrix as input to find the 14 estimated principal components. The eigenvalues for each component are

Component	1	2	3	4	5	6	7
Eigenvalue	3.876	3.159	1.379	1.234	1.102	0.968	0.730

Component	8	9	10	11	12	13	14
Eigenvalue	0.535	0.486	0.270	0.141	0.079	0.022	0.018

Since the total variance $V = 14$ (which is the sum of the eigenvalues), the first component explains $100(3.876)/14 = 27.7\%$ of the total variance; the second component explains $100(3.159)/14 = 22.6\%$ of the variance, and so forth. Hence, the cumulative proportions of total variance explained by each component are

Component	1	2	3	4	5	6	7
Cumulative proportion	0.28	0.50	0.60	0.69	0.77	0.84	0.89

Component	8	9	10	11	12	13	14
Cumulative proportion	0.93	0.96	0.98	0.99	1.00	1.00	1.00

For the sake of illustration, we suppose that 0.77 is a reasonable proportion of the total variance to explain; so the first $q = 5$ principal components sufficiently represent the original variables. The coefficients a_{ij}, $i = 1, \ldots, 5, j = 1, \ldots, 14$, are given in Table 5.6.1. Hence the first principal component is $Y_1 = -0.0206(\text{age}) + \cdots - 0.4423(\text{hematocrit})$. Similarly, the other components are read in the same way.

To interpret these components we use Comment 5.6.1.4 to determine which variables contribute most to each component. Since the correlation between X_i and Y_j is $\alpha_{ji}[V(Y_j)]^{1/2}$, $i = 1, \ldots, 14, j = 1, \ldots, 5$, we arbitrarily pick those variables yielding a correlation ≥ 0.4 in absolute value. For example, $[V(Y_1)]^{1/2} = (3.876)^{1/2} = 1.97$, so that $X_{13} = $ hemoglobin has an absolute correlation of $|1.97(-0.4467)| = 0.88$ with Y_1. The seven variables circled in column 1 all have absolute correlations greater than 0.4 with Y_1. Similarly, the six variables circled in column 2 have absolute correlations greater than 0.4 with Y_2, and so forth. Inspecting the circled variables for each component suggests an interpretation of each component. For example, Y_1 could be termed a blood characteristics component, Y_2 a pressure and flow component, Y_3 an aging component, Y_4 a urine output component, and Y_5 a red cell index component.

5.7 *Factor Analysis*

In the preceding section the principal component technique was shown to be one method for describing the dependence structure of a set of p response variables $X_1, ..., X_p$ having a multivariate distribution with mean $\mu^{p \times 1} = (\mu_1, ..., \mu_p)'$ and covariance matrix $\Sigma^{p \times p} = (\sigma_{ij})$. It was seen that the principal components $Y_1, ..., Y_p$ are expressible as linear combinations

$$Y_1 = \sum_{j=1}^{p} \alpha_{1j} X_j,$$
$$\vdots$$
$$Y_p = \sum_{j=1}^{p} \alpha_{pj} X_j,$$

(5.7.1)

of the response variables $X_1, X_2, ..., X_p$, are uncorrelated, and are ordered by their variances $V(Y_i)$, $i = 1, ..., p$. Furthermore, the total variance V is preserved under the transformation, that is,

$$V = \sum_{i=1}^{p} \sigma_{ii} = \sum_{i=1}^{p} V(Y_i).$$

(5.7.2)

Note that Eq. (5.7.1) can be inverted so that each response variable can be expressed as a linear combination of the principal components, that is

$$X_1 = \sum_{j=1}^{p} \beta_{1j} Y_j,$$
$$\vdots$$
$$X_p = \sum_{j=1}^{p} \beta_{pj} Y_j,$$

(5.7.3a)

for some constants β_{ij}, $i, j = 1, ..., p$. Furthermore, it can be shown that $\beta_{ij} = \alpha_{ji}$ for $i, j = 1, ..., p$, so that

$$X_1 = \sum_{j=1}^{p} \alpha_{ji} Y_j,$$
$$\vdots$$
$$X_p = \sum_{j=1}^{p} \alpha_{jp} Y_j.$$

(5.7.3b)

From this set of equations, called the *principal component model*, it follows that

$$\sigma_{ij} = \sum_{k=1}^{p} \alpha_{ki} V(Y_k) \alpha_{kj} \quad \text{for} \quad i \neq j$$

(5.7.4)

and

$$\sigma_{ii} = \sum_{k=1}^{p} \alpha_{ki}^2 V(Y_k) \quad \text{for} \quad i, j = 1, ..., p.$$

(5.7.5)

These two equations represent the actual restructuring or "factoring" of the variances and covariances of the response variables as functions of the α_{ij} and the variances of the principal components.

This section discusses a more general approach toward linear representation of the response variables. This is accomplished by the *factor model*

$$X_1 = \sum_{j=1}^{m} \lambda_{1j} F_j + e_1, \quad \cdots, \quad X_p = \sum_{j=1}^{m} \lambda_{pj} F_j + e_p, \tag{5.7.6}$$

where λ_{ij} are constants and m is generally much smaller than p. The variables $F_1, ..., F_m$ are called *common* (*primary* or *latent*) *factors*, since they are common to all p response variables. They are assumed to be uncorrelated with unit variances. The variables $e_1, ..., e_p$ are called *specific* (or *unique*) *factors*, since they are unique to each response variable. They are assumed to be uncorrelated with

$$V(e_i) = \tau_i, \quad i = 1, ..., p, \tag{5.7.7}$$

called the *specific variance* or *specificity* of the ith response. The variables F_i and e_j are assumed to be uncorrelated, $i = 1, ..., m, j = 1, ..., p$. The constants λ_{ij} are called the *factor loadings*.

We can now write the new restructuring of the variances and covariances of the response variables as

$$\sigma_{ij} = \lambda_{i1} \lambda_{j1} + \cdots + \lambda_{im} \lambda_{jm} \quad \text{for} \quad i \neq j \tag{5.7.8}$$

and

$$\sigma_{ii} = \lambda_{i1}^2 + \cdots + \lambda_{im}^2 + \tau_i \quad \text{for} \quad i,j = 1, ..., p. \tag{5.7.9}$$

These equations are the analogs of Eqs. (5.7.4) and (5.7.5), respectively, The quantity $\sum_{j=1}^{m} \lambda_{ij}^2$ is called the *communality* of the ith response, and it is seen that the communality of the ith response equals the variance of the ith response minus the specificity of the ith response, $i = 1, ..., p$.

Thus, the p components of the principal component model may be viewed as p common factors explaining the dependence structure of the p response variables, while the $m < p$ common factors of the factor model explain most of the dependence structure with the residual accounted for by the specific factors. In other words, the principal component model attributes all of the total variance to the p common factors, while the factor model splits the variance of each response variable into two parts—the variance due to the common factors (the communality) and the variance due to the sampling variation of each response variable (the specificity).

The technique of factor analysis concerns itself with estimating the factor loadings λ_{ij} and the specific variances τ_i, $i = 1, ..., p, j = 1, ..., m$. It is also concerned with methods for evaluating the common factors for an individual as a function of his responses. These values are called *factor scores*.

Once the factor loadings have been obtained, the major burden of the analyst is to make the "best" interpretation of the common factors. This involves the technique of factor *rotation*, which, due to its subjectivity, is the most controversial part of factor analysis.

Since factor analysis has become a separate discipline, this section makes no attempt to give a complete treatment of the subject. Comprehensive treatments may be found in Harman (1967) and Thurstone (1945). Section 5.7.1 discusses the *principal factor method* of obtaining the factor loadings. Although the *maximum likelihood method* is more theoretically appealing, the principal factor method is used most often in packaged factor analysis programs. A good treatment of the maximum likelihood solution may be found in Morrison (1967). Section 5.7.2 discusses some of the many techniques of factor rotation, and Section 5.7.3 discusses estimation of the factor scores.

COMMENT 5.7.1 ∗ The models given by Eqs. (5.7.3) and (5.7.6) take on compact form when written in matrix notation. The principal components are given by

$$\mathbf{Y} = \mathbf{AX},$$

where

$$\mathbf{Y}^{p \times 1} = (Y_1, \ldots, Y_p)', \qquad \mathbf{X}^{p \times 1} = (X_1, \ldots, X_p)', \qquad \text{and} \qquad \mathbf{A}^{p \times p} = (\alpha_{ij}).$$

Thus, the principal component model is

$$\mathbf{X} = \mathbf{BY},$$

where $\mathbf{B}^{p \times p} = \mathbf{A}^{-1} = \mathbf{A}'$, since $\mathbf{A}$ is an orthogonal matrix. The covariance matrix $\mathbf{\Sigma}^{p \times p}$ can then be written as

$$\mathbf{\Sigma} = \mathbf{A'VA},$$

where

$$\mathbf{V}^{p \times p} = \begin{bmatrix} V(Y_1) & 0 & \cdots & 0 \\ 0 & V(Y_2) & \cdots & 0 \\ \vdots & \vdots & \vdots & \vdots \\ 0 & 0 & \cdots & V(Y_p) \end{bmatrix}.$$

The factor model is

$$\mathbf{X} = \mathbf{\Lambda F} + \mathbf{e},$$

where

$$\mathbf{\Lambda}^{p \times p} = (\lambda_{ij}), \qquad \mathbf{F}^{m \times 1} = (F_1, \ldots, F_m)', \qquad \text{and} \qquad \mathbf{e}^{p \times 1} = (e_1, \ldots, e_p)'.$$

The covariance matrix Σ can then be written as

$$\Sigma = \Lambda\Lambda' + T,$$

where

$$T^{p \times p} = \begin{bmatrix} \tau_1 & 0 & \cdots & 0 \\ 0 & \tau_2 & \cdots & 0 \\ \vdots & \vdots & \vdots & \vdots \\ 0 & 0 & \cdots & \tau_p \end{bmatrix}.*$$

5.7.1 The Principal Factor Solution

Contrary to previous sections which solve the problem first in terms of the population parameters and then in terms of the sample estimates, in this section we assume from the onset that we have a random sample of responses $\mathbf{x}_1^{p \times 1}, \ldots, \mathbf{x}_n^{p \times 1}$ from a multivariate distribution with mean $\boldsymbol{\mu}^{p \times 1} = (\mu_1, \ldots, \mu_p)'$ and covariance matrix $\Sigma^{p \times p} = (\sigma_{ij})$. Let $\mathbf{S}^{p \times p} = (s_{ij})$ be the sample covariance matrix and $\mathbf{R}^{p \times p} = (r_{ij})$ be the sample correlation matrix, where $r_{ij} = s_{ij}/(s_{ii}s_{jj})^{1/2}$, $i,j = 1, \ldots, p$. The first problem of factor analysis is to find from $\mathbf{S}$ or $\mathbf{R}$ the estimated factor loadings l_{ij} of λ_{ij} and the estimated specific variances t_i of τ_i, $i = 1, \ldots, p, j = 1, \ldots, m$. It should be noted that in most applications $\mathbf{R}$ is preferred to $\mathbf{S}$ since experimenters prefer working with standardized variables (see Comment 5.6.1.3).

The most theoretically appealing method for finding these estimates is that of maximum likelihood. This method, however, is computationally difficult and has not received much usage. Other methods have been described which can be used on a desk calculator; the best known and most frequently used for this purpose is the *centroid method*. Other techniques include the *group centroid method*, the *multiple group method*, the *rank reduction method*, the *orthogonalization method*, *Jacobi-type methods*, and *order reduction methods*. For a description of all of these methods see Horst (1965).

Since the advent of the digital computer, the most commonly used technique has been the *principal factor solution* which can be applied to either the sample covariance or correlation matrix. To obtain this solution we first extract the p estimated principal components

$$Y_i = \sum_{j=1}^{p} a_{ij} X_j, \qquad i = 1, \ldots, p. \tag{5.7.10}$$

Recall that the p principal components are uncorrelated and that the variance $V(Y_i)$ of the ith principal component is the ith largest eigenvalue of the sample covariance or correlation matrix, $i = 1, \ldots, p$. Also recall that $\mathbf{a}_i = (a_{i1}, \ldots, a_{ip})'$ is the eigenvector corresponding to the ith principal component. In terms of the response variables, we have

$$X_i = \sum_{j=1}^{p} a_{ji} Y_j, \qquad i = 1, \dots, p. \tag{5.7.11}$$

The principal factor solution chooses as the m common factors the first m principal components scaled as

$$F_j = \frac{Y_j}{[V(Y_j)]^{\frac{1}{2}}}, \qquad j = 1, \dots, m. \tag{5.7.12}$$

The estimated factor loadings are

$$l_{ij} = a_{ji}[V(Y_j)]^{\frac{1}{2}}, \qquad i = 1, \dots, p, \quad j = 1, \dots, m, \tag{5.7.13}$$

and the estimated specific factors are

$$e_i = \sum_{j=m+1}^{p} a_{ji} Y_j, \qquad i = 1, \dots, p. \tag{5.7.14}$$

Hence, we have the *estimated factor model*

$$X_i = \sum_{j=1}^{m} l_{ij} F_j + e_i, \qquad i = 1, \dots, p. \tag{5.7.15}$$

In this estimated model, each common factor has unit variance, and the factors are uncorrelated. Furthermore, the common factors are uncorrelated with the specific factors. However, note that the covariance between e_i and e_k is

$$\text{cov}(e_i, e_k) = \sum_{j=m+1}^{p} a_{ji} a_{jk} V(Y_j) \qquad \text{for} \quad i, k = 1, \dots, p, \quad i \neq k. \tag{5.7.16}$$

Since these covariances are not necessarily zero, this constitutes a violation of the original assumptions of the model.

The communality h_i^2 and the specificity t_i of X_i are estimated by

$$h_i^2 = \sum_{j=1}^{m} l_{ji}^2 = \sum_{j=1}^{m} a_{ji}^2 V(Y_j) \tag{5.7.17}$$

and

$$t_i = \sum_{j=m+1}^{p} a_{ji}^2 V(Y_j), \tag{5.7.18}$$

respectively, for $i = 1, \dots, p$.

There exist packaged computer programs for solving this problem. Some of the inputs and options to these programs include (1) specification of the number of factors to be extracted, (2) the form of the matrix which is to be factor analyzed, (3) estimates of the communalities and the maximum number of iterations to determine the communalities. Other input options are discussed in the next sections. The details of the above inputs are:

(1)   The number of factors to be extracted is specified either by an integer m or a constant c. In the latter case the number of factors m is taken to be the number of eigenvalues greater than c.

(2)   The matrix to be factor-analyzed may be (a) the covariance matrix, (b) the covariance matrix about the origin, that is, the matrix of sums of squares and crossproducts, (c) the correlation matrix, (d) the correlation matrix about the origin, that is, the correlation matrix obtained from (b), or (e) the matrix of factor loadings.

(3)   Recall that in the principal component model, the total variance due to the common factors (that is, principal components) is preserved. In the factor analysis model, it is sometimes desirable to obtain estimates of the common factors which preserve the total communality $\sum_{i=1}^{p} h_i^2$, that is, the total variance due to the common factors. In fact, this is usually encouraged in applications in psychology and educational measurement. For this reason the user may specify initial estimates of each communality and the maximum number of iterations to achieve convergence of the total communality. These estimates, which are substituted for the diagonal elements of the matrix to be factor-analyzed, may be (a) squared multiple correlation coefficients in the case of a correlation matrix, or variances due to regression in the case of a covariance matrix, (b) the maximum absolute row value, or (c) estimates obtained from a preliminary analysis. Obtaining the estimated factor loadings and the new communalities from this altered matrix constitutes an *iteration*. For the next iteration, the new communalities are substituted for the diagonal elements in the original matrix; then the loadings and new communalities are obtained. This process is repeated until the maximum number of iterations is reached or the maximum change in the communality estimates is less than a small specified level. Note that the user may leave the diagonal elements of the original matrix unchanged but may still specify the number of iterations to achieve convergence of the total communality.

COMMENTS 5.7.2   1.   In deciding on the number m of factors, the user is guided by several criteria:

(a) The physical structure may suggest a given number of intrinsic factors.

(b) When the unaltered correlation matrix is used, some factor analysts recommend taking as many factors as there are eigenvalues $\geqslant 1$.

(c) As in principal component analysis, choose the number of factors which explain a specified proportion of the total variance or total communality.

2.   In deciding on diagonal elements for the matrix, statistical considerations would suggest leaving those elements unchanged. However, when total communality is to be preserved, the most commonly used estimates are the squared multiple correlations for **R** or the variance due to regression for **S**.

3. It should be remembered that the extracted factors are different for all the choices of the matrix to be factor-analyzed.

4. It should be noted that the correlation between X_i and F_j is

$$\text{corr}(X_i, F_j) = l_{ij} \quad \text{for} \quad i = 1, ..., p, \quad j = 1, ..., m.$$

Hence, to interpret each factor we look for variables with large loadings in absolute value. These are the variables that are highly correlated with the factor.

EXAMPLE 5.7.1   Data on $n = 113$ critically ill patients was collected on admission to an intensive care unit. The variables were $p = 13$ initial measurements of arterial and venous pressures, blood flow, heart rate and volumes (see Table 5.7.1). BMDX72 was used with the sample correlation matrix (given in Table 5.7.1) as input to find the principal factor solution. The following cases were considered.

EXAMPLE 5.7.1a   In this case the diagonal elements of the correlation matrix were unaltered and only one iteration was specified. The eigenvalues for each principal component are

Component	1	2	3	4	5	6	7
Eigenvalue	3.875	2.980	1.269	1.233	1.095	0.766	0.711

Component	8	9	10	11	12	13
Eigenvalue	0.507	0.290	0.150	0.084	0.023	0.019

and the cumulative proportions of total variance explained by each component are

Component	1	2	3	4	5	6	7
Cumulative proportion	0.30	0.53	0.62	0.72	0.80	0.86	0.92

Component	8	9	10	11	12	13
Cumulative proportion	0.96	0.98	0.99	0.99	1.00	1.00

The choice for m—the number of common factors—was motivated by the fact that the factors should correspond to blood pressures, volumes, and constituents. Hence, choosing $m = 3$ we obtain as the estimated factor loadings Table A.

Hence, $l_{11} = 0.21$ is the correlation of systolic pressure with the first factor, $l_{12} = 0.88$ is the correlation of the same variable with the second factor, and so forth. To interpret the factors, we examine the loadings which are greater in absolute value than some cutoff value $r = 0.4$, say. These loadings are circled in Table A. The first factor is essentially a weighted average of 8 of the 13 response variables; the second factor involves arterial pressures and blood flow; the third factor includes heart rate, appearance time and plasma volume index. These three factors are not easily interpreted, and, as will be seen in the next section, factor rotation will ameliorate this situation.

TABLE 5.7.1. *Correlation Matrix for Example 5.7.1*

Variable	SP 1	MAP 2	HR 3	DP 4	MVP 5	L(CI) 6	L(AT) 7	L(MCT) 8	UO 9	L(PVI) 10	L(RCI) 11	Hgb 12	Hct 13
Systolic pressure	1.00												
Mean arterial pressure	0.90	1.00											
Heart rate	−0.10	−0.07	1.00										
Diastolic pressure	0.81	0.95	0.00	1.00									
Mean venous pressure	−0.03	−0.07	0.05	−0.13	1.00								
Log cardiac index	0.12	0.03	−0.05	−0.07	−0.05	1.00							
Log appearance time	−0.13	−0.11	−0.15	−0.04	−0.01	−0.49	1.00						
Log mean circulation time	−0.17	−0.11	0.02	−0.00	0.14	−0.68	0.84	1.00					
Urine output	0.13	0.15	−0.12	0.12	−0.23	0.09	−0.21	−0.18	1.00				
Log plasma volume index	−0.08	−0.17	−0.13	−0.27	0.13	0.54	−0.16	−0.28	0.04	1.00			
Log red cell index	0.09	0.11	−0.02	0.14	−0.06	−0.11	0.20	0.21	−0.05	0.04	1.00		
Hemoglobin	0.09	0.21	0.09	0.33	−0.09	−0.48	0.39	0.47	−0.07	−0.49	0.38	1.00	
Hematocrit	0.09	0.21	0.06	0.32	−0.08	−0.48	0.40	0.49	−0.09	−0.50	0.39	0.97	1.00

TABLE A

Variable	1	2	3
1 SP	0.21	0.88 ②	−0.22
2 MAP	0.33	0.90 ①	−0.13
3 HR	0.05	−0.08	0.59 ①
4 DP	0.45 ⑦	0.83 ③	−0.04
5 MVP	−0.07	−0.18	−0.35
6 L(CI)	−0.70 ④	0.33	−0.10
7 L(AT)	0.61 ⑥	−0.44 ⑤	−0.42 ③
8 L(MCT)	0.71 ③	−0.48 ④	−0.26
9 UO	−0.13	0.31	0.18
10 L(PVI)	−0.61 ⑤	−0.03	−0.52 ②
11 L(RCI)	0.40 ⑧	0.03	−0.32
12 Hgb	0.87 ②	−0.00	0.15
13 Hct	0.88 ①	−0.01	0.13

EXAMPLE 5.7.1b   As suggested in Comment 5.7.2.1, this case chooses the factors to correspond to eigenvalues ≥ 1. From the eigenvalues listed in Example 5.7.1a, we see that $m = 5$. The first three factors are the same as in the previous example. The loadings for the 4th and 5th factor are given in Table B. Using a cutoff of $r = 0.4$, the 4th factor is most highly correlated with heart rate, venous pressure, and urine output, and factor 5 is most correlated with red-cell index. Except for the fifth factor, it is still difficult to interpret the remaining factors.

	TABLE B				TABLE C		
Variable	4	5		Variable	$m = 3$	$m = 5$
1 SP	0.15	−0.09		1 SP	0.87	0.90
2 MAP	0.14	−0.09		2 MAP	0.95	0.97
3 HR	0.48 ③	0.33		3 HR	0.37	0.71
4 DP	0.13	−0.07		4 DP	0.91	0.93
5 MVP	0.71 ①	−0.03		5 MVP	0.17	0.67
6 L(CI)	−0.06	0.34		6 L(CI)	0.62	0.74
7 L(AT)	−0.12	−0.20		7 L(AT)	0.76	0.81
8 L(MCT)	0.05	−0.21		8 L(MCT)	0.81	0.86
9 UO	−0.59 ②	−0.22		9 UO	0.15	0.55
10 L(PVI)	−0.07	0.31		10 L(PVI)	0.66	0.76
11 L(RCI)	−0.23	0.69 ①		11 L(RCI)	0.27	0.80
12 Hgb	−0.09	0.26		12 Hgb	0.79	0.87
13 Hct	−0.08	0.26		13 Hct	0.80	0.87

The estimated communalities for these two cases are given in Table C. It is noteworthy that when $m = 3$, variables 5, 9, and 11 have communalities less than 0.3, while when $m = 5$, all communalities are greater than 0.5. This verifies the facts that (1) the communality (that is, the variance explained by common factors) increases with m; and (2) the influence of the common factors varies from one response variable to another.

EXAMPLE 5.7.1c   This example demonstrates the effect of altering the diagonal elements of the sample correlation matrix. The squared multiple correlation coefficient of X_i on the remaining variables was substituted for the ith diagonal element, $i = 1, ..., 13$. Again, only one iteration was specified, and the number of factors extracted was $m = 3$. Since $\mathbf{R}$ is altered, the eigenvalues, the cumulative proportions of total variance, and the factor loadings are all different from the above two cases. These loadings, along with the squared multiple correlation coefficients and estimated communalities, are given in Table D.

TABLE D

Variable	Factor 1	Factor 2	Factor 3	Multiple R^2	Estimated communalities
1 SP	0.24	0.85 ②	0.23	0.85	0.84
2 MAP	0.37	0.89 ①	0.17	0.96	0.98
3 HR	0.03	−0.06	−0.23	0.22	0.06
4 DP	0.49 ⑦	0.81 ③	0.10	0.94	0.92
5 MVP	−0.06	−0.13	0.20	0.28	0.06
6 L(CI)	−0.64 ④	0.32	−0.08	0.63	0.53
7 L(AT)	0.58 ⑤	−0.45 ⑤	0.48 ①	0.81	0.78
8 L(MCT)	0.68 ③	−0.50 ④	0.40 ②	0.86	0.89
9 UO	−0.09	0.23	−0.08	0.18	0.07
10 L(PVI)	−0.55 ⑥	0.00	0.22	0.50	0.36
11 L(RCI)	0.43 ⑧	0.00	−0.02	0.26	0.11
12 Hgh	0.88 ②	−0.05	−0.36	0.96	0.93
13 Hct	0.89 ①	−0.06	−0.35	0.96	0.93

Using the same cutoff of $r = 0.4$, we see that the first factor is essentially a weighted average of the same eight variables as in the other cases; the second factor involves arterial pressures and blood flow, while the third factor is most correlated with blood flow. The first two factors are similar to those of the above two cases, while the third factor is different. The estimated communalities are, in general, smaller than those obtained using the unaltered correlation matrix.

5.7.2 Factor Rotation

After the factor loadings have been extracted, the next step in the analysis is the interpretation of each factor. To aid in the interpretation, we exploit the *indeterminacy of the factor solution* whereby we can find new common factors $F_1^{(R)} ..., F_m^{(R)}$, which are linear combinations of the old common factors $F_1, ..., F_m$, and which are uncorrelated with unit variances. Thus, the new set of factors also satisfies the factor model and, furthermore, there are an infinite number of such sets. The process of obtaining a set of new factors is called an *orthogonal factor rotation* and the rotated factor model can then be written as

$$X_i = \sum_{j=1}^{m} c_{ij} F_j^{(R)} + e_i, \qquad i = 1, ..., p, \qquad (5.7.19)$$

where the constants c_{ij} are the rotated factor loadings. It should be noted that after an orthogonal factor rotation has been made, the communality of each response variable X_i remains the same, that is,

$$h_i^2 = \sum_{j=1}^{m} c_{ij}^2 = \sum_{j=1}^{m} l_{ij}^2, \qquad i = 1, ..., p. \qquad (5.7.20)$$

The constants c_{ij} may be expressed as

$$c_{ij} = \sum_{k=1}^{m} l_{ik} q_{kj}, \qquad i = 1, ..., p, \quad j = 1, ..., m, \qquad (5.7.21)$$

where q_{kj} are constants, $k = 1, ..., m, j = 1, ..., m$. Now, to aid in the interpretation of the factors, these constants q_{kj} are chosen so that the resulting rotated loadings c_{ij} are *simple*. Roughly speaking, the loadings are simple if many of them are nearly zero, while a few of them are relatively large. The goal, then, is to represent each response variable with relatively large loadings on one or at most a few factors and nearly zero loadings on the remaining factors (Thurstone, 1945). If simple loadings are obtained, then the job of interpreting the factors becomes easier (recall from Comment 5.7.2.4 that a loading is the correlation between the response variable and the factor).

Factor analysts have devised many methods of orthogonal rotations—both graphical and analytical—in order to obtain simple loadings. An excellent review of all of these methods is found in Harman (1967). Analytical methods attempt to achieve simple loadings by minimizing a function of the c_{ij} called an *objective function*. The objective function for the commonly used orthogonal rotation techniques can be represented as

$$G = \sum_{k=1}^{m} \sum_{\substack{j=1 \\ j \neq k}}^{m} \left[\sum_{i=1}^{p} c_{ik}^2 c_{ik}^2 - \frac{\gamma}{p} \left(\sum_{i=1}^{p} c_{ij}^2 \right) \left(\sum_{i=1}^{p} c_{ik}^2 \right) \right], \qquad (5.7.22)$$

where the usual range of γ is from 0 to 1 for orthogonal rotations.

When $\gamma = 0$, the rotation resulting from minimizing G is called a *quartimax* rotation. In this case it can be shown that minimizing G is equivalent to maximizing

$$\frac{1}{pm} \sum_{j=1}^{m} \sum_{i=1}^{p} (c_{ij}^2 - \overline{c_{..}^2})^2, \qquad (5.7.23)$$

where

$$\overline{c_{..}^2} = \frac{1}{pm} \sum_{j=1}^{m} \sum_{i=1}^{p} c_{ij}^2. \tag{5.7.24}$$

Equation (5.7.23) is the variance of the squares of all the factor loadings. Thus, the quartimax method maximizes the variation among all squared factor loadings, that is, it chooses factor loadings which are sufficiently diverse in magnitude in order to produce maximum variation. This has the effect of increasing large loadings and decreasing small loadings, thus linking a given factor with as few variables as possible.

When $\gamma = 1$, the rotation is called a *varimax* rotation, and it is this rotation which is most used in applications. In this case, it can be shown that minimizing G is equivalent to maximizing

$$\frac{1}{p} \sum_{j=1}^{m} \sum_{i=1}^{p} (c_{ij}^2 - \overline{c_{.j}^2})^2, \tag{5.7.25}$$

where

$$\overline{c_{.j}^2} = \frac{1}{p} \sum_{i=1}^{p} c_{ij}^2, \qquad j = 1, \dots, m. \tag{5.7.26}$$

Equation (5.7.25) is the sum of the variances of the squared factor loadings within each column. Thus, the varimax method makes the squares of the loadings of each factor as varied as possible. This also has the effect of increasing large loadings and decreasing small loadings. In this method, we obtain simple loadings for each factor separately, while in the quartimax method, this is done for all factors simultaneously.

So far we have only considered orthogonal rotations of the common factors. Some factor analysts have argued that orthogonality of the factors is not as important a criterion as that of obtaining simple loadings. Hence, they have relaxed the condition of obtaining uncorrelated factors and instead look for new correlated factors $F_1^{(R)}, \dots, F_m^{(R)}$ which are linear combinations of the old factors $F_1, \dots, F_m$ and which have unit variances. Thus, the new set of factors does not satisfy the factor model of Eq. (5.7.6). The process of obtaining such a set of new factors is called an *oblique rotation*. The rotated factor model can still be represented by Eq. (5.7.19) with the constants c_{ij}, $i = 1, \dots, p, j = 1, \dots, m$, represented by Eq. (5.7.21). Since the rotated factors can be correlated, we now have a wider range for the constants q_{kj}, $k, j = 1, \dots, m$. In turn, we have more choices for the constants c_{ij}.

Analytical methods attempting to achieve simple loadings by searching for oblique rotations which minimize G [Eq. (5.7.22)] are called *direct oblimin methods* (see Jennrich and Sampson, 1966, for the mathematical details). Harman

(1967, p. 336) suggests that the range of γ should now be from $-\infty$ to 0. The more negative γ is, the more correlated the factors will be. When $\gamma = 0$, we have the *direct quartimin* method which is the oblique analog of the quartimax method. However, since the factors are not required to be uncorrelated, this procedure is not the same as maximizing the variance of the squared factor loadings [see Eq. (5.7.23)].

Historically the direct oblimin methods were not the first oblique factor rotations to be developed. To review this historical development, we first define several new concepts:

(1)   Given a set of rotated factors $F_1^{(R)}, ..., F_m^{(R)}$, the $p \times m$ matrix of correlations between the response variables $X_1, ..., X_p$ and these factors is called the *factor structure*. It should be noted that if the rotated factors are uncorrelated, then the factor structure is the same as the matrix of factor loadings.

(2)   Corresponding to each rotated factor $F_i^{(R)}, i = 1, ..., m$, we can find a factor G_i which is uncorrelated with the remaining rotated factors $F_j^{(R)}$ $j = 1, ..., m, j \neq i$. The factors $G_1, ..., G_m$ are called *reference factors* and these factors are said to be *bi-orthogonal* to $F_1^{(R)}, ..., F_m^{(R)}$ (Thurstone, 1945). It should be noted that if the rotated factors are uncorrelated, then $G_i = F_i^{(R)}, i = 1, ..., m$.

(3)   The $p \times m$ matrix whose ij element v_{ij} is the correlation between the response variable X_i and the reference factor $G_j, i = 1, ..., p, j = 1, ..., m$, is called the *reference factor structure* (or, simply, the *reference structure*). If the rotated factors are uncorrelated then the reference factor structure is the same as the factor structure.

Now, historically, orthogonal rotations attempted to find simple factor structure (sometimes called simple structure), while oblique rotations attempted to find simple reference structure. Thus, an oblique rotation attempts to minimize the objective function

$$G = \sum_{k=1}^{m} \sum_{\substack{j=1 \\ j \neq k}}^{m} \left[\sum_{i=1}^{p} v_{ij}^2 v_{ik}^2 - \frac{\gamma}{p} \left(\sum_{i=1}^{p} v_i^2 \right) \left(\sum_{i=1}^{p} v_{ik}^2 \right) \right], \qquad (5.7.27)$$

where $v_{ij} = \text{corr}(X_i, G_j)$ and γ ranges from 0 to 1. Analytical methods attempting to achieve simple reference structure are called *indirect oblimin methods*, or simply, *oblimin methods*. The direct and indirect oblimin methods are related by $v_{ij} = d_j c_{ij}$, where d_j is a constant, $i = 1, ..., p, j = 1, ..., m$.

When $\gamma = 0$, we have the *(indirect) quartimin method*; when $\gamma = \frac{1}{2}$ we have the *(indirect) bi-quartimin method*; and when $\gamma = 1$, we have the *(indirect) covarimin method*. The rotated axes are most oblique when $\gamma = 0$ and least oblique when $\gamma = 1$ (see Harman, 1967, p. 326).

COMMENTS 5.7.3   1. A sophisticated packaged factor analysis program will permit the user to choose the kind of rotations that he desires. His options may include (a) no rotation, (b) orthogonal rotations, (c) direct oblimin rotations (oblique rotations for simple loadings), or (d) (indirect) oblimin rotations (oblique rotations for simple reference structure). He may also specify the value of γ for the objective function G, as well as the maximum number of rotations made. Rotations will be performed until the maximum number is reached or until the change in G relative to its original value is less than some specified value.

2. As if Gesund is not yet confused, a further refinement is that the loadings c_{ij} can be "normalized" by replacing them with c_{ij}/h_i, $i = 1, ..., p, j = 1, ..., m$. This is called *Kaiser normalization* and has the property that each response variable will contribute an amount commensurate with its communality (Kaiser, 1958).

EXAMPLE 5.7.1 (*continued*)   The factors obtained for the three cases are rotated in order to aid in their interpretation. The maximum number of rotations was set at 50. The results are as follows.

EXAMPLE 5.7.1a   The three factors were rotated using varimax rotation. The rotated factor loadings are given in the accompanying table. The factors are indeed easier to interpret. In particular, $F_1^{(R)}$ involves flow measurements and the last three variables, and could be called a blood flow and composition factor. Factor 2 is highly correlated with the three arterial pressure variables and could be called an arterial pressure factor. Finally, $F_3^{(R)}$ could be interpreted as a blood volume factor.

		Factor		
Variable		1	2	3
1	SP	−0.08	0.93 ②	−0.08
2	MAP	−0.04	0.97 ①	0.06
3	HR	−0.22	−0.19	0.53 ④
4	DP	0.04	0.93 ③	0.21
5	MVP	0.20	−0.12	−0.34
6	L(CI)	−0.63 ③	0.08	−0.48 ⑤
7	L(AT)	0.86 ②	−0.11	−0.02
8	L(MCT)	0.88 ①	−0.14	0.17
9	UO	−0.32	0.21	0.08
10	L(PVI)	−0.20	−0.13	−0.78 ①
11	L(RCI)	0.46 ⑥	0.23	−0.05
12	Hgb	0.60 ⑤	0.26	0.60 ②
13	Hct	0.61 ④	0.26	0.59 ③

EXAMPLE 5.7.1b   The five factors were rotated using varimax rotation. The first three columns of rotated factor loadings are no longer the same as the three columns of loadings of Example 5.7.1a. The rotated factor loadings are given in the accompanying table. The interpretation is $F_1^{(R)}$ = blood flow factor, $F_2^{(R)}$ = arterial pressure factor, $F_3^{(R)}$ = heart rate and plasma factor, $F_4^{(R)}$ = urine output factor, and $F_5^{(R)}$ = blood composition factor.

		Factor				
Variable		1	2	3	4	5
1	SP	−0.11	0.94 ③	−0.09	0.02	0.03
2	MAP	−0.01	0.98 ①	−0.00	−0.04	0.04
3	HR	−0.08	−0.10	0.81 ①	0.17	0.04
4	DP	0.10	0.95 ②	0.09	−0.08	0.09
5	MVP	0.03	−0.00	−0.00	0.81 ①	−0.14
6	L(CI)	−0.85 ②	0.02	−0.14	0.01	0.07
7	L(AT)	0.78 ③	−0.13	−0.36	0.14	0.19
8	L(MCT)	0.88 ①	−0.12	−0.14	0.21	0.14
9	UO	−0.15	0.14	−0.19	−0.67 ②	−0.14
10	L(PVI)	−0.61 ⑥	−0.21	−0.49 ②	0.27	0.18
11	L(RCI)	0.08	0.07	−0.09	0.02	0.88 ①
12	Hgb	0.64 ⑤	0.21	0.32	−0.16	0.54 ②
13	Hct	0.65 ④	0.21	0.30	−0.15	0.54 ③

EXAMPLE 3.7.1c   The three factors were rotated obliquely using the direct quartimin method. The rotated factor loadings are given in the accompanying

		Factor		
Variable		1	2	3
1	SP	−0.13	0.94 ②	0.01
2	MAP	0.00	0.99 ①	−0.01
3	HR	0.27	−0.17	−0.21
4	DP	0.17	0.91 ③	−0.01
5	MVP	−0.24	−0.02	0.24
6	L(CI)	−0.36	0.06	−0.48 ③
7	L(AT)	−0.08	0.02	0.92 ①
8	L(MCT)	0.07	−0.02	0.91 ②
9	UP	0.00	0.13	−0.22
10	L(PVI)	−0.59 ③	−0.04	0.01
11	L(RCI)	0.25	0.09	0.11
12	Hgb	0.96 ①	0.01	0.01
13	Hct	0.95 ②	0.01	0.03

table. The interpretation is $F_1^{(R)} = $ blood composition factor, $F_2^{(R)} = $ arterial pressure factor, $F_3^{(R)} = $ blood flow factor. Note that interpretation is more obvious since we had more freedom in rotations. The factor correlation matrix is

| | | Factor | | |
		1	2	3
	1	1.00	0.22	0.49
Factor	2	0.22	1.00	−0.12
	3	0.49	−0.12	1.00

Note that factors $F_1^{(R)}$ and $F_3^{(R)}$ are most highly correlated. Factors $F_2^{(R)}$ and $F_3^{(R)}$ are least and negatively correlated.

5.7.3 Factor Scores

In many applications it is desirable to know the value of each factor for a given response vector $\mathbf{x} = (x_1, \ldots, x_p)'$. These values of the factors are called *factor scores*. For example, in educational measurements, it has been shown that there are two common factors—verbal and quantitative—which explain mental aptitude. For a given set of test scores $x_1, \ldots, x_p$, it would be desirable to evaluate the student's factor scores to summarize his mental ability. There is no unique way of doing this, but the customary method is to use regression techniques. Treating the factors as dependent variables and the response variables as the independent variables, it is possible to write an equation of the form

$$\hat{F}_j = \sum_{i=1}^{p} b_{ij} z_i, \qquad j = 1, \ldots, m, \tag{5.7.28}$$

where $\hat{F}_j$ is the jth estimated factor score, z_i is the ith standard score for the ith response variable, that is,

$$z_i = \frac{x_i - \bar{x}_i}{s_i}, \qquad i = 1, \ldots, p,$$

and b_{ij} are the estimated regression coefficients, sometimes called *factor score coefficients*. It should be recalled from Chapter 3 that b_{ij} are functions of the response variables and the correlations between the common factors and the response variables. (These quantities have already been obtained.) Packaged programs usually offer the option of obtaining factor score coefficients and factor scores for each individual in the sample.

Problems

Notes: 1. Data Set A is the data in Example 1.4.1, Tables 1.4.1 and 1.4.2. Data Set B is the data in Example 1.4.2, Tables 1.4.3 and 1.4.4.
2. In Data Set A, all continuous variables may be assumed normal, except for CI, AT, MCT, and PVI, whose logarithms may be assumed normal. In Data Set B all continuous variables may be assumed normal, except for systolic and diastolic pressures (1950 and 1962) whose logarithms may be assumed normal.

Section 5.1

5.1.1. (A class exercise)   Collect data on the height, weight, and age of each male student in the class and perform an analysis of outliers. Interpret the results.

5.1.2.   Repeat Problem 5.1.1 for female students in the class. Notice that the outliers detected may be due to "lying."

Section 5.2

5.2.1. (Data Set A)   (a) Using initial data on all patients, test whether the population mean vector of $X_1 = $ SP, $X_2 = $ HR, $X_3 = $ DP, $X_4 = $ MVP is equal to the mean of healthy individuals defined as $\mu_1 = 120$, $\mu_2 = 80$, $\mu_3 = 80$, $\mu_4 = 3$.
   (b) Using multiple confidence intervals, determine which means differ significantly from the healthy values.

5.2.2. (Data Set A)   Repeat Problem 5.2.1 using the final data for patients who survived.

5.2.3. (Data Set A)   Repeat Problem 5.2.1 using the final data for patients who died.

5.2.4. (Data Set A)   Comment on the results of Problems 5.2.1, 5.2.2, and 5.2.3.

5.2.5. (Data Set A)   (a) Defining $X_1, ..., X_4$ as in Problem 5.2.1, test whether the mean initial vectors are the same for survivors and nonsurvivors.
   (b) Make 90% multiple confidence intervals for all component mean differences.

5.2.6. (Data Set A)   Repeat Problem 5.2.5 for final values. Compare the results of these two problems.

Section 5.3

5.3.1. (Data Set B)   Define $W_1 = $ those who survived past 1968, and $W_2 = $ others in the population.

   (a) Estimate the Bayes classification procedure based on $X_1 = $ age, $X_2 = $ log systolic, $X_3 = $ log diastolic pressures, and $X_4 = $ serum cholesterol (1950). Use estimates of the prior probabilities based on the data.

(b) Compute D^2.

(c) Estimate the probabilities of misclassification by two different methods.

(d) Test the hypothesis that the mean vectors of W_1 and W_2 are equal.

5.3.2. (Data Set B)   Do Problem 5.3.1 using $X_1, ..., X_4$ as defined there and $X_5 = \log SP(1962)$, $X_6 = \log DP(1962)$, and $X_6 = SER-CH(1962)$. Also

(e) Graph the posterior probability that an individual will survive past 1968 as a function of the discriminant function z.

(f) Does including X_5, X_6, X_7 improve discrimination? [Hint: see Comment 5.5.1.2c.]

Section 5.4

5.4.1 (Data Set A)   Use initial data. Define $W_1, ..., W_6$ as in Example 5.4.1 and let $X_1 = MAP$, $X_2 = MVP$, $X_3 = \log_{10} CI$, $X_4 = UO$, $X_5 = \log_{10} PVI$, $X_6 = Hgb$, $X_7 = Hct$.

(a) Use a packaged program to derive a classification procedure, assuming equal prior probabilities and costs of misclassification.

(b) Suppose a patient has the following data: $X_1 = 70$, $X_2 = 10$, $X_3 = 0.3$, $X_4 = 10$, $X_5 = 1.5$, $X_6 = 10$, $X_7 = 30$. Where would you classify him?

(c) Estimate some probabilities of misclassification which might be of interest.

(d) Test the hypothesis that the six mean vectors are equal.

(e) Compare the results with those of Example 5.4.1.

Section 5.5

5.5.1. (Data Set A)   Do Problem 5.4.1 using a stepwise discriminant analysis program. Also,

(f) Obtain a "best" set of variables for classification.

Section 5.6

5.6.1. (Data Set A)   In Example 5.6.1, we obtained the principal components based on the sample correlation matrix of *initial* measurements on 14 variables for *all* patients. Compare these results with your principal component analysis on:

(a) the *initial* measurements on the same variables for patients who *survived*;

(b) the *final* measurements on the same patients for patients who *survived*.

Also compare (a) and (b).

5.6.2. (Data Set B)   Perform a principal component analysis on the variables $X_1 = age$, $X_2 = SP(1950)$, $X_3 = DP(1950)$, $X_4 = height(1950)$, $X_5 = SER-CH$ (1950), $X_6 = weight(1950)$, $X_7 = SP(1962)$, $X_8 = DP(1962)$, $X_9 = SER-CH(1962)$, $X_{10} = weight(1962)$. Include an interpretation of the principal components accounting for approximately 70% of the total variation.

Section 5.7

Note: Problems for this section depend on the particular factor analysis program or programs available to the reader and the options allowed. Thus many problems may be formulated which use all or parts of Data Sets A and B. Two problems, which allow many variations, are included here.

5.7.1. (Data Set A)   In Example 5.7.1 we discussed various factor analyses of *initial* data on 14 variables for *all* patients. Perform similar analyses on the same variables and compare the results as follows:

   (a) Same as Example 5.7.1(a); use *initial* data on patients who *survived*;
   (b) Same as Example 5.7.1(b); use *initial* data on patients who *died*;
   (c) Same as Example 5.7.1(c); use *final* data on patients who *survived*.

5.7.2. (Data Set B)   Define $X_1, ..., X_{10}$ as in Problem 5.6.2. In parts (a)–(f), use the correlation matrix and always *interpret* the factors.

   (a) Perform a principal factor analysis with no rotation using three factors.
   (b) Repeat (a) using an orthogonal rotation.
   (c) Repeat (a) using an oblique rotation.
   (d) Perform a principal factor analysis, with no rotation, using as many factors as there are eigenvalues $\geqslant 1$.
   (e) Repeat (d) using an orthogonal rotation.
   (f) Repeat (d) using an oblique rotation.

APPENDIX I
REVIEW OF FUNDAMENTAL CONCEPTS

In this appendix we present an overview of some of the more important concepts in probability and statistics. We present this material in a verbal and non-mathematical style by motivating concepts and methods in the context of applications. We attempt to maintain accuracy of presentation without being mathematically rigorous.

This appendix is in no way meant to be complete. A complete discourse would require proofs of all statements, more theoretical examples, and additional details. Instead, one purpose is to summarize the concepts needed to develop the material in the book. Other purposes are to establish the notation used in this book and to present the distributions which are commonly used in statistical applications. We also define some concepts relating to multivariate distributions.

Readers interested in a more detailed development of the material should refer to the following references (this list includes some popular text books and is in no way meant to be complete). For an elementary treatment of statistics which does not require knowledge of calculus, see Dixon and Massey (1969), Dunn (1967), and Snedecor and Cochran (1967). For an intermediate treatment which requires knowledge of calculus, see Brownlee (1965), Hoel (1963), Hogg and Craig (1965), Lindgren (1968), and Mood and Graybill (1963). For advanced treatment of statistical theory including multivariate analysis, see Anderson (1958), Cramér (1946), Dempster (1969), Kendall and Stuart (1967, 1968, 1969), Morrison (1967), and Rao (1965). For probability theory see Feller (1966, 1968), Fisz (1963), Loève (1963), and Parzen (1960). For a decision theoretic approach to statistics see Ferguson (1967) and Lehmann (1959).

In this appendix, Section I.1 discusses the concepts of probability theory;

Section I.2 presents the common univariate distributions; Section I.3 discusses samples and sampling distributions; while Sections I.4–I.5 discuss statistical inference. Section I.6 defines vectors of observations and presents the multivariate normal distribution.

I.1 Concepts of Probability Theory

There are many approaches to the study of probability theory. Probability theory is a branch of mathematics dealing with random phenomena which the mathematician studies from an axiomatic point of view, see for example Feller (1966, 1968). The statistician, on the other hand, is interested in probability theory as a tool for developing statistical theory and methodology. We introduce the ideas and concepts of probability theory in an intuitive manner, maintaining accuracy but not mathematical rigor. The ideas are developed through the use of examples. The protagonist of one of our examples, a conscientious practicing physician named Dr. Thucidides A. Gesund, is faced with uncertain decisions and learns the role of probability and statistics in arriving at reasonable conclusions. Section I.1.1 defines the concept of population (or universe), Section I.1.2 introduces random variables, Section I.1.3 presents a discussion of probability appropriate for studying random variables, Section I.1.4 defines the distribution of a random variable, and Section I.1.5 presents the concept of the expected value of a random variable or a function of a random variable. In Section I.1.6 these ideas are generalized to more than one random variable.

I.1.1 Populations

A *population* (or *universe*) W may be considered as the total collection of objects w of interest for a given problem. The objects may be persons, animals, manufactured items, agricultural plots, and so forth. Each object is called an *element* (or *individual*) of the population, and the appropriate measurement made on each element is called an *observation*. Frequently, the problem of interest involves an *experiment* in which each element is subjected to a treatment. In this case, each element is called an *experimental unit*.

EXAMPLE I.1.1   A drug manufacturer has developed a new drug for treating hypertension, that is, elevated arterial blood pressure. Dr. Gesund is interested in evaluating the effect of this drug on his hypersensitive patients. His experiment consists of measuring the systolic blood pressure (mm Hg) before giving the drug, administering the drug, measuring the systolic blood pressure (mm Hg) after a two-week period, and then calculating the change in pressure. His objective is to decide from this difference whether or not the drug is effective in reducing arterial blood pressure.

In this experimental situation, the population of interest is all individuals with hypertension who take the proposed drug. The treatment is the administration of the drug, the experimental unit is the patient, and an observation is the change in systolic blood pressure in a two-week period for a given patient.

EXAMPLE I.1.2   In studying the income of families in the United States for a given year, each family is an element of the population of families in the United States and an observation is defined as the income in cents of a given family.

EXAMPLE I.1.3   A wholesale dealer of toys is negotiating a contract with a toy manufacturer. He is interested in knowing the percentage of defective toys in the population of all toys produced by the manufacturer. Each toy is an element of the population, and an observation consists of determining whether or not a given toy is defective.

In Example I.1.1, the population is hypothetical, since there is no way of determining each element of the population at the time of the experiment. Furthermore, the population is conceivably *infinite*, since it includes all possible individuals who will take the drug. On the other hand, the population of Example I.1.2 is an actual (and *finite*) population, since it is theoretically possible to determine the income of all families in the U.S. for a given year. The population of Example I.1.3 could be either hypothetical or actual. If the manufacturer is negotiating the sale of a given lot of toys which have already been manufactured, then the population is actual. On the other hand, if he is negotiating the sale of toys to be produced over a period of five years, then the population is hypothetical. Since the population is not always easily definable, the researcher is urged to give careful thought to this question so as not to overgeneralize the conclusions of his research.

Most problems in this book are of the type in which the population is hypothetical. Problems dealing with actual populations belong to the area of statistics known as *survey sampling* discussed, for example, in Cochran (1953).

I.1.2 Random Variables

A *random variable* X is a function which assigns a numerical value $X(w)$ to each element w in the population W. In this book, random variables are usually denoted by capital letters $X, Y, Z, \ldots$. The actual value x, say, assumed by X for a given individual w is called a *realization* of X and represents our observation. Realizations are denoted by small letters $x, y, z, \ldots$. Sometimes a random variable will be simply called a "*variable*."

In Example I.1.1, $X(w) = $ change in systolic blood pressure (mm Hg) for patient w over a two-week period of treatment. In Example I.1.2, $Y(w) = $ income in cents of family w in the given year, and in Example I.1.3, we may define

$$Z(w) = \begin{cases} 1, & \text{if toy } w \text{ is defective,} \\ 0, & \text{if toy } w \text{ is nondefective.} \end{cases}$$

The functions $X(w)$, $Y(w)$, and $Z(w)$ define the random variables X, Y, and Z, respectively. The choice of 0 and 1 for $Z(w)$ is arbitrary, and any two different numbers may be substituted for them.

The set of all possible distinct realizations of a random variable is called the *sample space S*. In Examples I.1.1–I.1.3, the sample spaces are the real line, the set of nonnegative integers, and the set $S = \{0, 1\}$, respectively.

Any subset E of the sample space S is called an *event*. Events are usually denoted by $E, E_1, E_2, \ldots$. In Example I.1.3 there are four possible events: $E_1 = \{0\}$, $E_2 = \{1\}$, $E_3 = \{0, 1\}$, and $E_4 = $ the empty set. The event E_1 is the subset of nondefective toys, E_2 is the subset of defective toys, E_3 is the sample space of all toys, and E_4 is the "empty" or null set of no members. In Example I.1.2 the event $E = \{0, 1, \ldots, 300,000\}$ denotes the subset of families with income less than or equal to \$3000. In Example I.1.1 the event $E = \{x \mid 19 \leqslant x \leqslant 48\}$, which is read "the set of x such that x is no smaller than 19 and no larger than 48," denotes the subset of patients with change of systolic pressure (mm Hg) in the interval $19 \leqslant x \leqslant 48$.

The concepts of population, random variable, and sample space are represented graphically in Fig. I.1.1 for Example I.1.1.

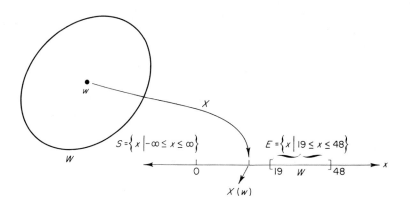

FIG. I.1.1. *Representation of population (W), random variable ($X = $ change in systolic pressure), and sample space ($S = $ real line: $-\infty < x < \infty$) for Example I.1.1.*

If the sample space S consists of a finite number of values, it is called *discrete*.[†]
A random variable with a discrete sample space is called a *discrete random variable*. The random variable Z of Example I.1.3 is discrete, since the sample space consists of only two values. Such a discrete random variable is called *dichotomous*. The random variable Y of Example I.1.2 is also discrete.[‡] The random variable X of Example I.1.1, is not discrete but is called *continuous*. This last concept will be discussed in Section I.1.4.

I.1.3 Probability

The basic concept underlying most statistical theory is that of the *probability* associated with a random variable. In the case of a discrete random variable X, the probability that X assumes the value x is the proportion p_x of individuals in the population possessing the value x. We write this relationship as $\Pr(X = x) = p_x$. [Some texts use the notation $P(X = x)$ or $P(x)$.] For the dichotomous variable of Example I.1.3, we have $\Pr(Z = 1) = p$ and $\Pr(Z = 0) = 1 - p$, where p is the proportion of defective items in the population. For the discrete random variable of Example I.1.2, let p_i be the proportion of families in the population with an annual income of i cents. Then $\Pr(Y = i) = p_i$, $i = 0, 1, \ldots, T$, where T is the maximum income (in cents) for a family in the given year. It is obvious that $\Pr(Y = i) = 0$ if $i < 0$ or $i > T$.

For a discrete random variable X let $E = \{x_1, \ldots, x_n\}$ be an event in S. Then the probability that X takes any value x_i in E is the sum of the probabilities p_{x_i} associated with each x_i, $i = 1, \ldots, n$. In symbols we write

$$\Pr(X \text{ in } E) = \sum_{i=1}^{n} p_{x_i} = \sum_{x_i \text{ in } E} p_{x_i}.^{§} \tag{I.1.1}$$

[Other texts write the left side of this equation as $\Pr(X \in E)$ or $\Pr(E)$.] It is obvious then that

$$\Pr(X \text{ in } S) = \sum_{x_i \text{ in } S} p_{x_i} = 1. \tag{I.1.2}$$

In Example I.1.2 let $E = \{0, 1, \ldots, 300{,}000\}$. Then the probability $\Pr(Y \text{ in } E)$ that a family has an annual income of \$3000 or less [written as $\Pr(0 \leqslant Y \leqslant 300{,}000) = \Pr(Y \leqslant 300{,}000)$] is

$$\Pr(Y \leqslant 300{,}000) = \sum_{i=0}^{300{,}000} p_i.$$

---

† Discrete sample spaces also include countably infinite sets.
‡ The set of nonnegative integers is countably infinite.
§ This notation means "sum of p_{x_i} for all x_i in E."

Similarly, the probability $\Pr(Y$ not in $E)$ that a family has an income over $3000, written as $\Pr(300{,}000 < Y < T) = \Pr(Y > 300{,}000)$, is

$$\Pr(Y > 300{,}000) = \sum_{i=300{,}001}^{T} p_i.$$

Since $\Pr(Y \leqslant T) = 1$, we also have

$$\Pr(Y > 300{,}000) = 1 - \sum_{i=0}^{300{,}000} p_i.$$

For *any* random variable X, the probability that X takes a value in a given event E is the proportion of individuals in the population with values in E. Hence, for Example I.1.1, $\Pr(X$ in $E)$ where $E = \{x \mid 19 \leqslant x \leqslant 48\}$, or $\Pr(19 \leqslant X \leqslant 48)$, denotes the proportion of individuals in the population with systolic pressure change in the interval $19 \leqslant x \leqslant 48$.

For any random variable we can write a useful relationship for nonoverlapping (*disjoint*) events $E_1, E_2, ..., E_k$. The probability that X belongs to any one of these events (the *union* of $E_1, ..., E_k$) is the sum of the probabilities associated with each event E_i, $i = 1, ..., k$. Symbolically, we have

$$\Pr\big(X \text{ in } (E_1 \cup E_2 \cup \cdots \cup E_k)\big) = \sum_{i=1}^{k} \Pr(E_i), \qquad (I.1.3)$$

where E_i are disjoint, $i = 1, ..., k$. The notation $\cup$ is a symbol for the union of events, and the expression "X in $(E_1 \cup E_2 \cdots \cup E_k)$" is read "$X$ in E_1 or E_2 or $\cdots$ or E_k."

I.1.4 Distribution of a Random Variable

The *distribution* of a random variable X is a means of describing the probability structure of the population in terms of realizations of X. In the case of a discrete variable, the distribution, called the *discrete distribution*, may be specified by listing the values $p_x = \Pr(X = x)$ for each x in the sample space S. In many cases, we can write a mathematical function $p(x)$ which relates p_x to x. The function $p(x)$ is called the *probability function* of the discrete random variable. Probability functions are characterized by constants called *parameters*. A parameter is any characteristic of the population.

For Example I.1.3 the discrete distribution is given by

z	p_z
0	$1-p$
1	p

where p is the proportion of defective items in the population. This can also be expressed by a probability function as

$$p(z) = \begin{cases} p^z(1-p)^{1-z}, & \text{for} \quad z = 0, 1, \\ 0, & \text{otherwise.} \end{cases}$$

This distribution is characterized by the single parameter p. For Example I.1.2 the distribution is given by

y	p_y
0	p_0
1	p_1
$\vdots$	$\vdots$
T	p_T

where p_i is the proportion of families with an income of i cents in the given year.

For any random variable X, the *cumulative distribution function* $F(x)$, sometimes abbreviated as cdf, is defined as

$$F(x) = \Pr(X \leqslant x). \tag{I.1.4}$$

In the case of a discrete random variable X with probability function $p(x)$, we have from Eq. (I.1.1)

$$F(x) = \sum_{u \leqslant x} p(u). \tag{I.1.5}$$

For Example I.1.3 the cdf is

$$F(z) = \begin{cases} 0, & \text{if} \quad z < 0, \\ 1-p, & \text{if} \quad 0 \leqslant z < 1, \\ 1, & \text{if} \quad z \geqslant 1, \end{cases}$$

and is graphed in Fig. I.1.2a. For Example I.1.2, the cdf is

$$F(y) = \begin{cases} 0, & \text{if} \quad y < 0, \\ \sum_{i=0}^{k} p_i, & \text{if} \quad K \leqslant y < K+1, \quad K = 0, ..., T-1, \\ 1, & \text{if} \quad y \geqslant T, \end{cases}$$

and is graphed in Fig. I.1.2b.

Note that in both figures the cdf has a "jump" or a "step" at different values of the random variable. If $F(x)$ is a continuous[†] function of x, that is, the graph of

---

† The precise definition of continuous functions can be found in Rudin (1964).

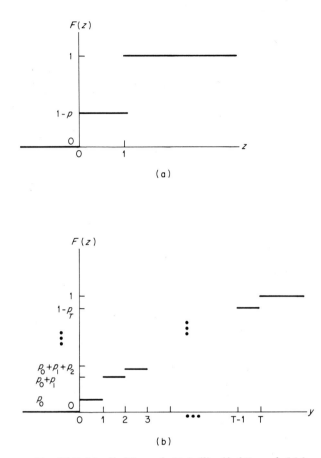

FIG. I.1.2. (a) *cdf of Example I.1.3;* (b) *cdf of Example I.1.2.*

$F(x)$ contains no jumps, then the random variable X is called a *continuous random variable* and its distribution is called a *continuous distribution*.

An important concept related to a continuous random variable is that of a *density* (or *frequency function*). The density[†] $f(x)$ of a continuous random variable X is a nonnegative function defined such that $F(x)$ is the area to the left of x under the graph of $f(x)$. This graph is called the *frequency curve*. This is a continuous analog of Eq. (I.1.5). Densities (and hence cdf's) are also characterized by parameters.

[†] In terms of differential calculus, the density $f(x)$ is the derivative of $F(x)$ with respect to x, that is

$$f(x) = \frac{dF(x)}{dx}.$$

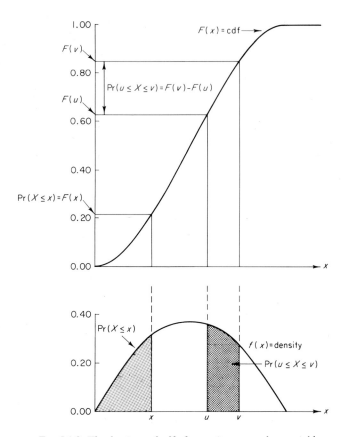

FIG. I.1.3. *The density and cdf of a continuous random variable.*

Figure I.1.3 illustrates the density and cdf of a continuous random variable. In this figure the shaded area to the left of x under the graph of $f(x)$ is the probability that $X \leqslant x$. This is equal to $F(x)$ on the upper graph. It is also noted from this figure that the area under $f(x)$ between u and v may be expressed in terms of F as

$$\Pr(u \leqslant X \leqslant v) = F(v) - F(u). \tag{I.1.6}$$

Some of the common discrete and continuous distributions are discussed in Section I.2.

COMMENTS I.1.1   1. The probability function $p(x)$ of a discrete random variable has the following properties:

(a) $0 \leqslant p(x) \leqslant 1$, for all x;

(b) $\sum_{x=-\infty}^{\infty} p(x) = 1$, that is, the sum of $p(x)$ over all possible values of x is equal to unity;

(c) $\Pr(u \leqslant X \leqslant v) = \sum_{x=u}^{v} p(x)$;

(d) $F(x) = \sum_{u \leqslant x} p(u)$.

*2. The density $f(x)$ of a continuous random variable has the following properties:

(a) $f(x) \geqslant 0$ for all x;

(b) $\int_{-\infty}^{\infty} f(x)\, dx = 1$;

(c) $\Pr(u \leqslant X \leqslant v) = \int_{u}^{v} f(x)\, dx$;

(d) $F(x) = \int_{-\infty}^{x} f(u)\, du$. Hence, $dF(x)/dx = f(x)$.*

3. For any random variable X the cumulative distribution function $F(x)$ has the following properties:

(a) $F(-\infty) = 0$, $F(\infty) = 1$;

(b) $F(x)$ is a nondecreasing function of x;

(c) $\Pr(u \leqslant X \leqslant v) = F(v) - F(u)$.

4. For a continuous random variable X the following relationship holds:

$$\Pr(X \leqslant x) = \Pr(X < x).$$

I.1.5 Expectation

The *expected value* $E(X)$ of a random variable X may be intuitively expressed as the average of the realizations $X(w)$ for all w in the population W. To motivate the general definition of expected value, we first restrict our attention to a finite population. We number the individuals in the population sequentially, so that $w = 1, 2, \ldots, N$. The expected value (*mean* or *average*) of X (also written as μ) is

$$\mu = E(X) = \frac{1}{N} \sum_{i=1}^{N} X(i). \tag{I.1.7}$$

If we denote the distinct elements of S by $x_1, \ldots, x_K$ ($K \leqslant N$), then we can express μ as

$$\mu = E(X) = \frac{\sum_{k=1}^{K} x_k n_k}{N}, \tag{I.1.8}$$

where n_k is the number of elements of W with value x_k. But $(n_k/N) = p_{x_k}$ by definition, and, therefore, we can write

$$\mu = \sum_{k=1}^{K} x_k p_{x_k}. \tag{I.1.9}$$

This definition is applicable to discrete random variables whether the population is finite or infinite. However, when W is infinite, there is no standard way to empirically obtain the probabilities p_{x_k} unless a probability function $p(x)$ describes the distribution. In this case we have

$$\mu = \sum_{k=1}^{K} x_k p(x_k), \tag{I.1.10}$$

where K may be infinite.

In Example I.1.2 we have

$$E(Y) = \sum_{i=0}^{T} ip_i$$

and in Example I.1.3 we have

$$E(Z) = 0(1-p) + 1(p) = p.$$

Equation I.1.10 can be extended to the case of a continuous random variable X except that summation is replaced by integration and the probability function $p(x)$ is replaced by a density function $f(x)$ (see starred Comment I.1.2.1).

The concept of expected value extends to any function $g(X)$ of X. The *expected value $E(g(X))$ of a function of the random variable X* is the average value of $g(X(w))$ for all w in W. Thus, for a discrete random variable, we have from Eq. (I.1.9),

$$E(g(X)) = \sum_{k=1}^{K} g(x_k) p_{x_k} \tag{I.1.11}$$

and this generalizes in the same way to the continuous case.

Special functions $g(X)$ are X^i and $[X-E(X)]^i$ for $i \geqslant 1$. The expected values $E(X^i)$ and $E[X-E(X)]^i$ are called the *ith moment about zero* and the *ith moment about the mean* (or *ith central moment*), respectively. The second central moment is known as the *variance* and is denoted by σ^2 or $V(X)$. The positive square root is called the *standard deviation* σ or $[V(x)]^{1/2}$.

Note that $V(X)$ may be expressed as

$$V(X) = E[X-E(X)]^2 = E(X^2) - [E(X)]^2. \tag{I.1.12}$$

Thus, in Example I.1.2, the variance is

$$V(Y) = \sum_{i=0}^{T} [i-E(Y)]^2 p_i$$

$$= \sum_{i=0}^{T} i^2 p_i - \left[\sum_{i=0}^{T} ip_i \right]^2,$$

and in Example I.1.3,

$$V(Z) = (0-p)^2(1-p) + (1-p)^2 p$$
$$= p(1-p),$$

since $E(Z) = p$.

COMMENTS I.1.2   *1.   For a continuous random variable X with density $f(x)$, the expected value of X is

$$E(X) = \int_{-\infty}^{\infty} xf(x)\,dx.$$

The expected value of a function $g(X)$ is

$$E(g(X)) = \int_{-\infty}^{\infty} g(x)f(x)\,dx.$$

In particular, the variance is

$$\sigma^2 = E(X - E(X))^2 = \int_{-\infty}^{\infty} (x - E(X))^2 f(x)\,dx.*$$

2.   The population mean μ is a *measure of central tendency*. In a physical sense, μ is the center of gravity of the graph of the probability function or density. Other measures of central tendency are the population *median* and the population *mode*. The population median is any value M such that

$$\Pr(X \leqslant M) \geqslant \tfrac{1}{2} \qquad \text{and} \qquad \Pr(X \geqslant M) \geqslant \tfrac{1}{2}.$$

The population median is not necessarily unique. For example, consider the following distribution:

x	p_x
0	$\frac{1}{8}$
1	$\frac{3}{8}$
2	$\frac{3}{8}$
3	$\frac{1}{8}$

The median is any value M such that $1 \leqslant M \leqslant 2$, since, for $M = 1$,

$$\Pr(X \leqslant M) = \tfrac{1}{2}, \qquad \Pr(X \geqslant M) = \tfrac{7}{8};$$

for $M = 2$,

$$\Pr(X \leqslant M) = \tfrac{7}{8}, \qquad \Pr(X \geqslant M) = \tfrac{1}{2};$$

and for $1 < M < 2$,

$$\Pr(X \leqslant M) = \tfrac{1}{2} = \Pr(X \geqslant M).$$

Since the median is not unique in this case, the convention is to choose the middle value $M = (1+2)/2 = \frac{3}{2}$. On the other hand, in the distribution

x	p_x
0	$\frac{1}{4}$
1	$\frac{1}{2}$
2	$\frac{1}{4}$

the median is the unique value $M = 1$, since it is the only value satisfying the definition. If X is a continuous random variable, then M is chosen so that

$$\Pr(X \leqslant M) = \Pr(X \geqslant M) = \tfrac{1}{2}.$$

The *population mode* is defined as any value of X for which the probability function or density is maximum. Thus, for the last example given above, the mode is 1. Such a distribution is called *unimodal*. For the first example above, there are two modes: 1 and 2. Such a distribution is called *bimodal*.

In comparing the three measures of centrality, we note that they are all equal for symmetric unimodal distributions. The mean has the most appealing theoretical properties. The median is sometimes difficult to calculate, especially if the ordering of the realizations is necessary. However, it may be a more meaningful measure of centrality for an asymmetric (or skewed) distribution such as that of Example I.1.2. The mode is particularly useful for gambling and decision making applications of probability theory.

3.   The variance is a *measure of dispersion* (or *variation*). The standard deviation measures the spread of the distribution in the same units used to measure the realizations of the random variable. Another measure of dispersion is the population *mean absolute deviation* defined as the expected value of the absolute difference between the random variable X and its mean. Sometimes, the median is substituted for the mean in this definition. The mean absolute deviation is intuitively appealing since it measures the "average deviation." However, the variance is mathematically more tractable and is therefore used in most applications.

4.   The following relationships hold for the expected value and the variance:

(a) $E(a+bX) = a+bE(X)$, where a and b are constants. Multiplying a random variable X by a constant b (that is, changing its scale) changes the scale of the average value by the same quantity. Similarly, adding a constant a to a random variable X (that is, changing its origin) accordingly shifts the average value by the same amount.

(b) $V(a+bX) = b^2 V(X)$, where a and b are constants. Multiplying a random variable X by a constant b multiplies the variance by the square of the constant (that is, multiplies the standard deviation by the absolute value of the constant). However, adding a constant a to a random variable does not change the variance.

5. The population mean, median, mode, variance, and higher moments are characteristics of the population, and, hence, they are parameters. Some of these parameters (or functions of them) may appear in the probability function or density.

I.1.6 Several Random Variables

In many cases we measure several characteristics of an element w of the population W. This necessitates defining several random variables $X_1, X_2, ..., X_k$ (k finite > 1). Each random variable X_i is a function which assigns a numerical value $X_i(w)$ to each element w in W, $i = 1, ..., k$. The actual value x_i assumed by X_i for a given w is a *realization* of X_i, $i = 1, ..., k$.

EXAMPLE I.1.1a   Dr. Gesund measures both systolic and diastolic blood pressures for each patient. Let $X_1(w)$ = change in systolic pressure (mm Hg), and $X_2(w)$ = change in diastolic blood pressure (mm Hg), both measured on patient w. These functions define the random variables X_1 and X_2, respectively.

EXAMPLE I.1.2a   In the study of family income in the U. S., let $Y_1(w)$ = income in cents of family w, $Y_2(w)$ = size of family w, and $Y_3(w)$ = age in years of the head of the household of family w. These functions define the random variables Y_1, Y_2, and Y_3, respectively.

EXAMPLE I.1.3a   The wholesale dealer of toys defines $Z_1(w) = 1$ or 0 depending on whether or not toy w is defective, and $Z_2(w)$ = suggested retail price in cents of toy w. These functions define the random variables Z_1 and Z_2, respectively.

We may also view measurements on the same characteristic on k individuals from W as k random variables. Let $w_1, w_2, ..., w_k$ be k individuals from W, and let X be a random variable. Then we define the k random variables $X_1, X_2, ..., X_k$ as

$$X_1(w_1, ..., w_k) = X(w_1),$$
$$X_2(w_1, ..., w_k) = X(w_2),$$
$$\vdots$$
$$X_k(w_1, ..., w_k) = X(w_k).$$

EXAMPLE I.1.1b   Dr. Gesund measures the change in systolic blood pressure (mm Hg) on k patients. Hence, $X_i(w_1, ..., w_k) = X(w_i)$ = change in systolic blood pressure for patient w_i, $i = 1, ..., k$.

EXAMPLE I.1.2b   We find the income in cents for k families. Hence, $Y_i(w_1, ..., w_k) = Y(w_i) =$ income of family w_i, $i = 1, ..., k$.

EXAMPLE I.1.3b   The wholesale dealer determines which of k toys is defective. Hence, $Z_i(w_1, ..., w_k) = Z(w_i) = 1$ or 0 depending on whether or not toy w_i is defective, $i = 1, ..., k$.

We now discuss the concepts underlying the probability structure of the population in terms of the realizations of several random variables, that is, the *joint distribution* of several random variables. For either definition of $X_1, ..., X_k$, we may represent the k realizations $x_1, ..., x_k$ as a k-*tuple* $(x_1, ..., x_k)$. The set of all possible distinct k-tuples is then the *sample space S*. For X_1, X_2 in Example I.1.1a, $S =$ the ordinary plane; for Y_1, Y_2, and Y_3 in Example I.1.2a, $S =$ all possible triples of nonnegative integers; and for Z_1, Z_2 of Example I.1.3a, $S =$ all possible pairs (z_1, z_2), where $z_1 = 1$ or 0 and $z_2 =$ a positive integer. For Example I.1.1b, $S =$ all k-tuples $(x_1, ..., x_k)$ with x_i real, that is, S is the k-dimensional Euclidean space. For Example I.1.2b, $S =$ all k-tuples $(y_1, ..., y_k)$, where y_i is a nonnegative integer. For example I.1.3b, $S =$ all possible sequences of length k consisting of 0's and 1's.

As before, any subset E of S is an *event*. For example, in Example I.1.1a, $E =$ the first quadrant $= \{x_1, x_2 | x_1 \geqslant 0$ and $x_2 \geqslant 0\}$ denotes the subset of non-negative change of both systolic and diastolic pressures. In Example I.1.2a, $E = \{y_1, y_2, y_3 | y_2 = 3, y_3 = 42\}$ denotes the subset of incomes of families of size 3 in which the head of the household is 42. In Example I.1.3a, $E = \{z_1, z_2 | z_1 = 1\}$ denotes the subset of prices of defective toys.

In order to define the *probability of an event* $\Pr(E)$, we distinguish between the two ways of defining k random variables. When $X_1, ..., X_k$ represent k characteristics on the same individual, then $\Pr(E)$ is the proportion of individuals in the population with k-tuples in the event E. When $X_1, ..., X_k$ are measures of the same characteristic on k individuals, it is necessary to construct a modified population $G = \{(w_1, ..., w_k) | w_i$ is in $W, i = 1, ..., k\}$. Then $\Pr(E)$ is the proportion of elements of G producing a k-tuple in E. For either definition, we can generalize the cdf of a random variable X to the *joint cdf* of the random variables $X_1, ..., X_k$ by defining

$$F(x_1, ..., x_k) = \Pr(X_1 \leqslant x_1, ..., X_k \leqslant x_k), \qquad (\text{I.1.13})$$

that is, the joint cdf is the probability of the event $E = \{X_1 \leqslant x_1$ and $X_2 \leqslant x_2, ...$ and $X_k \leqslant x_k\}$. If the X_i are all discrete, then it may be possible to express the probability of the event $E = \{X_1 = x_1, ..., X_k = x_k\}$ as a *joint probability function* $p(x_1, ..., x_k)$. If the X_i are all continuous, then it may be possible to extend the concept of the density of a single random variable to that of the *joint density* $f(x_1, ..., x_k)$ of the k random variables. Probabilities of events may then be obtained by integration (see Comment I.1.3.2). However, we can write [analogous to Eq. (I.1.6)] the relation

$$\Pr(u_1 \leqslant X_1 \leqslant v_1, \ldots, u_k \leqslant X_k \leqslant v_k) = F(v_1, \ldots, v_k) - F(u_1, \ldots, u_k).$$

$$(\text{I.1.14})$$

The treatment of k random variables of which some are continuous and some are discrete is beyond the scope of this book. References may be found in Afifi and Elashoff (1969).

Other distributions related to the joint distribution of $X_1, \ldots, X_k$ need to be discussed. The distribution of each X_i is called the *marginal distribution* of X_i, $i = 1, \ldots, k$. This is the same as the distribution of X_i when considered separately. The corresponding probability function (or density) is called the *marginal probability function* (or *marginal density*). The joint distribution of a subset of m of the random variables, $1 \leqslant m < k$, with the remaining $k - m$ random variables fixed at specified values is called a *conditional distribution* and is derived as follows: Rearrange the random variables so that $X_{m+1}, \ldots, X_k$ are the variables whose values are held fixed at $X_{m+1} = x_{m+1}, \ldots, X_k = x_k$. Then the distribution of $X_1, \ldots, X_m$ in the subpopulation for which $X_{m+1}, \ldots, X_k$ are held fixed is called the *conditional distribution of $X_1, \ldots, X_m$ given $X_{m+1} = x_{m+1}, \cdots, X_k = x_k$*. Again, the conditional distribution may be discrete or continuous, and the corresponding probability function (or density) is called the *conditional probability function* (or *conditional density*).

For Example I.1.3a, there are two marginal distributions—the marginal distribution Z_1, specifying the distribution of defective and nondefective toys, and the marginal distribution of Z_2, specifying the distribution of toy prices. If we were interested in the distribution of the price of defective toys, we fix $Z_1 = 1$ and examine the distribution of Z_2 for the resulting subpopulation. This is the conditional distribution of Z_2 given $Z_1 = 1$.

We now define *statistical independence* of two random variables X_1 and X_2. The random variables X_1 and X_2 are said to be *statistically independent* if a realization of X_1 does not affect the realization of X_2, and conversely. In other words, the distribution of X_1 for a given value of $X_2 = x_2$ is the same for all values of x_2, and conversely. This leads to the fact that X_1 and X_2 are statistically independent if the conditional distribution of X_1 given $X_2 = x_2$ is identical with the marginal distribution of X_1 for all values x_2. Similarly, X_1 and X_2 are statistically independent if the conditional distribution of X_2 given $X_1 = x_1$ is identical with the marginal distribution of X_2 for all values x_1. It may be shown that the alternative definitions of statistical independence are:

(a) $p(x_1, x_2) = p_1(x_1) p_2(x_2)$ for all x_1 and x_2, for discrete random variables; $f(x_1, x_2) = f_1(x_1) f_2(x_2)$ for all x_1 and x_2, for continuous random variables. That is, the joint probability function (density) is the product of the two marginal probability functions (densities).

(b) $F(x_1, x_2) = F_1(x_1) F_2(x_2)$, for all x_1 and x_2, for discrete or continuous random variables. That is, the joint cdf is equal to the product of the two marginal cdf's.

(c) $\Pr(X_1 \text{ in } E_1 \text{ and } X_2 \text{ in } E_2) = \Pr(X_1 \text{ in } E_1) \Pr(X_2 \text{ in } E_2)$ for all events E_1 and E_2.

We will usually refer to statistical independence of random variables as simply *independence* of random variables. Two random variables which are not independent are called *dependent*.

The k random variables $X_1, \ldots, X_k$ are *mutually (statistically) independent* if and only if $F(x_1, \ldots, x_k) = F_1(x_1) \cdots F_k(x_k)$ for all values of $x_1, \ldots, x_k$ and for discrete or continuous random variables. Expressions (a) and (c) may also be extended to k random variables.

For examples, the random variables $X_1, \ldots, X_k$ of Example I.1.1b may be shown to be mutually independent random variables. The same is true for Y_i and Z_i, $i = 1, \ldots, k$ of Examples I.1.2b and I.1.3b, respectively. In Example I.1.1a changes in systolic and diastolic pressure should be dependent, since diastolic pressure is necessarily less than systolic pressure, that is, the value of systolic pressure is an upper limit on diastolic pressure. In Example I.1.2a one would also expect dependence between age and income, and income and family size. In general, it is not safe to assume the independence of random variables defined on the same individual. On the other hand, measurements made on different elements of the population are more likely to be independent.

COMMENTS I.1.3   1.   The joint probability function $p(x_1, \ldots, x_k)$ of the discrete random variables $X_1, \ldots, X_k$ has the following properties:

(a) $0 \leqslant p(x_1, \ldots, x_k) \leqslant 1$ for all $x_1, \ldots, x_k$.

(b) $\sum_{x_1} \cdots \sum_{x_k} p(x_1, \ldots, x_k) = 1.$[†]

(c) $\Pr(u_1 \leqslant X_1 \leqslant v_1, \ldots, u_k \leqslant X_k \leqslant v_k) = \sum_{x_1 = u_1}^{v_1} \cdots \sum_{x_k = u_k}^{v_k} p(x_1, \ldots, x_k).$

(d) $F(x_1, \ldots, x_k) = \sum_{u_1 \leqslant x_1} \cdots \sum_{u_k \leqslant x_k} p(u_1, \ldots, u_k).$

(e) $p(x_i) = \sum_{x_1} \cdots \sum_{x_{i-1}} \sum_{x_{i+1}} \sum_{x_k} p(x_1, \ldots, x_k)$ is the marginal probability function of x_i.

*2.   The joint density $f(x_1, \ldots, x_k)$ of the continuous random variables of $X_1, \ldots, X_k$ has the following properties:

(a) $f(x_1, \ldots, x_k) \geqslant 0$ for all $x_1, \ldots, x_k$.

(b) $\int_{-\infty}^{\infty} \cdots \int_{-\infty}^{\infty} f(x_1, \ldots, x_k) \, dx_1 \cdots dx_k = 1.$

(c) $\Pr(u_1 \leqslant X_1 \leqslant v_1, \ldots, u_k \leqslant X_k \leqslant v_k) = \int_{u_k}^{v_k} \cdots \int_{u_1}^{v_1} f(x_1, \ldots, x_k) \, dx_1 \cdots dx_k.$

(d) $F(x_1, \ldots, x_k) = \int_{-\infty}^{x_k} \cdots \int_{-\infty}^{x_1} f(u_1, \ldots, u_k) \, du_1 \cdots du_k.$

(e) $f(x_i) = \int_{-\infty}^{\infty} \cdots \int_{-\infty}^{\infty} f(x_1, \ldots, x_k) \, dx_1 \cdots dx_{i-1} \, x_{i+1} \cdots dx_k$ is the marginal density of x_i.*

† The notation $\sum_{x_i}$ denotes summation over all possible values of x_i.

I.2 Common Univariate Distributions

In this section, we discuss some commonly used univariate distributions, that is, distributions of a single random variable. In particular, we discuss those distributions which are utilized in this book. For each distribution, the probability function or density is given, and appropriate applications are discussed. A summary table (Table I.2.1) at the end of the section lists each distribution along with its mean and variance.

I.2.1 The Binomial Distribution

Let $X_1, ..., X_n$ be n independent dichotomous random variables, each taking value 1 with probability p or value 0 with probability $1-p$. Let

$$X = \sum_{i=1}^{n} X_i. \tag{I.2.1}$$

Then X is a random variable with sample space $S = \{0, 1, 2, ..., n\}$. The distribution of X is called the *binomial distribution*. Its probability function $p(i) = \Pr(X = i)$ is denoted by $b_n(i, p)$ and is given by

$$b_n(i, p) = \binom{n}{i} p^i (1-p)^{n-i} \quad \text{for} \quad i = 0, ..., n, \tag{I.2.2}$$

where

$$\binom{n}{i} = \frac{n!}{i!(n-i)!} \tag{I.2.3}$$

and

$$\begin{cases} k! = 1 \cdot 2 \cdot 3 \cdots (k-1)k, & k \geqslant 1, \\ 0! = 1. \end{cases} \tag{I.2.4}$$

The quantity $k!$ is read "k factorial" and $\binom{n}{i}$, called the *binomial coefficient*, is read "n combinations i." Tables of factorials can be found in mathematical handbooks such as Burington (1965).

Table 1 of Appendix II lists values of the binomial probability for $n \leqslant 10$ and various values of p. For example, from this table, the probability that $X = 3$ with $n = 10$ and $p = 0.5$ is

$$b_{10}(3, 0.5) = 0.1172.$$

This can be verified directly from Eqs. (I.2.2)–(I.2.4) as follows:

$$\binom{n}{i} = \binom{10}{3} = \frac{10!}{3!\,7!} = \frac{10 \cdot 9 \cdot 8}{3 \cdot 2 \cdot 1} = 120,$$

$$p^i = (\tfrac{1}{2})^3 = \tfrac{1}{8}, \quad \text{and} \quad (1-p)^{n-i} = (\tfrac{1}{2})^7 = \tfrac{1}{128}.$$

Hence,

$$b_{10}(3,0.5) = 120(\tfrac{1}{8})(\tfrac{1}{128}) \doteq 0.1172.$$

The interpretation of $b_n(i,p)$ is that it is the probability of obtaining i 1's in n independent trials, where the probability of obtaining a 1 on any given trial is p. Thus, in Example I.1.3, if the probability that any toy is defective is $p = \tfrac{1}{2}$,[†] then the probability that exactly 3 out of 10 toys are defective is $b_{10}(3,0.5) = 0.1172$.

Table 1 in Appendix II may also be used to compute the cumulative distribution function. For example, if X is binomial with $p = 0.5$ and $n = 10$, then the probability of no more than three 1's is

$$\Pr(X \leqslant 3) = \sum_{i=0}^{3} b_{10}(i,0.5) = 0.0010 + 0.0098 + 0.0439 + 0.1172$$

$$= 0.1719.$$

For values of n and p not in this table, the computer is extremely useful for calculating the binomial probabilities.

Sometimes it is more convenient to consider the *proportion* of 1's rather than the number of 1's. Thus, we define the derived random variable

$$Y = \frac{X}{n} = \sum_{i=1}^{n} \frac{X_i}{n}, \tag{I.2.5}$$

which has probability function

$$\Pr\left(Y = \frac{i}{n}\right) = b_n(i,p), \qquad i = 0, \ldots, n. \tag{I.2.6}$$

Its mean and variance are p and $p(1-p)/n$, respectively.

I.2.2 The Poisson Distribution

Let X be a random variable with sample space $S = \{0, 1, 2, \ldots\}$. Then X has the *Poisson distribution with parameter λ* if

$$p(i) = \Pr(X = i) = \frac{e^{-\lambda} \lambda^i}{i!}, \qquad i = 0, 1, \ldots, \tag{I.2.7}$$

† Dr. Gesund sold his stock in this company after learning this fact.

where $i!$ is given by Eq. (I.2.4) and e is a constant approximately equal to 2.7183. Tables of $e^{-\lambda}$ can be found in mathematical handbooks such as Burrington (1965).

This distribution characterizes phenomena which occur randomly in time. As examples, the number of particles emitted from radioactive material per unit of time, the number of telephone calls per minute at a telephone exchange under steady state conditions may reasonably be assumed to follow the Poisson distribution. In all these examples the average rate per unit of time is the parameter λ, and the probability of i occurrences in the unit of time is given by Eq. (I.2.7). Furthermore, the distribution of the length of time between successive occurrences follows an exponential distribution discussed in Section I.2.4.

As an example, suppose that the average rate of incoming calls at steady state is 4/min at a telephone exchange between the hours of 10–11 a.m. Then the probability that there are no more than 3 calls between 10:00 and 10:01 is

$$\Pr(X \leqslant 3) = \Pr(X = 0) + \cdots + \Pr(X = 3)$$

$$= \sum_{i=0}^{3} \frac{e^{-4} 4^i}{i!}$$

$$= e^{-4}(\tfrac{1}{1} + \tfrac{4}{1} + \tfrac{16}{2} + \tfrac{32}{3}) = 0.433.$$

I.2.3 Uniform Distributions

The simplest continuous distribution is called the *uniform distribution* or *rectangular distribution*. The random variable X is said to be uniformly distributed over the interval $[a, b]$ if its density function is

$$f(x) = \begin{cases} \dfrac{1}{b-a}, & a \leqslant x \leqslant b, \\ 0, & \text{otherwise.} \end{cases} \tag{I.2.8}$$

The cdf is given by

$$F(x) = \begin{cases} 0, & x < a, \\ \dfrac{x-a}{b-a}, & a \leqslant x \leqslant b, \\ 1, & x > b. \end{cases} \tag{I.2.9}$$

This distribution is sometimes denoted by $U(a, b)$. If x is a realization of X, then x is said to be *randomly selected* from the interval $[a, b]$.

COMMENTS I.2.1   1.   If X is $U(a, b)$, then the random variable

$$Z = \frac{X-a}{b-a}$$

is uniformly distributed over the unit interval $[0, 1]$, that is, Z is $U(0, 1)$. This distribution is called the *standardized uniform distribution*.

2.   Computer programs exist which select random numbers z from the interval $[0, 1]$. To select x randomly from the interval $[a, b]$, the program selects z randomly from $[0, 1]$ and then calculates

$$x = (b-a)z + a.$$

Methods for selecting z may be found in the bibliography of Martin (1968).

3.   An important discrete distribution is called the *equal probability distribution* which has the probability function

$$p(i) = \Pr(X = i) = \frac{1}{k}, \qquad i = 1, ..., k,$$

where k is a positive integer. If x is a realization of a random variable X distributed according to this distribution, then x is said to be *randomly selected* from the integers $1, 2, ..., k$. The computer is useful in performing this operation by first randomly selecting a value z from $U(0, 1)$, calculating $y = kz + 1$, and then calculating $x = $ the largest integer contained in y. For example, if $k = 10$, and the randomly selected $z = 0.561$, then $y = 6.61$ and $x = 6$. Hence, the 6th value is randomly selected from the set of integers $1, 2, ..., 10$.

4.   Numbers selected randomly from the interval $[0, 1]$ may be used to select random realizations of a random variable with known distribution. This is discussed in Section 1.6.

I.2.4 The Exponential Distribution

The continuous random variable X is said to have an *exponential distribution with parameter* θ if it has the density function

$$f(x) = \theta e^{-\theta x}, \qquad x \geqslant 0, \quad \theta > 0. \tag{I.2.10}$$

The cdf is

$$F(x) = 1 - e^{-\theta x}. \tag{I.2.11}$$

If the distribution of the number of occurrences per unit of time of a particular phenomenon is Poisson with parameter λ, then the distribution of the length of time between successive occurrences is exponential with parameter $\theta = \lambda$.

I.2.5 The Normal Distribution

The most commonly used distribution in statistical applications is the *normal* (or *Gaussian*) *distribution*. The continuous random variable X is said to have a *normal distribution with parameters* μ *and* σ^2 if it has the density function

$$f(x) = \frac{1}{\sqrt{2\pi}\,\sigma} \exp -\frac{1}{2}\left(\frac{x-\mu}{\sigma}\right)^2, \quad -\infty < x < \infty, \quad \sigma > 0, \quad -\infty < \mu < \infty.$$

(I.2.12)

As noted in Table I.2.1, the mean is μ and the variance is σ^2 (standard deviation σ).

Since the cdf does not have a closed form, cumulative frequencies may be obtained by numerical integration (see Ralston and Wilf, 1960). This distribution is usually denoted by $N(\mu, \sigma^2)$.

COMMENTS I.2.2   1.   The density given in Eq. (I.2.12) is symmetric about μ, bell shaped (see Fig. I.2.1) and has the properties:

(a) The area between $\mu \pm \sigma$ is approximately 0.68 (that is, approximately 68% of the individuals in the population have values of X within one standard deviation of the mean);

(b) The area between $\mu \pm 2\sigma$ is approximately 0.95 (i.e., about 95% of the individuals in the population have values of X within two standard deviations of the mean);

(c) The third central moment is zero, and the fourth central moment is $3\sigma^4$.

The parameter μ determines the central location of the distribution and the parameter σ determines the shape. As σ becomes smaller, the spread becomes smaller and the peak becomes higher. As σ becomes larger, the density becomes flatter with less peakedness (see Fig. I.2.1).

2.   If X is $N(\mu, \sigma^2)$, then the transformed variable Z defined by

$$Z = \frac{X-\mu}{\sigma}$$

is $N(0, 1)$, that is, Z has mean $\mu = 0$ and standard deviation $\sigma = 1$. The density of Z is

$$f(z) = \frac{1}{\sqrt{2\pi}} \exp\left(-\frac{z^2}{2}\right)$$

and its distribution is called the *standard normal distribution*. If Z is $N(0, 1)$, then we sometimes denote $\Pr(Z \leqslant z)$ by $\Phi(z)$. The density of Z is denoted by $\phi(z)$. Tables of $\Phi(z)$ are given in Table 2 (Appendix II). For example,

$$\Pr(z \leqslant -1.0) = 0.1587,$$

$$\Pr(z \geqslant 1.0) = 1 - \Pr(z \leqslant 1.0) = 1 - 0.8413 = 0.1587,$$

$$\Pr(-1 \leqslant z \leqslant 2) = \Pr(z \leqslant 2) - \Pr(z \leqslant -1) = 0.9773 - 0.1587 = 0.8186.$$

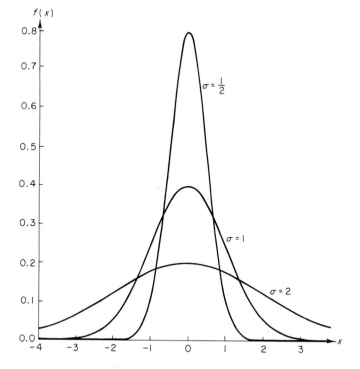

Fɪɢ. I.2.1. *Density of the normal distribution with $\mu = 0$ and three choices of σ^2.*

3.   If X is $N(\mu, \sigma^2)$, then areas to the left of a given value of x can be obtained by transforming X to Z and then using Table 2 (Appendix II). For example, if $\mu = 30$ and $\sigma = 20$, then to calculate $\Pr(X \leqslant 25)$, we have

$$\Pr(X \leqslant 25) = \Pr\left(Z \leqslant \frac{25 - 30}{20}\right) = \Pr(Z \leqslant -0.25) = 0.4013.$$

To calculate $\Pr(X \geqslant 51)$, we have

$$\Pr(X \geqslant 51) = \Pr\left(Z \geqslant \frac{51 - 30}{20}\right) = \Pr(Z \geqslant 1.05) = 1.0 - 0.8531 = 0.1469.$$

Finally,

$$\Pr(25 \leqslant X \leqslant 51) = \Pr(X \leqslant 51) - \Pr(X \leqslant 25)$$

$$= \Pr(Z \leqslant 1.05) - \Pr(Z \leqslant -0.25)$$

$$= 0.8531 - 0.4013 = 0.4518.$$

4.   If X is $N(\mu, \sigma^2)$, then for constants a and b, $Y = a + bX$ is $N(a + b\mu, b^2\sigma^2)$.

5. If X_1 is $N(\mu_1, \sigma_1{}^2)$, X_2 is $N(\mu_2, \sigma_2{}^2)$, ..., and X_k is $N(\mu_k, \sigma_k{}^2)$, and $X_1, ..., X_k$ are mutually independent, then for constants $a, b_1, ..., b_k$, $Y = a + \sum_{i=1}^k b_i X_i$ is also normally distributed with mean $a + \sum_{i=1}^k b_i \mu_i$ and variance $\sum_{i=1}^k b_i{}^2 \sigma_i{}^2$. That is, a linear combination of independent normal variables is also normal. A more general result is given in Section I.6.

6. Many observed phenomena follow an approximate normal distribution. For this reason, the bulk of classical statistical theory assumes normality of the random variable under consideration. As will be shown later, a second reason for assuming normality is a consequence of the central limit theorem, and a third reason is that some useful statistical theories are not highly dependent on this assumption.

EXAMPLE I.1.1 (*continued*)   In order to make probability statements about X = change in systolic pressure due to the drug, Dr. Gesund assumes that X is normally distributed with mean $\mu = 30$ mm Hg and standard deviation $\sigma = 20$ mm Hg. Using Comment I.2.2.3 he can then calculate probabilities of interest. Since μ and σ here are the same as those used in that comment, he can conclude (if the assumption of normality is valid) that:

(a) 40.13% of his patients using this drug will show a reduction of systolic blood pressure not larger than 25 mm Hg;

(b) 14.69% of his patients will show a reduction greater than or equal to 51 mm Hg; and

(c) 45.18% of his patients will show a reduction in the range of 25 to 51 mm Hg.

Note that since these three possibilities exhaust the sample space, the total percent is 100%.

I.2.6 The Chi Square (χ^2) Distribution

If $Z_1, Z_2, ..., Z_v$ are mutually independent $N(0, 1)$ random variables, where v is a positive integer, the variable U defined by

$$U = \sum_{i=1}^{v} Z_i{}^2 \tag{I.2.13}$$

has a *chi square* (χ^2) *distribution with parameter* v. This parameter is called the number of *degrees of freedom*. The density of U is

$$f(u) = \frac{u^{(v/2)-1} e^{-u/2}}{2^{v/2} \left[\frac{v}{2} - 1 \right]!}, \qquad u > 0, \quad v \text{ integer} \geqslant 1. \tag{I.2.14}$$

The cdf does not have a closed form, and the distribution of U is denoted by $\chi^2(v)$.

COMMENTS I.2.3   1.   The density $f(u)$, which depends on a single parameter v, has a long right tail when v is small and tends to be symmetric as v increases (see Fig. I.2.2).

2.   We now define the percentile of any distribution. For any random variable X, the *qth percentile* of the distribution of X is the value $x_{q/100}$ defined by

$$\Pr(X \leqslant x_{q/100}) = q/100.$$

Selected percentiles of the χ^2 distribution for a range of degrees of freedom v are given in Table 3 (Appendix II). For example, the 90th percentile of the χ^2 distribution with $v = 9$ is $\chi^2_{0.90}(9) = 14.7$. This implies that 90% of the individuals in the population have the value of the random variable $\leqslant 14.7$. Similarly, the 5th percentile of the χ^2 distribution with $v = 15$ is $\chi^2_{0.05}(15) = 7.26$.

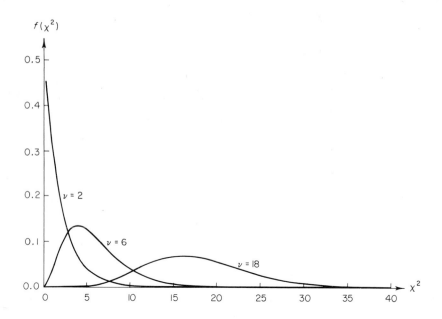

FIG. I.2.2. *Density of the chi square distribution with three choices of degrees of freedom v.*

I.2.7 The Student's t Distribution

If Z is $N(0, 1)$ and U is $\chi^2(v)$, where Z and U are independent, then the random variable T defined by

$$T = \frac{Z}{\sqrt{U/v}} \tag{I.2.15a}$$

has a *Student's t distribution with parameter v*. The parameter v is the number of degrees of freedom. The density of T is

$$f(t) = \frac{\left(\frac{v-1}{2}\right)!}{\sqrt{v\pi}\left(\frac{v-2}{2}\right)!\left[1+\frac{t^2}{v}\right]^{(v+1)/2}}, \quad -\infty < t < \infty, \quad v \text{ integer} \geqslant 1.$$

(I.2.15b)

The cdf does not have a closed form, and the distribution of T is denoted by $t(v)$.

COMMENTS I.2.4   1.   This density, which is symmetric, is generally flatter and more spread out than the normal density (see Fig. I.2.3). As $v \to \infty$, the density of the Student's t distribution approaches that of $N(0, 1)$.
  2.   Upper percentiles of the t distribution are tabulated in Table 5 (Appendix II). For example, the 95th percentile of t with $v = 10$ is $t_{0.95}(10) = 1.812$. Due to symmetry we obtain lower percentiles from the relation

$$t_{q/100}(v) = -t_{1-(q/100)}(v).$$

Therefore, the 5th percentile of t with $v = 10$ is $t_{0.05}(10) = -1.812$.

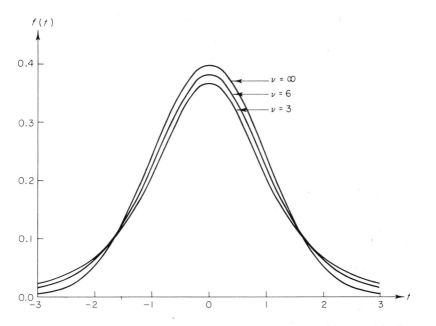

FIG. I.2.3. *Density of the Student's t distribution with three choices of degrees of freedom v.*

I.2.8 The F Distribution

If U is $\chi^2(v_1)$ and V is $\chi^2(v_2)$, and U and V are independent, then

$$W = \frac{U/v_1}{V/v_2} \tag{I.2.16}$$

has a *F distribution with parameters v_1 and v_2*. The parameters v_1 and v_2 are called the numerator and the denominator degrees of freedom, respectively. The density of W is

$$f(w) = \frac{\left(\dfrac{v_1+v_2-2}{2}\right)!}{\left(\dfrac{v_1-2}{2}\right)!\left(\dfrac{v_2-2}{2}\right)!}\left(\frac{v_1}{v_2}\right)^{v_1/2}\frac{w^{(v_1-2)/2}}{\left(1+\dfrac{v_1 w}{v_2}\right)^{(v_1+v_2)/2}},$$

$$w > 0, \quad v_1, v_2 \text{ integers} \geqslant 1. \tag{I.2.17}$$

The cdf does not have a closed form, and the distribution of W is denoted by $F(v_1, v_2)$.

COMMENTS I.2.5.   1.   This density has a long right tail and tends to be less skewed as both v_1 and v_2 increase (see Fig. I.2.4).

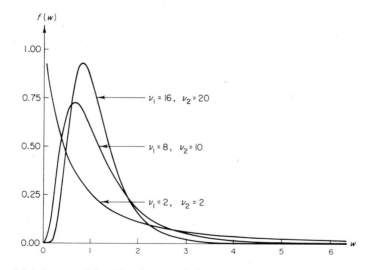

FIG. I.2.4. *Density of the F distribution with three choices of degrees of freedom v_1, v_2.*

2. Upper percentiles of the F distribution are tabulated in Table 6 (Appendix II). For example, the 97.5th percentile of F with $v_1 = 5$ and $v_2 = 19$ is $F_{0.975}(5, 19) = 3.33$. Lower percentiles may be obtained from the relationship

$$F_{q/100}(v_1, v_2) = \frac{1}{F_{1-(q/100)}(v_2, v_1)}.$$

Therefore, the 5th percentile of F with $v_1 = 4$, and $v_2 = 15$ is

$$F_{0.05}(4, 15) = \frac{1}{F_{0.95}(15, 4)} = \frac{1}{5.86}$$

$$= 0.171.$$

I.2.9 Summary

Table I.2.1 summarizes the eight distributions discussed in this section. For each distribution, its type, probability function or density, mean and variance, and relevant table in Appendix II are given.

I.3 Samples from a Population

The major objective of statistical analysis is to ascertain some of the properties of the population under consideration. If the population is finite, then a survey of each individual (provided that this is feasible) is the best procedure. However, most problems of interest involve either infinite populations or populations which are finite but difficult to survey. Hence, the best procedure is to carefully select from the population a subset of n elements, called a *sample of size n*, to examine its properties, and then to generalize these results to the population. The generalization to the population is called *statistical inference*. In this section we first discuss the question of selecting a sample of size n from which generalizations to the population are justifiable, then we expand the meaning of statistical inference, and discuss the concept of sampling distributions.

I.3.1 Random Samples

The basic requirement of a sample is that it be representative of the population. Although it is difficult to define what is meant by "representative," the usual method is to obtain a sample which is random. A *simple random sample* of size n

TABLE I.2.1. *Summary of Commonly Used Univariate Distributions*

Distribution	Type	Probability function or density	Mean	Variance	Table in Appendix II
Binomial	Discrete	$\binom{n}{i} p^i (1-p)^{n-i}$	np	$np(1-p)$	1
Poisson	Discrete	$\dfrac{e^{-\lambda}\lambda^i}{i!}$	λ	λ	—
Uniform (Rectangular)	Continuous	$\dfrac{1}{b-a}$	$\dfrac{a+b}{2}$	$\dfrac{(b-a)^2}{12}$	—
Exponential	Continuous	$\theta\,e^{-\theta x}$	$\dfrac{1}{\theta}$	$\dfrac{1}{\theta^2}$	—
Normal	Continuous	$\dfrac{1}{\sqrt{2\pi}\,\sigma}\exp-\left\{\dfrac{1}{2}\left(\dfrac{x-\mu}{\sigma}\right)^2\right\}$	μ	σ^2	2
$\chi^2(v)$	Continuous	$\dfrac{(\chi^2)^{(v/2)-1}\exp-\chi^2}{2^{v/2}\left[\dfrac{v}{2}-1\right]!}$	v	$2v$	3
$t(v)$	Continuous	$\dfrac{\left(\dfrac{v-1}{2}\right)!}{\sqrt{v\pi}\,\left(\dfrac{v-2}{2}\right)!}\left[1+\dfrac{t^2}{v}\right]^{(v+1)/2}$	0	$\dfrac{v}{v-2}$	5
$F(v_1,v_2)$	Continuous	$\dfrac{\left(\dfrac{v_1+v_2-2}{2}\right)!}{\left(\dfrac{v_1-2}{2}\right)!\left(\dfrac{v_2-2}{2}\right)!}\left(\dfrac{v_1}{v_2}\right)^{v_1/2}\dfrac{F^{v_1-2}}{\left(1+\dfrac{v_1 F}{v_2}\right)^{(v_1+v_2)/2}}$	$\dfrac{v_2}{v_2-2}$	$\dfrac{2v_2^2(v_1+v_2-2)}{v_1(v_2-2)^2(v_2-4)}$	6

is a sample so taken that every conceivable sample of size n has the same probability of being drawn from the population. In order to satisfy this definition, each member of the sample should be replaced into the population before selecting the next member. This is called *sampling with replacement*. Another random sample (not simple) is one in which each selected member is not returned to the population, and hence, it may not appear more than once in the sample. This is called *sampling without replacement*. If the population is infinite, then either sampling with or without replacement produces a simple random sample. If the population is finite and large relative to the sample size, then sampling without replacement approximates a simple random sample. If the population is finite and the sample size is a significant proportion of the population size, then the differences between the two methods become apparent.

Formally, a random sample of size n is a set of realizations of n independent identically distributed random variables. These random variables represent measurements of the same characteristics on n individuals as defined in Section I.1.6. Intuitively, this means that each individual in the population has the same probability of appearing in the sample and that the choice of any member of the sample is not affected by other choices. The main advantages of random sampling are that the effects of uncontrolled factors may be eliminated, and that most theoretical results are more easily derived assuming random sampling. Other methods of sampling are discussed in detail in Cochran (1953).

The standard procedures of obtaining a random sample from a finite population are discussed in most elementary texts, for example, Dixon and Massey (1969). Section I.2.3 discussed computer methods for obtaining a random sample. Practically speaking, there is no standard way of obtaining a simple random sample from an infinite population. Hence, the researcher is forced to restrict attention to a finite subset of the population. Infinite populations arise in experimental situations such as that in Example I.1.1. Dr. Gesund, out of necessity, must restrict his hypothetical infinite population to an actual finite population of patients taking the drug during the period of his research. Furthermore, out of convenience, he restricts this population to the subpopulation of patients who live in the vicinity of his office. From this subpopulation he may select a random sample of size n.

I.3.2 Sampling Distributions

In this section we develop the notion of a *sampling distribution*. To motivate this concept, we first examine the components of statistical inference. Statistical inference may be viewed as a method for making statements concerning unknown parameters of the population under study. Such statements may be divided into two major categories—*estimation* and the *testing of hypotheses*. The first category deals with obtaining an estimate of a given parameter either by (a) computing a single estimate (called a *point estimate*) from a sample, or (b) computing an interval (called a *confidence interval*) which hopefully includes the true value of the parameter. These methods of estimation are discussed in Section I.4. The second category of statistical inference, which deals with testing the validity of a statement about the parameter(s), called a *statistical hypothesis*, is discussed in Section I.5.

For purposes of statistical inference, we assume that we have a random sample $x_1, ..., x_n$ in which the x_i are realizations of independent, identically distributed, random variables X_i. We then calculate some function $g(x_1, ..., x_n)$ of the random sample called a *statistic*. Repeating this procedure for every possible sample of size n we obtain the *sampling population* of g. The distribution of this population is called the *sampling distribution of the statistic g* (see Fig. I.3.1). Examples of sampling distributions will be discussed in the following sections.

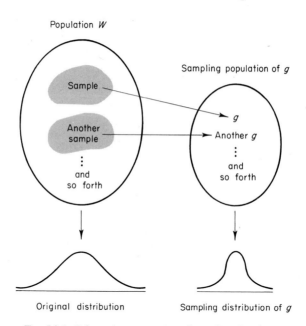

FIG. I.3.1. *Schematic presentation of sampling distribution.*

EXAMPLE I.1.1 (*continued*)   Dr. Gesund now knows that the random variable X, defined by $X(w) =$ change in systolic pressure for patient w, is a continuous random variable. Furthermore, he assumes that X is normally distributed with mean μ and variance σ^2. He then selects a random sample of 9 patients in the vicinity of his office and treats them with the drug. From the observations $x_1, ..., x_9$, he is now prepared to estimate the parameter μ or to test a hypothesis concerning it.

I.4 Estimation of Population Parameters

We assume that we have a random sample $x_1, ..., x_n$ of realizations of the random variables $X_1, ..., X_n$ from a population with density (or probability function) of the form $f(x; \theta_1, ..., \theta_k)$. The function is now written in this new form to represent either a density or a probability function and to indicate that it involves k parameters $\theta_1, ..., \theta_k$, some of which may be known. It is desired to estimate one or more of the θ_j from our sample. Each function $g(X_1, ..., X_n)$ which we choose to estimate a given parameter is called a (*point*) *estimator*, and the numerical value $g(x_1, ..., x_n)$ which it assumes from our sample is called a (*point*) *estimate*. Since each estimator is itself a random variable, we can study its sampling distribution in order to examine its properties. The desirable properties of estimators are discussed in Section I.4.1, and methods of obtaining them are discussed in Section I.4.2. Confidence intervals are discussed in Section I.4.3. For a theoretical presentation and examples see, for example, Lindgren (1968).

I.4.1 Properties of Point Estimators

In this section, we denote a given parameter to be estimated by θ, and an estimator by $\hat{\theta} = g(X_1, ..., X_n)$. One desired property of an estimator is that of *unbiasedness*. An estimator $\hat{\theta}$ is said to be *unbiased* if

$$E(\hat{\theta}) = \theta \quad \text{for all} \quad \theta. \tag{I.4.1}$$

This implies that the sampling distribution of $\hat{\theta}$ is centered at the parameter θ. That is, an unbiased estimator $\hat{\theta}$ is equal to θ on the average.

In some problems, it may be possible to devise many unbiased estimators. The intuitive procedure for choosing among these estimators is to pick one that has the least variability. An unbiased estimator $\hat{\theta}$ of θ with minimum variance among all unbiased estimators of θ is said to be *efficient*. If $V(\hat{\theta})$ is this minimum variance and $V(\hat{\theta}_a)$ is the variance of any other unbiased estimator $\hat{\theta}_a$ of θ, then the efficiency of $\hat{\theta}_a$ is

$$\text{Efficiency} = \frac{V(\hat{\theta})}{V(\hat{\theta}_a)}. \tag{I.4.2}$$

This is no smaller than 0 and no larger than 1. An efficient estimator is sometimes called a *minimum variance unbiased estimator*.

Sometimes an estimator becomes efficient as the sample size n increases. The limiting efficiency of an estimator as the sample size increases indefinitely is called the *asymptotic efficiency*. When the asymptotic efficiency is 1, the estimator is called *asymptotically efficient*.

Another desirable property of an estimator $\hat{\theta}$ is that of consistency. Formally, an estimator $\hat{\theta}$ is said to be *consistent* if, for any positive quantity ε,

$$\Pr\{|\hat{\theta}-\theta| < \varepsilon\} \to 1 \qquad \text{as} \quad n \to \infty. \tag{I.4.3}$$

Intuitively, this implies that as the sample size n increases, our estimator gets closer to the true value of the parameter.

I.4.2 Methods of Estimation

There is no single, best procedure for estimating the parameters $\theta_1, \theta_2, ..., \theta_k$ of interest. In this section some of the standard theoretical methods of estimation are introduced. The oldest modern method of obtaining point estimators is the *method of moments*. In brief, to estimate k parameters $\theta_1, ..., \theta_k$ by this method, we equate the first k population moments to the first k sample moments (sample moments are defined in Section 2.2.1). The solutions of the k equations provide estimators of the parameters. In general, these estimators are consistent.

The most popular method of estimation is the method of *maximum likelihood*. To define this method, we define the *likelihood function* as

$$L = \prod_{i=1}^{n} f(x_i; \theta_1, ..., \theta_k), \tag{I.4.4}$$

where $\prod$ denotes multiplication of the factors $f(x_i; \theta_1, ..., \theta_k)$. The maximum likelihood technique consists in determining the values $\hat{\theta}_1, ..., \hat{\theta}_k$ which maximize L with respect to $\theta_1, ..., \theta_k$. The resulting estimators $\hat{\theta}_1, ..., \hat{\theta}_k$, called *maximum likelihood estimators*, have the properties of being consistent, asymptotically normal, and asymptotically efficient under some general conditions. They are frequently biased, however. The computer is extremely useful in solving numerically for the estimates for a given sample.

A third method is that of *minimum χ^2*. For this method, we assume that the sample space is partitioned into c mutually exclusive and exhaustive categories. Let n_i be the observed number of sample values in the ith category, $i = 1, ..., c$. Hence, $\sum_{i=1}^{c} n_i = n$. Furthermore, let $p_i(\theta_1, ..., \theta_k)$ be the probability of falling in the ith category, $i = 1, ..., c$. These probabilities are obtained from the assumed probability function or density, and are functions of the parameters. Then the minimum χ^2 method of estimation consists in determining the values of $\theta_1, ..., \theta_k$ which minimize

$$\chi^2 = \sum_{i=1}^{c} \frac{[n_i - np_i(\theta_1, ..., \theta_k)]^2}{np_i(\theta_1, ..., \theta_k)}. \tag{I.4.5}$$

The computer is also useful for numerically calculating the solution.

Finally, a popular method of estimation is that of *least squares*. This technique and its properties are discussed in detail in Chapters 3 and 4.

I.4.3 Confidence Interval for a Parameter

After a point estimator $\hat{\theta}$ of a parameter θ is obtained, it is desirable to have an indication of the reliability of this estimator. This may be accomplished by computing the standard deviation of the sampling distribution of $\hat{\theta}$. This quantity is called the *standard error of the estimate* and is a measure of its variability. Another approach is to construct a *confidence interval*. Associated with the interval is the probability that the interval includes the true unknown value of θ. This probability is a measure of our confidence in the interval including the parameter, hence the name.

More precisely, we preselect a quantity $\alpha, 0 < \alpha < 1$, and find two numbers $a(\hat{\theta})$ and $b(\hat{\theta})$ depending on the estimator $\hat{\theta}$, such that

$$\Pr\{a(\hat{\theta}) \leqslant \theta \leqslant b(\hat{\theta})\} = 1 - \alpha. \tag{I.4.6}$$

The interval $[a(\hat{\theta}), b(\hat{\theta})]$ is then called a $100(1-\alpha)\%$ *confidence interval for θ.*[†] The probability that this interval includes the true value of θ is equal to $1-\alpha$ (the *confidence level*). The values of $a(\hat{\theta})$ and $b(\hat{\theta})$ depend on the sampling distribution of $\hat{\theta}$ and are called the *confidence limits* for θ. These limits are random variables, which vary from sample to sample. Of the confidence intervals based on all possible samples of size n, we expect $100(1-\alpha)\%$ of them to contain the true value of θ. Commonly used values for α are 0.10, 0.05, and 0.01, corresponding to 90%, 95%, and 99% confidence intervals, respectively. For fixed n, the larger the confidence level, the larger the confidence interval. Furthermore, for fixed α, the length of the confidence interval decreases as n increases.

COMMENTS I.4.1   1.   If X is $N(\mu, \sigma^2)$, then the maximum likelihood estimators for μ and σ^2 are

$$\hat{\mu} = \frac{1}{n} \sum_{i=1}^{n} x_i = \bar{x},$$

the *sample mean*, and

$$\hat{\sigma}^2 = \frac{1}{n} \sum_{i=1}^{n} (x_i - \bar{x})^2,$$

respectively. The estimator $\bar{x}$ is unbiased, consistent, and efficient. Its sampling distribution is also normal with mean μ and variance σ^2/n, that is, $\bar{x}$ is $N(\mu, \sigma^2/n)$. The estimator $\hat{\sigma}^2$ is biased but consistent. The sampling distribution of the quantity $n\hat{\sigma}^2/\sigma^2$ is χ^2 with $n-1$ degrees of freedom. Hence, the mean of $\hat{\sigma}^2$ is $(n-1)\sigma^2/n$, and the variance is $2(n-1)\sigma^4/n^2$.

[†] Confidence intervals are written in other forms. For example, $a(\hat{\theta}) \leqslant \theta \leqslant b(\hat{\theta})$. Also, if $b(\hat{\theta}) = \hat{\theta} + c$ and $a(\hat{\theta}) = \hat{\theta} - c$ for some constant c, then we may write the interval as $\hat{\theta} \pm c$.

The unbiased form of the maximum likelihood estimator for σ^2 is

$$s^2 = \frac{1}{n-1} \sum_{i=1}^{n} (x_i - \bar{x})^2,$$

the *sample variance*. The sampling distribution of $(n-1)s^2/\sigma^2$ is $\chi^2(n-1)$.

The quantity s is the usual estimator for the standard deviation σ. This estimator is biased. The standard error of the mean $\bar{x}$ is $\sigma/\sqrt{n}$ and is, therefore, estimated by $s/\sqrt{n}$.

2.   An important concept utilized throughout the book is the number of degrees of freedom of a sum of squares. In general, if s^2 is an unbiased estimator of σ^2 such that vs^2/σ^2 has a $\chi^2(v)$ distribution, then we say that s^2 *has v degrees of freedom*.

3.   An important theoretical result concerning sampling distributions is the *central limit theorem*. Various forms of this theorem are given in Feller (1968); here we state one of its simplest forms.

If $X_1, \ldots, X_n$ are independent identically distributed random variables with mean μ and finite variance σ^2, then as $n \to \infty$, the distribution of

$$\sqrt{n}\left(\frac{\bar{X} - \mu}{\sigma}\right) \to N(0, 1), \qquad \text{where} \quad \bar{X} = \frac{1}{n} \sum_{i=1}^{n} X_i.$$

An important application of this theorem is that if we obtain a random sample of size n from a population with finite variance, then regardless of the distribution of our random variable X, the distribution of the sample mean $\bar{X}$ is approximately $N(\mu, \sigma^2/n)$ for large n. Other theoretical implications of this theorem are given in Chapter 2.

4.   If X has a binomial distribution with parameters n and p, then the maximum likelihood estimator for p is

$$\hat{p} = \bar{x} = \frac{1}{n} \sum_{i=1}^{n} x_i.$$

This estimator is unbiased. It follows from the central limit theorem that for large n, the sampling distribution of $\hat{p}$ is approximately normal with mean p and variance $p(1-p)/n$, that is, $\hat{p}$ is approximately $N(p, p(1-p)/n)$ for large n.

5.   A $100(1-\alpha)\%$ confidence interval for the mean μ of a normal distribution with known σ is

$$\left[\bar{x} - z_{1-(\alpha/2)} \frac{\sigma}{\sqrt{n}}, \; \bar{x} + z_{1-(\alpha/2)} \frac{\sigma}{\sqrt{n}} \right],$$

where $z_{1-(\alpha/2)}$ is the $100(1-(\alpha/2))$th percentile of $N(0, 1)$.

EXAMPLE I.1.1 (*continued*)   The 9 sample realizations in mm Hg collected by Dr. Gesund are $-10, -5, 0, 25, 30, 35, 45, 50$ and 55. The maximum likelihood estimate for μ is $\bar{x} = \frac{1}{9}[-10-5+\cdots+55] = 25$, and the maximum likelihood estimate for σ^2 is $\hat{\sigma}^2 = \frac{1}{9}[(-10-25)^2 + \cdots + (55-25)^2] = 4800/9 = 533.3$. The unbiased estimate for σ^2 is $s^2 = 4800/8 = 600$, and the usual estimate for the standard deviation is $s = \sqrt{600} = 24.5$. Finally, the estimate of the standard error of the mean $\bar{x}$ is $s/\sqrt{n} = 24.5/3 = 8.15$.

Assuming that σ is known and is equal to $\sigma = 20$ mm Hg (see example following Section I.2.5), then a 95% confidence interval for μ is

$$\left[25 - 1.96\left(\frac{20}{\sqrt{9}}\right), \ 25 + 1.96\left(\frac{20}{\sqrt{9}}\right)\right] = (11.9, 38.1).$$

Hence, the probability is 0.95 that this interval includes the true value of μ.

I.5 Testing of Hypotheses

In many scientific investigations it is possible to formulate the problem as a hypothesis to be confirmed or rejected. Thus, the theory to be investigated is the basis for a statistical hypothesis. A *statistical hypothesis* is a statement either about the values of one or more of the parameters of a given distribution or about the form of the distribution itself. Hence, a statistical hypothesis is a statement concerning the population described by the distribution.

In Example I.1.1, Dr. Gesund is interested in determining whether the proposed drug does reduce arterial pressure in hypertensive patients. He thus formulates a hypothesis which states that "the average reduction in blood pressure is greater than zero," that is, the drug has a positive effect in reducing hypertension. He learns from statistical theory that this hypothesis is tested by first formulating another hypothesis, the hypothesis of "no average change due to the drug." If he lets μ be the population mean of the distribution of the two-week reduction in systolic pressure, then he can write this last hypothesis in the form $H_0: \mu = 0$. The hypothesis of interest to Dr. Gesund, however, is $H_1: \mu > 0$. His concern then is to reach a decision confirming either H_0 or H_1 based on a sample of patients. The hypothesis H_0 is called the *null hypothesis*; it is the "hypothesis of no change." The "hypothesis of interest" H_1 is called the *alternative hypothesis*. Most statistical hypothesis testing problems may be formulated so that the null and alternative hypotheses are defined in this manner.

A *statistical test of a hypothesis* is a procedure for determining whether the null hypothesis should be accepted or rejected. The reason for emphasizing the null hypothesis is that H_0 is usually viewed as a statement which is more meaningful if it is rejected. This is based on the idea that a theory should be rejected if evidence exists which contradicts it, but should not necessarily be accepted if evidence cannot be found to contradict it.

Without any theoretical justification, Gesund (MD) argues that to test his hypothesis concerning μ, he looks at the sample mean change in systolic pressure $\bar{x}$. He decides that if $\bar{x}$ exceeds a certain value, called the *critical value*, he will reject H_0 and accept H_1; if $\bar{x}$ does not exceed the critical value, he will not reject H_0. (This will be shown later to be a theoretically valid procedure.) For the sake of simplicity of expression, "not rejecting H_0" will be written simply as "accepting H_0." It must be kept in mind, however, that his decision, that is, rejecting or accepting H_0, is based on his sample observations, and that his decision is necessarily subject to error.

In general, there are two types of errors associated with a decision. If H_0 is in fact true, and the decision is to reject H_0, then an error, called the *Type I error* has been committed. On the other hand, if H_1 is in fact true, and the decision is to accept H_0, then an error, called the *Type II error* has been committed. These errors appear in Table I.5.1 along with the probabilities of making each decision

TABLE I.5.1. *Two Types of Error in Making a Statistical Decision*

	H_0 true	H_0 false
Reject H_0	Type I error Probability $= \alpha$	Correct decision Probability $= 1 - \beta$
Accept H_0	Correct decision Probability $= 1 - \alpha$	Type II error Probability $= \beta$

given the true situation. The probability of committing a Type I error is denoted by α, while the probability of committing a Type II error is denoted by β. These probabilities can be expressed as

$$\alpha = \Pr\{\text{reject } H_0 | H_0 \text{ true}\}, \quad \text{and} \quad \beta = \Pr\{\text{accept } H_0 | H_0 \text{ false}\}.$$

$$(\text{I.5.1})$$

(The slash "|" is read "given.") The statistical problem is to find the decision procedure which in some way minimizes the probability of committing either of these errors, that is, minimizes α and β.

In Example I.1.1, let $\bar{x}_c$ be the critical value of $\bar{x}$. To calculate the probabilities α and β associated with his decision, Dr. Gesund examines the sampling distribution of $\bar{x}$ *under H_0* (that is, when H_0 is true), and also the sampling distribution of $\bar{x}$ under H_1. Since H_1 includes any and all values of $\mu > 0$, he restricts his attention to a particular value $\mu = \mu_1 > 0$, say. These distributions are shown in Fig. I.5.1. The *null and alternative distributions* are the sampling distributions

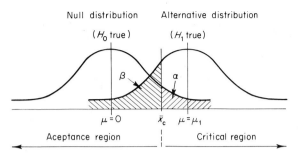

Null distribution   Alternative distribution

$(H_0 \text{ true})$ $(H_1 \text{ true})$

FIG. I.5.1. *Probabilities associated with the test H_0: $\mu = 0$, against H_1: $\mu = \mu_1 > 0$.*

of $\bar{x}$ under H_0 and H_1, respectively. Since his concern is to minimize α and β, he looks for an $\bar{x}_c$ which achieves this goal. Examination of Fig. I.5.1 shows that as $\bar{x}_c$ moves to the right, α decreases but β increases. Similarly, as $\bar{x}_c$ moves to the left, β decreases but α increases. The usual solution to this dilemma is to fix α at some small value and hope that β is also small. This fixed value of α is called the *significance level*. Usual values for α are $\alpha = 0.10, 0.05$, and 0.01. For a fixed value of α, the "goodness" of the test procedure is measured by the probability of rejecting H_0 when H_1 is true. This probability, called the *power* of the test, is usually denoted by π and is expressed by the relationship

$$\pi = 1 - \beta = \Pr\{\text{rejecting } H_0 | H_1 \text{ true}\}. \qquad (\text{I.5.2})$$

It should be noted that the power is a function of the specific alternative value of the parameter. A "good" test for a fixed α is one which has large power. Sometimes it is possible to find a test which is "best" in that it has minimum β among *all* tests possessing significance level α. Equivalently, a best test is one which has maximum power π among all tests possessing significance level α.

Fixing α determines the critical value $\bar{x}_c$. The *critical region* is then the subset of the sample space corresponding to the rejection of H_0. The complementary region, corresponding to acceptance of H_0, is called the *region of acceptance*. For Example I.1.1, the critical region is $\bar{x} \geqslant \bar{x}_c$, and the region of acceptance is $\bar{x} < \bar{x}_c$ (see Fig. I.5.1).

I.5.1 A Procedure for Constructing a Test of a Hypotheses

In general, a test of a statistical hypothesis is equivalent to identifying the subset of the sample space which corresponds to the critical region for a fixed significance level α. There may be many tests which achieve the same α, but the objective is to find one which maximizes the power. Although there is no standard procedure for determining the most powerful test, one procedure, the *likelihood ratio procedure*, is often very useful. We now discuss this procedure.

Let $x_1, ..., x_n$ be a random sample from a population with density (or probability function) $f(x; \theta_1, ..., \theta_k)$. The *likelihood ratio* λ is defined as

$$\lambda = \frac{\text{Max}_{H_0} L(\theta_1, ..., \theta_k)}{\text{Max } L(\theta_1, ..., \theta_k)}. \qquad (\text{I.5.3})$$

The denominator is the maximum value of the likelihood function $L(\theta_1, ..., \theta_k)$ [given by Eq. (I.4.4)] over all possible values of the parameters $\theta_1, ..., \theta_k$. The numerator is the maximum value of $L(\theta_1, ..., \theta_k)$ over all values of the parameters which are possible under H_0. Note that λ is a random variable, since it is a function of $X_1, ..., X_n$. Since H_0 represents a restriction on the values of the parameters, λ necessarily satisfies the inequality $0 \leqslant \lambda \leqslant 1$. Intuitively, if λ is close to 1, we would be inclined to accept H_0. Thus, the *likelihood ratio test procedure* consists of rejecting H_0 if $0 \leqslant \lambda \leqslant \lambda_c$, where λ_c is chosen so that

$$\Pr(\lambda \leqslant \lambda_c | H_0) = \alpha, \qquad (\text{I.5.4})$$

that is, λ_c is determined from the distribution of λ under H_0 so that the critical region has probability α under H_0. It is useful to note that

$$\text{Max } L(\theta_1, ..., \theta_k) = L(\hat{\theta}_1, ..., \hat{\theta}_k), \qquad (\text{I.5.5})$$

where $\hat{\theta}_i$ is the maximum likelihood estimate of θ_i, $i = 1, ..., k$. Similarly,

$$\underset{H_0}{\text{Max }} L(\theta_1, ..., \theta_k) = L(\hat{\theta}_1^{(0)}, ..., \hat{\theta}_k^{(0)}), \qquad (\text{I.5.6})$$

where $\hat{\theta}_i^{(0)}$ is either the value of θ_i specified by H_0 or its maximum likelihood estimate under H_0, $i = 1, ..., k$. Since the value of λ_c cannot be determined unless the null distribution of λ is known, it is sometimes necessary to resort to the asymptotic distribution of λ. Under H_0 the distribution of $-2 \ln \lambda$ approaches the $\chi^2(\nu)$ distribution as $n \to \infty$. The number of degrees of freedom ν = the number of independent parameters specified by H_0.

COMMENTS I.5.1   1.   Assume that X is $N(\mu, \sigma^2)$ with σ^2 known and, let $x_1, ..., x_n$ be a random sample from this distribution. Consider testing $H_0: \mu = \mu_0$ against the *one-sided alternative* $H_1: \mu > \mu_0$ at significance level α. The critical region of the best test is given by $\bar{x} \geqslant \bar{x}_u$, where $\bar{x} = (1/n) \sum_{i=1}^{n} x_i$ is the sample mean, and $\bar{x}_u$ is chosen such that $\Pr(\bar{x} \geqslant \bar{x}_u | H_0) = \alpha$. Since from Comment I.4.1.1 we know that $\bar{x}$ is $N(\mu_0, \sigma^2/n)$ under H_0, then $\bar{x}_u = \mu_0 + z_{1-\alpha}(\sigma/\sqrt{n})$, where $z_{1-\alpha}$ is the $100(1-\alpha)$th percentile of the $N(0, 1)$ distribution. The critical region is therefore the right tail $(\bar{x} \geqslant \bar{x}_u)$ (see Fig. I.5.2a).

Similarly, to test $H_0: \mu = \mu_0$ against the other one-sided alternative $H_1: \mu < \mu_0$, the best test gives the left tail as the critical region, namely $\bar{x} \leqslant \bar{x}_l = \mu_0 + z_\alpha(\sigma/\sqrt{n})$ (see Fig. I.5.2b). These two tests are called *one-sided tests*.

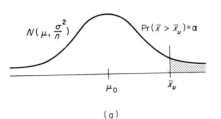

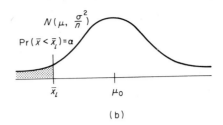

(a)

(b)

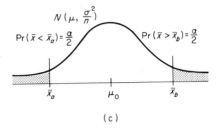

FIG. I.5.2. *Critical regions for testing H_0: $\mu = \mu_0$, with σ^2 known. (a) Alternative H_1: $\mu > \mu_0$; (b) Alternative H_1: $\mu < \mu_0$; (c) Alternative H_1: $\mu \neq \mu_0$.*

(c)

Finally, to test H_0: $\mu = \mu_0$ against the *two-sided alternative* H_1: $\mu \neq \mu_0$, we obtain as the critical region of the likelihood ratio test the two tails

$$\bar{x} \leqslant \bar{x}_a = \mu_0 + z_{\alpha/2}\frac{\sigma}{\sqrt{n}} \quad \text{and} \quad \bar{x} \geqslant \bar{x}_b = \mu_0 + z_{1-(\alpha/2)}\frac{\sigma}{\sqrt{n}}$$

(see Fig. I.5.2c). This test is called a *two-sided test*. We note that the two one-sided tests have the property that they have maximum power against any value of μ which is possible under H_1. A test which is best for all alternatives is called a *uniformly most powerful test*.

2. The power of each test in Comment I.5.1.1 is easy to calculate. For the alternative H_1: $\mu = \mu_1 > \mu_0$, we have

$$\pi = \Pr\left(Z \leqslant z_\alpha + \frac{(\mu_1 - \mu_0)}{\sigma}\sqrt{n}\right),$$

where Z is $N(0, 1)$. For the alternative H_1: $\mu = \mu_1 < \mu_0$, we have

$$\pi = \Pr\left(Z \leqslant z_\alpha - \frac{(\mu_1 - \mu_0)}{\sigma}\sqrt{n}\right).$$

Finally, for the alternative H_1: $\mu = \mu_1 \neq \mu_0$, we have

$$\pi = \Pr\left(Z \leqslant z_{\alpha/2} - \frac{(\mu_1 - \mu_0)}{\sigma}\sqrt{n}\right) + \Pr\left(Z \leqslant z_{\alpha/2} + \frac{(\mu_1 - \mu_0)}{\sigma}\sqrt{n}\right).$$

3. Each alternative in Comment I.5.1.2 specifies the value of μ, that is, $\mu = \mu_1$. A hypothesis which specifies the values of each parameter is called a

simple hypothesis. If the hypothesis does not specify the values of some of the parameters, then it is called a *composite* hypothesis. Each alternative in Comment I.5.1.1 is composite.

4. It must be kept in mind that the conclusion of any statistical test of hypothesis is subject to error. In particular, accepting a null hypothesis H_0 should not lead to the conclusion that H_0 is correct. In either case, the result of the test should be considered only as one of the factors affecting a final decision. Other factors are the accumulated experiences and intuition of the researcher.

I.5.2 The Concept of P Value

Most critical regions of tests of hypotheses are expressed in terms of some statistic g called the *test statistic*. The test statistic is usually chosen so that its null distribution is available in tabulated form, for example, g has an $N(0, 1)$, χ^2, t, or F distribution. The critical region of the test can then be expressed in terms of the test statistic g. Depending on H_0 and H_1, the critical region expressed in terms of g takes one of the forms: (a) $g \leqslant g_l$, (b) $g \geqslant g_u$, or (c) $g \leqslant g_a$ and $g \geqslant g_b$, where g_l, g_u, or g_a and g_b are chosen from the table of the distribution of g so that under H_0,

$$\Pr(g \leqslant g_l) = \alpha, \qquad \Pr(g \geqslant g_u) = \alpha, \qquad \text{or}$$

$$\Pr(g \leqslant g_a) = \Pr(g \geqslant g_b) = \alpha/2. \tag{I.5.7}$$

Cases (a) and (b) represent *one-sided critical regions*, while case (c) is a *two-sided critical region*.

The procedure of performing the test consists of computing g from the sample and checking whether or not it falls in the appropriate critical region for g. If it does, we reject H_0; if it does not, we accept H_0.

Let g_0 be the computed value of g from the sample. An equivalent procedure (and one that we use in the book) is to calculate the probability that the test statistic assumes the value g_0 or a more extreme value than g_0 under H_0. (Extreme values are determined by the critical region.) This probability is called the *P value*, and is denoted by P in this book. If P is less than α, then H_0 is rejected at level α, otherwise H_0 is accepted. For cases (a), (b), and (c), we have under H_0,

$$P = \Pr(g \leqslant g_0), \qquad P = \Pr(g \geqslant g_0), \qquad \text{or}$$

$$P = 2 \operatorname{Min}[\Pr(g \geqslant g_0), \Pr(g \leqslant g_0)], \tag{I.5.8}$$

respectively. The last equality doubles the smaller of the quantities $\Pr(g \geqslant g_0)$ or $\Pr(g \leqslant g_0)$.

COMMENTS I.5.2   1. As in Comment I.5.1.1, let X be $N(\mu, \sigma^2)$ with σ^2 known. Then the test statistic

$$z_0 = \frac{\bar{x} - \mu_0}{\sigma}\sqrt{n}$$

has a $N(0, 1)$ distribution under H_0 and is used to test $H_0: \mu = \mu_0$ against a one- or two-sided alternative. If the alternative is $H_1: \mu > \mu_0$, then the critical region for z is $z \geqslant z_{1-\alpha}$. The P value is the area under the $N(0, 1)$ frequency curve to the right of z_0 (see Fig. I.5.3a). If the alternative is $H_1: \mu < \mu_0$, then the critical region is $z \leqslant z_\alpha$, and the P value is the area under the $N(0, 1)$ frequency curve to the left of z_0 (see Fig. I.5.3b). Finally, if $H_1: \mu \neq \mu_0$, then the critical regions are $z \leqslant z_{\alpha/2}$ and $z \geqslant z_{1-(\alpha/2)}$, so that the P value is double the area to the right of the absolute value of z_0 under the $N(0, 1)$ frequency curve (see Fig. I.5.3c). In all three cases, if the P value is less than α, then H_0 is rejected at the significance level α. Another way of expressing rejection of H_0 is to say that the test is *statistically significant* at level α. If H_0 is accepted, the test is *statistically nonsignificant*.

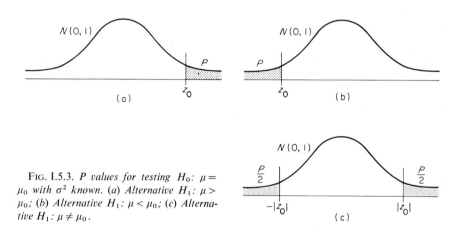

Fig. I.5.3. *P values for testing* $H_0: \mu = \mu_0$ *with* σ^2 *known.* (a) *Alternative* $H_1: \mu > \mu_0$; (b) *Alternative* $H_1: \mu < \mu_0$; (c) *Alternative* $H_1: \mu \neq \mu_0$.

2. A $100(1-\alpha)\%$ confidence interval for a parameter θ may be used for testing $H_0: \theta = \theta_0$ against the two-sided alternative $H_1: \theta \neq \theta_0$ at significance level α. In this case, we accept H_0 if the interval includes the value of θ_0, otherwise we reject H_0.

3. A $100(1-\alpha)\%$ confidence interval for μ with known σ^2 was given in Comment I.4.1.5 as

$$\left[\bar{x} - z_{1-(\alpha/2)}\frac{\sigma}{\sqrt{n}}, \; \bar{x} + z_{1-(\alpha/2)}\frac{\sigma}{\sqrt{n}} \right].$$

This interval can be used for testing $H_0: \mu = \mu_0$ against $H_1: \mu \neq \mu_0$ at significance level α. If the interval includes μ_0, we accept H_0, otherwise we reject H_0.

EXAMPLE I.1.1 (*continued*)   From the sample of 9 patients, Dr. Gesund has calculated $\bar{x} = 25$. Assuming for the sake of illustration that $\sigma = 20$, he can now test the hypothesis $H_0: \mu = \mu_0 = 0$ against $H_1: \mu > 0$ at significance level $\alpha = 0.05$. The critical value is $\bar{x}_u = 0 + z_{0.95}(20/\sqrt{9}) = 1.645(6.67) = 10.97$ (see Comment I.5.1.1). Since $25 > 10.97$, H_0 is rejected, and he asserts that there is a significant positive change in systolic blood pressure due to the drug at $\alpha = 0.05$. If $H_1: \mu = \mu_1 = 30$, then the power of this test (see Comment I.5.1.2) is

$$\pi = \Pr\left(Z < -1.645 + \frac{(30-0)}{20}\sqrt{9}\right) = \Pr(Z < 2.86) > 0.998.$$

Hence, the probability is greater than 0.99 that the null hypothesis has been correctly rejected, if μ is actually 30.

As another method for testing H_0, we can use the test statistic of Comment I.5.2.1. Thus,

$$z_0 = \frac{(25-0)}{20}\sqrt{9} = 3.75.$$

Since the critical region is $z > z_{0.95} = 1.645$, H_0 is rejected.

As a third possibility, we calculate from Table 1 (Appendix II) $P = \Pr(z > 3.75) < 0.0001$. Since this P value is less than $\alpha = 0.05$, we reject H_0.

Finally, we can test H_0 against the two-sided alternative $H_1: \mu \neq 0$ at $\alpha = 0.05$ using the 95% confidence interval for μ. Since the interval $25 \pm 1.96(20/3) = (11.9, 38.1)$ does not include 0, we reject H_0.

I.6 The Multivariate Normal Distribution

In this section we discuss the theoretical framework underlying multiple measurements, that is, the theory of more than one random variable defined on an individual of the population. To do this, we define a vector and a matrix of random variables, and in particular, the mean vector and the covariance matrix. We then define the most commonly used joint distribution in statistical applications—the multivariate normal distribution. Applications of this distribution appear in Chapters 3–5.

I.6.1 Random Vectors and Matrices

The definitions of vectors and matrices given in this section are for the special case where the components are random variables or realizations of random variables.

In many statistical applications, the researcher measures $k > 1$ characteristics on each individual w in the population W. As discussed in Section I.1.6, we define k random variables $X_1, X_2, ..., X_k$ to correspond to these characteristics. It is useful to represent the k random variables as a *random vector*, that is, an ordered k-tuple arranged in a column such as

$$\mathbf{X} = \begin{bmatrix} X_1 \\ X_2 \\ \vdots \\ X_k \end{bmatrix}. \tag{I.6.1}$$

Each X_i is called a *component* of the random vector. We usually represent random vectors by boldface capital letters $\mathbf{X}, \mathbf{Y}, \mathbf{Z}, ...$.

A realization of the random vector $\mathbf{X}$ will be denoted by a *vector* of observations defined by

$$\mathbf{x} = \begin{bmatrix} x_1 \\ x_2 \\ \vdots \\ x_k \end{bmatrix}, \tag{I.6.2}$$

where the components of $\mathbf{x}$ are the realizations $x_1, ..., x_k$ of $X_1, ..., X_k$, respectively. Such a realization $\mathbf{x}$ is called a *multivariate observation*; if $k = 2$, it is called a *bivariate observation*. Vectors of observations will be represented by boldface small letters $\mathbf{x}, \mathbf{y}, \mathbf{z}, ...$.

Sometimes a random vector $\mathbf{X}$ will be superscripted as $\mathbf{X}^{k \times 1}$ to denote that k components are arranged in 1 column. The superscript will be deleted if the number of components is obvious. A vector of observations may also be superscripted in the same way.

Suppose we have a random sample of n individuals on whom we measure k characteristics. Let $x_{1j}, x_{2j}, ..., x_{kj}$ represent the k realizations for the jth individual, $j = 1, ..., n$. We can represent each set of k realizations by a vector

$$\mathbf{x}_j^{k \times 1} = \begin{bmatrix} x_{1j} \\ x_{2j} \\ \vdots \\ x_{kj} \end{bmatrix}, \qquad j = 1, ..., n,$$

and we can combine the n vectors into a rectangular array called a *matrix*,

$$\mathbf{x}^{k \times n} = \begin{bmatrix} x_{11} & x_{12} & \cdots & x_{1n} \\ x_{21} & x_{22} & \cdots & x_{2n} \\ \vdots & \vdots & \vdots & \vdots \\ x_{k1} & x_{k2} & \cdots & x_{kn} \end{bmatrix} = (x_{ij}). \tag{I.6.3}$$

The superscript indicates that there are k rows and n columns. Each x_{ij} is an *element* of the matrix. The second equality indicates that a typical element of the ith row and jth column is given by x_{ij}. Each element x_{ij} of the matrix is a realization of a univariate random variable X_{ij}, $i = 1, ..., k$, $j = 1, ..., n$. These $k \times n$ random variables may be represented in a *random matrix* defined by

$$\mathbf{X}^{k \times n} = \begin{bmatrix} X_{11} & X_{12} & \cdots & X_{1n} \\ X_{21} & X_{22} & \cdots & X_{2n} \\ \vdots & \vdots & \vdots & \vdots \\ X_{k1} & X_{k2} & \cdots & X_{kn} \end{bmatrix}. \tag{I.6.4}$$

I.6.2 The Mean Vector and Covariance Matrix of a Random Vector

Let $\mathbf{X}$ be a random vector with components $X_1, ..., X_k$. Then the components have a joint cdf denoted by

$$F(\mathbf{x}) = F(x_1, ..., x_k) = \Pr(X_1 \leqslant x_1, ..., X_k \leqslant x_k). \tag{I.6.5}$$

The moments of each component X_i may be obtained from the marginal distribution of X_i, and, in particular, we may obtain the expected value $\mu_i = E(X_i)$ of X_i, $i = 1, ..., k$. These k expected values may be represented by the *mean vector* $\boldsymbol{\mu}$ defined by

$$\boldsymbol{\mu} = E(\mathbf{X}) = \begin{bmatrix} \mu_1 \\ \mu_2 \\ \vdots \\ \mu_k \end{bmatrix}. \tag{I.6.6}$$

Similarly, the variances σ_i^2 of X_i may also be obtained from the marginal distribution of X_i, $i = 1, ..., k$. However, a new measure of variability, called the *covariance* σ_{ij} of X_i and X_j, may be calculated from the joint distribution of X_i and X_j. This quantity is defined by

$$\sigma_{ij} = \text{cov}(X_i, X_j) = E[(X_i - \mu_i)(X_j - \mu_j)] \tag{I.6.7}$$

for $i, j = 1, ..., k$. Note that $\sigma_{ij} = \sigma_{ji}$ and $\sigma_{ii} = \sigma_i^2$. If $\sigma_{ij} = 0$, then X_i and X_j are said to be *uncorrelated*; if $\sigma_{ij} > 0$, then, on the average, X_i and X_j increase (or decrease) together; if $\sigma_{ij} < 0$, then, on the average, X_i increases (decreases) when X_j decreases (increases).

The variances and covariances then constitute the elements of the *covariance matrix* $\boldsymbol{\Sigma}$ defined by

$$\Sigma^{k \times k} = \operatorname{cov}(\mathbf{X}) = \begin{bmatrix} \sigma_1^{\,2} & \sigma_{12} & \cdots & \sigma_{1k} \\ \sigma_{21} & \sigma_2^{\,2} & \cdots & \sigma_{2k} \\ \vdots & \vdots & \vdots & \vdots \\ \sigma_{k1} & \sigma_{k2} & \cdots & \sigma_k^{\,2} \end{bmatrix} \tag{I.6.8}$$

The covariance matrix is the generalization of the variance of a univariate random variable.

I.6.3 The Multivariate Normal Distribution

As discussed in Section I.1.6, if all of the components X_i of $\mathbf{X}$ are continuous random variables, then we may specify the (multivariate) distribution of $X_1, \ldots, X_k$ by the joint density $f(x_1, \ldots, x_k)$. The most commonly used multivariate distribution in statistical applications is the *multivariate (or k-variate) normal distribution*. It is characterized by its mean vector $\boldsymbol{\mu}$ and its covariance matrix $\boldsymbol{\Sigma}$, and its joint density is given in Comment I.6.1.1. If $\mathbf{X}$ has a multivariate normal distribution with mean vector $\boldsymbol{\mu}$ and covariance matrix $\boldsymbol{\Sigma}$, we say that $\mathbf{X}$ is $N(\boldsymbol{\mu}, \boldsymbol{\Sigma})$.

Some of the important properties of this distribution are:

(1) The marginal distribution of X_i is $N(\mu_i, \sigma_i^{\,2})$, where μ_i is the ith component of $\boldsymbol{\mu}$, and $\sigma_i^{\,2}$ is the element in the ith row and ith column (that is, the *i-th diagonal element*) of $\boldsymbol{\Sigma}$, $i = 1, \ldots, k$.

(2) More generally, we can define the *marginal distribution of a subset of l random variables* from $X_1, \ldots, X_k$, $1 < l < k$. Renumber the random variables so that this subset consists of the first l variables. Rearrange the mean vector and covariance matrix accordingly. Then if we define $\mathbf{X}_1$ by

$$\mathbf{X}_1^{l \times 1} = \begin{bmatrix} X_1 \\ \vdots \\ X_l \end{bmatrix}, \tag{I.6.9}$$

the marginal distribution of $\mathbf{X}_1$ is multivariate normal with mean

$$\boldsymbol{\mu}_1^{l \times 1} = \begin{bmatrix} \mu_1 \\ \vdots \\ \mu_l \end{bmatrix} \tag{I.6.10}$$

and covariance matrix

$$\boldsymbol{\Sigma}_{11}^{l \times l} = \begin{bmatrix} \sigma_1^{\,2} & \sigma_{12} & \cdots & \sigma_{1l} \\ \sigma_{21} & \sigma_2^{\,2} & \cdots & \sigma_{2l} \\ \vdots & \vdots & \vdots & \vdots \\ \sigma_{l1} & \sigma_{l2} & \cdots & \sigma_l^{\,2} \end{bmatrix}. \tag{I.6.11}$$

(3) The generalization of Comment I.2.2.5 is that for constants $a, b_1, \ldots, b_k$, the distribution of $Y = a + \sum_{i=1}^{k} b_i X_i$ is normal with mean $a + \sum_{i=1}^{k} b_i \mu_i$ and variance $\sum_{i=1}^{k} b_i^2 \sigma_i^2 + \sum_{i \neq j} \sum_{j=1}^{k} b_i b_j \sigma_{ij}$.

(4) If $\sigma_{ij} = 0$ for all $i \neq j$, that is, Σ is a *diagonal matrix*, then $X_1, \ldots, X_k$ are mutually independent. In particular, if X_i and $X_j (i \neq j)$ are uncorrelated, then they are also independent. This property is not necessarily possessed by other distributions.

(5) Let

$$\mathbf{X}_1 = \begin{bmatrix} X_1 \\ \vdots \\ X_l \end{bmatrix} \quad \text{and} \quad \mathbf{X}_2 = \begin{bmatrix} X_{l+1} \\ \vdots \\ X_k \end{bmatrix}.$$

Then the conditional distribution of $\mathbf{X}_1$ given

$$\mathbf{X}_2 = \mathbf{x}_2 = \begin{bmatrix} x_{l+1} \\ \vdots \\ x_k \end{bmatrix}$$

is also multivariate normal. The components of the mean vector of this conditional distribution are linear combinations of the components of $\mathbf{x}_2$, while the covariance matrix of this conditional distribution does not depend on $\mathbf{x}_2$ (see Comment I.6.1.4). This distribution plays an important role in linear regression (Chapter 3).

COMMENTS I.6.1   *1. A formal definition of the multivariate normal distribution is as follows. Let $Z_1, \ldots, Z_k$ be mutually independent $N(0, 1)$ random variables. Then $\mathbf{Z}^{k \times 1} = (Z_1, \ldots, Z_k)'^\dagger$ has the *standard spherical normal distribution* with density

$$f(\mathbf{z}) = (2\pi)^{-k/2} e^{-(1/2)\mathbf{z}'\mathbf{z}},$$

where $\mathbf{z} = (z_1, \ldots, z_k)'$. We write the distribution of $\mathbf{Z}$ as $N(\mathbf{0}, \mathbf{I})$, where $\mathbf{0}$ is the zero vector and $\mathbf{I}$ is the identity matrix. If $\mathbf{A}^{k \times k}$ is an arbitrary nonsingular matrix of constants, and $\boldsymbol{\mu}^{k \times 1}$ is a constant vector, then $\mathbf{X}^{k \times 1} = \mathbf{A}\mathbf{Z} + \boldsymbol{\mu}$ has a *multivariate (or k-variate) nondegenerate normal distribution*. Its density is

$$f(\mathbf{x}) = (2\pi)^{-k/2} |\Sigma|^{-1/2} \exp[-\tfrac{1}{2}(\mathbf{x} - \boldsymbol{\mu})' \Sigma^{-1} (\mathbf{x} - \boldsymbol{\mu})],$$

where $\mathbf{x} = (x_1, \ldots, x_k)'$, $\Sigma = \mathbf{A}\mathbf{A}'$, $|\Sigma|$ is the determinant of Σ, and Σ^{-1} is the inverse of Σ. Its mean vector is $\boldsymbol{\mu}$ and its covariance matrix is Σ. We say that $\mathbf{X}$ is $N(\boldsymbol{\mu}, \Sigma)$.

2. If $\mathbf{X}^{k \times 1}$ is $N(\boldsymbol{\mu}, \Sigma)$ and $\mathbf{B}^{m \times k}$ is a matrix of rank m, then $\mathbf{Y}^{m \times 1} = \mathbf{B}\mathbf{X}$ has an m-variate normal distribution.

† The prime denotes the transpose.

3. The region of the k-dimensional Euclidean space defined by $f(\mathbf{x}) = c$, c a constant, is an ellipsoid called the *ellipsoid of concentration*.

4. Let $\mathbf{X}^{k \times 1}$ be $N(\boldsymbol{\mu}, \boldsymbol{\Sigma})$ and let $\mathbf{X}_1 = (X_1, \ldots, X_l)'$ and $\mathbf{X}_2 = (X_{l+1}, \ldots, X_k)'$. Also, let

$$\boldsymbol{\mu}_1 = (\mu_1, \ldots, \mu_l)', \qquad \boldsymbol{\mu}_2 = (\mu_{l+1}, \ldots, \mu_k)',$$

$$\boldsymbol{\Sigma}_{11} = \begin{bmatrix} \sigma_1^2 & \cdots & \sigma_{1l} \\ \vdots & \vdots & \vdots \\ \sigma_{l1} & \cdots & \sigma_l^2 \end{bmatrix}, \qquad \boldsymbol{\Sigma}_{22} = \begin{bmatrix} \sigma_{l+1}^2 & \cdots & \sigma_{l+1,k} \\ \vdots & \vdots & \vdots \\ \sigma_{k,l+1} & \cdots & \sigma_k^2 \end{bmatrix}, \qquad \text{and}$$

$$\boldsymbol{\Sigma}_{12} = \boldsymbol{\Sigma}_{21}' = \begin{bmatrix} \sigma_{1,l+1} & \cdots & \sigma_{1k} \\ \vdots & \vdots & \vdots \\ \sigma_{l,l+1} & \cdots & \sigma_{lk} \end{bmatrix}.$$

Then $\mathbf{X}_1$ is $N(\boldsymbol{\mu}_1, \boldsymbol{\Sigma}_{11})$ and $\mathbf{X}_2$ is $N(\boldsymbol{\mu}_2, \boldsymbol{\Sigma}_{22})$. The conditional distribution of $\mathbf{X}_1$, given $\mathbf{X}_2 = \mathbf{x}_2 = (x_{l+1}, \ldots, x_k)'$ is

$$N\big(\boldsymbol{\mu}_1 + \boldsymbol{\Sigma}_{12} \boldsymbol{\Sigma}_{22}^{-1}(\mathbf{x}_2 - \boldsymbol{\mu}_2), \, \boldsymbol{\Sigma}_{11} - \boldsymbol{\Sigma}_{12} \boldsymbol{\Sigma}_{22}^{-1} \boldsymbol{\Sigma}_{21}\big).*$$

EXAMPLE I.1.1 (*continued*)   As in Example I.1.1a, Dr. Gesund measures $X_1(w) =$ change in systolic pressure (mm Hg) and $X_2(w) =$ change in diastolic pressure (mm Hg) for each patient w. These functions define the random variables X_1 and X_2, respectively, which can be expressed by the 2×1 random vector

$$\mathbf{X} = \begin{bmatrix} X_1 \\ X_2 \end{bmatrix}.$$

He assumes that $\mathbf{X}$ has a multivariate (or 2-variate) normal distribution called the *bivariate distribution* with mean vector $\boldsymbol{\mu}$ and covariance matrix $\boldsymbol{\Sigma}$, where

$$\boldsymbol{\mu} = \begin{bmatrix} \mu_1 \\ \mu_2 \end{bmatrix} \qquad \text{and} \qquad \boldsymbol{\Sigma} = \begin{bmatrix} \sigma_1^2 & \sigma_{12} \\ \sigma_{21} & \sigma_2^2 \end{bmatrix}.$$

The density of this distribution is

$$f(x_1, x_2) = \frac{1}{2\pi \sigma_1 \sigma_2 \sqrt{1 - \rho^2}}$$

$$\times \exp\left\{ \frac{-1}{2(1 - \rho^2)} \left[\left(\frac{x_1 - \mu_1}{\sigma_1} \right)^2 - 2\rho \frac{(x_1 - \mu_1)(x_2 - \mu_2)}{\sigma_1 \sigma_2} + \left(\frac{x_2 - \mu_2}{\sigma_2} \right)^2 \right] \right\},$$

where $\rho = \sigma_{12}/\sigma_1 \sigma_2$ is the *population correlation coefficient* (detailed discussion of ρ is in Section 3.1). This density has a true bell-shaped surface in the three-dimensional space with coordinate axes x_1, x_2, and $f(x_1, x_2)$. Probabilities are volumes above two-dimensional regions under this surface.

The marginal distribution of X_i is $N(\mu_i, \sigma_i^2)$, $i = 1, 2$, and if $\rho = 0$ (that is, X_1, X_2 are uncorrelated), then $f(x_1, x_2) = f_1(x_1)f_2(x_2)$, where $f_i(x_i)$ is the density of X_i, $i = 1$ or 2. This verifies that two uncorrelated normal random variables are also independent.

Finally, the conditional distribution of X_1 given $X_2 = x_2$ is normal with mean $\mu_1 + (\sigma_{12}/\sigma_2^2)(x_2 - \mu_2)$ and variance $\sigma_1^2 - \sigma_{12}^2/\sigma_2^2$. This conditional distribution gives rise to simple linear regression (Section 3.1).

APPENDIX II
STATISTICAL TABLES

TABLE 1. *Binomial Probabilities (Section I.2.1)*[a]

n	i	.01	.10	.20	.25	.30	.33	.40	.50
2	0	.9801	.8100	.6400	.5625	.4900	.4444	.3600	.2500
	1	.0198	.1800	.3200	.3750	.4200	.4444	.4800	.5000
	2	.0001	.0100	.0400	.0625	.0900	.1111	.1600	.2500
3	0	.9703	.7290	.5120	.4219	.3430	.2963	.2160	.1250
	1	.0294	.2430	.3840	.4219	.4410	.4444	.4320	.3750
	2	.0003	.0270	.0960	.1406	.1890	.2222	.2880	.3750
	3	.0000	.0010	.0080	.0156	.0270	.0370	.0640	.1250
4	0	.9606	.6561	.4096	.3164	.2401	.1975	.1296	.0625
	1	.0388	.2916	.4096	.4219	.4116	.3951	.3456	.2500
	2	.0006	.0486	.1536	.2109	.2646	.2963	.3456	.3750
	3	.0000	.0036	.0256	.0469	.0756	.0988	.1536	.2500
	4	.0000	.0001	.0016	.0039	.0081	.0123	.0256	.0625
5	0	.9510	.5905	.3277	.2373	.1681	.1317	.0778	.0312
	1	.0480	.3280	.4096	.3955	.3602	.3292	.2592	.1562
	2	.0010	.0729	.2048	.2637	.3087	.3292	.3456	.3125
	3	.0000	.0081	.0512	.0879	.1323	.1646	.2304	.3125
	4	.0000	.0004	.0064	.0146	.0284	.0412	.0768	.1562
	5	.0000	.0000	.0003	.0010	.0024	.0041	.0102	.0312

APPENDIX II

TABLE 1 (*continued*)[a]

n	i	.01	.10	.20	.25	.30	.33	.40	.50
6	0	.9415	.5314	.2621	.1780	.1176	.0878	.0467	.0156
	1	.0571	.3543	.3932	.3560	.3025	.2634	.1866	.0938
	2	.0014	.0984	.2458	.2966	.3241	.3292	.3110	.2344
	3	.0000	.0146	.0819	.1318	.1852	.2195	.2765	.3125
	4	.0000	.0012	.0154	.0330	.0595	.0823	.1382	.2344
	5	.0000	.0001	.0015	.0044	.0102	.0165	.0369	.0938
	6	.0000	.0000	.0001	.0002	.0007	.0014	.0041	.0156
7	0	.9321	.4783	.2097	.1335	.0824	.0585	.0280	.0078
	1	.0659	.3720	.3670	.3115	.2471	.2048	.1306	.0547
	2	.0020	.1240	.2753	.3115	.3177	.3073	.2613	.1641
	3	.0000	.0230	.1147	.1730	.2269	.2561	.2903	.2734
	4	.0000	.0026	.0287	.0577	.0972	.1280	.1935	.2734
	5	.0000	.0002	.0043	.0115	.0250	.0384	.0774	.1641
	6	.0000	.0000	.0004	.0013	.0036	.0064	.0172	.0547
	7	.0000	.0000	.0000	.0001	.0002	.0005	.0016	.0078
8	0	.9227	.4305	.1678	.1001	.0576	.0390	.0168	.0039
	1	.0746	.3826	.3355	.2670	.1977	.1561	.0896	.0312
	2	.0026	.1488	.2936	.3115	.2965	.2731	.2090	.1094
	3	.0001	.0331	.1468	.2076	.2541	.2731	.2787	.2188
	4	.0000	.0046	.0459	.0865	.1361	.1707	.2322	.2734
	5	.0000	.0004	.0092	.0231	.0467	.0683	.1239	.2188
	6	.0000	.0000	.0011	.0038	.0100	.0171	.0413	.1094
	7	.0000	.0000	.0001	.0004	.0012	.0024	.0079	.0312
	8	.0000	.0000	.0000	.0000	.0001	.0002	.0007	.0039
9	0	.9135	.3874	.1342	.0751	.0404	.0260	.0101	.0020
	1	.0830	.3874	.3020	.2253	.1556	.1171	.0605	.0176
	2	.0034	.1722	.3020	.3003	.2668	.2341	.1612	.0703
	3	.0001	.0446	.1762	.2336	.2668	.2731	.2508	.1641
	4	.0000	.0074	.0661	.1168	.1715	.2048	.2508	.2461
	5	.0000	.0008	.0165	.0389	.0735	.1024	.1672	.2461
	6	.0000	.0001	.0028	.0087	.0210	.0341	.0743	.1641
	7	.0000	.0000	.0003	.0012	.0039	.0073	.0212	.0703
	8	.0000	.0000	.0000	.0001	.0004	.0009	.0035	.0176
	9	.0000	.0000	.0000	.0000	.0000	.0001	.0003	.0020

TABLE 1 (*continued*)[a]

n	i	p .01	.10	.20	.25	.30	.33	.40	.50
10	0	.9044	.3487	.1074	.0563	.0282	.0173	.0060	.0010
	1	.0914	.3874	.2684	.1877	.1211	.0867	.0403	.0098
	2	.0042	.1937	.3020	.2816	.2335	.1951	.1209	.0439
	3	.0001	.0574	.2013	.2503	.2668	.2601	.2150	.1172
	4	.0000	.0112	.0881	.1460	.2001	.2276	.2508	.2051
	5	.0000	.0015	.0264	.0584	.1029	.1366	.2007	.2461
	6	.0000	.0001	.0055	.0162	.0368	.0569	.1115	.2051
	7	.0000	.0000	.0008	.0031	.0090	.0163	.0425	.1172
	8	.0000	.0000	.0001	.0004	.0014	.0030	.0106	.0439
	9	.0000	.0000	.0000	.0000	.0001	.0003	.0016	.0098
	10	.0000	.0000	.0000	.0000	.0000	.0000	.0001	.0010

[a] Since $b_n(i, p) = b_n(n - i, 1 - p)$, only $p \leqslant .5$ are given.

TABLE 2. *Cumulative* $N(0, 1)$ *(Section I.2.5)*[a]

z	·00	·01	·02	·03	·04	·05	·06	·07	·08	·09
− ·0	·5000	·4960	·4920	·4880	·4840	·4801	·4761	·4721	·4681	·4641
− ·1	·4602	·4562	·4522	·4483	·4443	·4404	·4364	·4325	·4286	·4247
− ·2	·4207	·4168	·4129	·4090	·4052	·4013	·3974	·3936	·3897	·3859
− ·3	·3821	·3783	·3745	·3707	·3669	·3632	·3594	·3557	·3520	·3483
− ·4	·3446	·3409	·3372	·3336	·3300	·3264	·3228	·3192	·3156	·3121
− ·5	·3085	·3050	·3015	·2981	·2946	·2912	·2877	·2843	·2810	·2776
− ·6	·2743	·2709	·2676	·2643	·2611	·2578	·2546	·2514	·2483	·2451
− ·7	·2420	·2389	·2358	·2327	·2297	·2266	·2236	·2206	·2177	·2148
− ·8	·2119	·2090	·2061	·2033	·2005	·1977	·1949	·1922	·1894	·1867
− ·9	·1841	·1814	·1788	·1762	·1736	·1711	·1685	·1660	·1635	·1611
−1·0	·1587	·1562	·1539	·1515	·1492	·1469	·1446	·1423	·1401	·1379
−1·1	·1357	·1335	·1314	·1292	·1271	·1251	·1230	·1210	·1190	·1170
−1·2	·1151	·1131	·1112	·1093	·1075	·1056	·1038	·1020	·1003	·09853
−1·3	·09680	·09510	·09342	·09176	·09012	·08851	·08691	·08534	·08379	·08226
−1·4	·08076	·07927	·07780	·07636	·07493	·07353	·07215	·07078	·06944	·06811
−1·5	·06681	·06552	·06426	·06301	·06178	·06057	·05938	·05821	·05705	·05592
−1·6	·05480	·05370	·05262	·05155	·05050	·04947	·04846	·04746	·04648	·04551
−1·7	·04457	·04363	·04272	·04182	·04093	·04006	·03920	·03836	·03754	·03673
−1·8	·03593	·03515	·03438	·03362	·03288	·03216	·03144	·03074	·03005	·02938
−1·9	·02872	·02807	·02743	·02680	·02619	·02559	·02500	·02442	·02385	·02330
−2·0	·02275	·02222	·02169	·02118	·02068	·02018	·01970	·01923	·01876	·01831
−2·1	·01786	·01743	·01700	·01659	·01618	·01578	·01539	·01500	·01463	·01426
−2·2	·01390	·01355	·01321	·01287	·01255	·01222	·01191	·01160	·01130	·01101
−2·3	·01072	·01044	·01017	·0^{2}9903	·0^{2}9642	·0^{2}9387	·0^{2}9137	·0^{2}8894	·0^{2}8656	·0^{2}8424
−2·4	·0^{2}8198	·0^{2}7976	·0^{2}7760	·0^{2}7549	·0^{2}7344	·0^{2}7143	·0^{2}6947	·0^{2}6756	·0^{2}6569	·0^{2}6387
−2·5	·0^{2}6210	·0^{2}6037	·0^{2}5868	·0^{2}5703	·0^{2}5543	·0^{2}5386	·0^{2}5234	·0^{2}5085	·0^{2}4940	·0^{2}4799
−2·6	·0^{2}4661	·0^{2}4527	·0^{2}4396	·0^{2}4269	·0^{2}4145	·0^{2}4025	·0^{2}3907	·0^{2}3793	·0^{2}3681	·0^{2}3573
−2·7	·0^{2}3467	·0^{2}3364	·0^{2}3264	·0^{2}3167	·0^{2}3072	·0^{2}2980	·0^{2}2890	·0^{2}2803	·0^{2}2718	·0^{2}2635
−2·8	·0^{2}2555	·0^{2}2477	·0^{2}2401	·0^{2}2327	·0^{2}2256	·0^{2}2186	·0^{2}2118	·0^{2}2052	·0^{2}1988	·0^{2}1926
−2·9	·0^{2}1866	·0^{2}1807	·0^{2}1750	·0^{2}1695	·0^{2}1641	·0^{2}1589	·0^{2}1538	·0^{2}1489	·0^{2}1441	·0^{2}1395
−3·0	·0^{2}1350	·0^{2}1306	·0^{2}1264	·0^{2}1223	·0^{2}1183	·0^{2}1144	·0^{2}1107	·0^{2}1070	·0^{2}1035	·0^{2}1001
−3·1	·0^{3}9676	·0^{3}9354	·0^{3}9043	·0^{3}8740	·0^{3}8447	·0^{3}8164	·0^{3}7888	·0^{3}7622	·0^{3}7364	·0^{3}7114
−3·2	·0^{3}6871	·0^{3}6637	·0^{3}6410	·0^{3}6190	·0^{3}5976	·0^{3}5770	·0^{3}5571	·0^{3}5377	·0^{3}5190	·0^{3}5009
−3·3	·0^{3}4834	·0^{3}4665	·0^{3}4501	·0^{3}4342	·0^{3}4189	·0^{3}4041	·0^{3}3897	·0^{3}3758	·0^{3}3624	·0^{3}3495
−3·4	·0^{3}3369	·0^{3}3248	·0^{3}3131	·0^{3}3018	·0^{3}2909	·0^{3}2803	·0^{3}2701	·0^{3}2602	·0^{3}2507	·0^{3}2415
−3·5	·0^{3}2326	·0^{3}2241	·0^{3}2158	·0^{3}2078	·0^{3}2001	·0^{3}1926	·0^{3}1854	·0^{3}1785	·0^{3}1718	·0^{3}1653
−3·6	·0^{3}1591	·0^{3}1531	·0^{3}1473	·0^{3}1417	·0^{3}1363	·0^{3}1311	·0^{3}1261	·0^{3}1213	·0^{3}1166	·0^{3}1121
−3·7	·0^{3}1078	·0^{3}1036	·0^{4}9961	·0^{4}9574	·0^{4}9201	·0^{4}8842	·0^{4}8496	·0^{4}8162	·0^{4}7841	·0^{4}7532
−3·8	·0^{4}7235	·0^{4}6948	·0^{4}6673	·0^{4}6407	·0^{4}6152	·0^{4}5906	·0^{4}5669	·0^{4}5442	·0^{4}5223	·0^{4}5012
−3·9	·0^{4}4810	·0^{4}4615	·0^{4}4427	·0^{4}4247	·0^{4}4074	·0^{4}3908	·0^{4}3747	·0^{4}3594	·0^{4}3446	·0^{4}3304

TABLE 2 *(continued)*[a]

z	·00	·01	·02	·03	·04	·05	·06	·07	·08	·09
·0	·5000	·5040	·5080	·5120	·5160	·5199	·5239	·5279	·5319	·5359
·1	·5398	·5438	·5478	·5517	·5557	·5596	·5636	·5675	·5714	·5753
·2	·5793	·5832	·5871	·5910	·5948	·5987	·6026	·6064	·6103	·6141
·3	·6179	·6217	·6255	·6293	·6331	·6368	·6406	·6443	·6480	·6517
·4	·6554	·6591	·6628	·6664	·6700	·6736	·6772	·6808	·6844	·6879
·5	·6915	·6950	·6985	·7019	·7054	·7088	·7123	·7157	·7190	·7224
·6	·7257	·7291	·7324	·7357	·7389	·7422	·7454	·7486	·7517	·7549
·7	·7580	·7611	·7642	·7673	·7703	·7734	·7764	·7794	·7823	·7852
·8	·7881	·7910	·7939	·7967	·7995	·8023	·8051	·8078	·8106	·8133
·9	·8159	·8186	·8212	·8238	·8264	·8289	·8315	·8340	·8365	·8389
1·0	·8413	·8438	·8461	·8485	·8508	·8531	·8554	·8577	·8599	·8621
1·1	·8643	·8665	·8686	·8708	·8729	·8749	·8770	·8790	·8810	·8830
1·2	·8849	·8869	·8888	·8907	·8925	·8944	·8962	·8980	·8997	·90147
1·3	·90320	·90490	·90658	·90824	·90988	·91149	·91309	·91466	·91621	·91774
1·4	·91924	·92073	·92220	·92364	·92507	·92647	·92785	·92922	·93056	·93189
1·5	·93319	·93448	·93574	·93699	·93822	·93943	·94062	·94179	·94295	·94408
1·6	·94520	·94630	·94738	·94845	·94950	·95053	·95154	·95254	·95352	·95449
1·7	·95543	·95637	·95728	·95818	·95907	·95994	·96080	·96164	·96246	·96327
1·8	·96407	·96485	·96562	·96638	·96712	·96784	·96856	·96926	·96995	·97062
1·9	·97128	·97193	·97257	·97320	·97381	·97441	·97500	·97558	·97615	·97670
2·0	·97725	·97778	·97831	·97882	·97932	·97982	·98030	·98077	·98124	·98169
2·1	·98214	·98257	·98300	·98341	·98382	·98422	·98461	·98500	·98537	·98574
2·2	·98610	·98645	·98679	·98713	·98745	·98778	·98809	·98840	·98870	·98899
2·3	·98928	·98956	·98983	$·9^2 0097$	$·9^2 0358$	$·9^2 0613$	$·9^2 0863$	$·9^2 1106$	$·9^2 1344$	$·9^2 1576$
2·4	$·9^2 1802$	$·9^2 2024$	$·9^2 2240$	$·9^2 2451$	$·9^2 2656$	$·9^2 2857$	$·9^2 3053$	$·9^2 3244$	$·9^2 3431$	$·9^2 3613$
2·5	$·9^2 3790$	$·9^2 3963$	$·9^2 4132$	$·9^2 4297$	$·9^2 4457$	$·9^2 4614$	$·9^2 4766$	$·9^2 4915$	$·9^2 5060$	$·9^2 5201$
2·6	$·9^2 5339$	$·9^2 5473$	$·9^2 5604$	$·9^2 5731$	$·9^2 5855$	$·9^2 5975$	$·9^2 6093$	$·9^2 6207$	$·9^2 6319$	$·9^2 6427$
2·7	$·9^2 6533$	$·9^2 6636$	$·9^2 6736$	$·9^2 6833$	$·9^2 6928$	$·9^2 7020$	$·9^2 7110$	$·9^2 7197$	$·9^2 7282$	$·9^2 7365$
2·8	$·9^2 7445$	$·9^2 7523$	$·9^2 7599$	$·9^2 7673$	$·9^2 7744$	$·9^2 7814$	$·9^2 7882$	$·9^2 7948$	$·9^2 8012$	$·9^2 8074$
2·9	$·9^2 8134$	$·9^2 8193$	$·9^2 8250$	$·9^2 8305$	$·9^2 8359$	$·9^2 8411$	$·9^2 8462$	$·9^2 8511$	$·9^2 8559$	$·9^2 8605$
3·0	$·9^2 8650$	$·9^2 8694$	$·9^2 8736$	$·9^2 8777$	$·9^2 8817$	$·9^2 8856$	$·9^2 8893$	$·9^2 8930$	$·9^2 8965$	$·9^2 8999$
3·1	$·9^3 0324$	$·9^3 0646$	$·9^3 0957$	$·9^3 1260$	$·9^3 1553$	$·9^3 1836$	$·9^3 2112$	$·9^3 2378$	$·9^3 2636$	$·9^3 2886$
3·2	$·9^3 3129$	$·9^3 3363$	$·9^3 3590$	$·9^3 3810$	$·9^3 4024$	$·9^3 4230$	$·9^3 4429$	$·9^3 4623$	$·9^3 4810$	$·9^3 4991$
3·3	$·9^3 5166$	$·9^3 5335$	$·9^3 5499$	$·9^3 5658$	$·9^3 5811$	$·9^3 5959$	$·9^3 6103$	$·9^3 6242$	$·9^3 6376$	$·9^3 6505$
3·4	$·9^3 6631$	$·9^3 6752$	$·9^3 6869$	$·9^3 6982$	$·9^3 7091$	$·9^3 7197$	$·9^3 7299$	$·9^3 7398$	$·9^3 7493$	$·9^3 7585$
3·5	$·9^3 7674$	$·9^3 7759$	$·9^3 7842$	$·9^3 7922$	$·9^3 7999$	$·9^3 8074$	$·9^3 8146$	$·9^3 8215$	$·9^3 8282$	$·9^3 8347$
3·6	$·9^3 8409$	$·9^3 8469$	$·9^3 8527$	$·9^3 8583$	$·9^3 8637$	$·9^3 8689$	$·9^3 8739$	$·9^3 8787$	$·9^3 8834$	$·9^3 8879$
3·7	$·9^3 8922$	$·9^3 8964$	$·9^4 0039$	$·9^4 0426$	$·9^4 0799$	$·9^4 1158$	$·9^4 1504$	$·9^4 1838$	$·9^4 2159$	$·9^4 2468$
3·8	$·9^4 2765$	$·9^4 3052$	$·9^4 3327$	$·9^4 3593$	$·9^4 3848$	$·9^4 4094$	$·9^4 4331$	$·9^4 4558$	$·9^4 4777$	$·9^4 4988$
3·9	$·9^4 5190$	$·9^4 5385$	$·9^4 5573$	$·9^4 5753$	$·9^4 5926$	$·9^4 6092$	$·9^4 6253$	$·9^4 6406$	$·9^4 6554$	$·9^4 6696$

[a] Abridged from Table II of A. Hald, "Statistical Tables and Formulas," 1952, Wiley, New York.

APPENDIX II

TABLE 3. *Percentiles of the χ^2 Distribution (Section I.2.6)*[a]

ν \ q	0.5	1	2.5	5	10	20	30	40
1	0.0⁴393	0.0³157	0.0³982	0.0²393	0.0158	0.0642	0.148	0.275
2	0.0100	0.0201	0.0506	0.103	0.211	0.446	0.713	1.02
3	0.0717	0.115	0.216	0.352	0.584	1.00	1.42	1.87
4	0.207	0.297	0.484	0.711	1.06	1.65	2.19	2.75
5	0.412	0.554	0.831	1.15	1.61	2.34	3.00	3.66
6	0.676	0.872	1.24	1.64	2.20	3.07	3.83	4.57
7	0.989	1.24	1.69	2.17	2.83	3.82	4.67	5.49
8	1.34	1.65	2.18	2.73	3.49	4.59	5.53	6.42
9	1.73	2.09	2.70	3.33	4.17	5.38	6.39	7.36
10	2.16	2.56	3.25	3.94	4.87	6.18	7.27	8.30
11	2.60	3.05	3.82	4.57	5.58	6.99	8.15	9.24
12	3.07	3.57	4.40	5.23	6.30	7.81	9.03	10.2
13	3.57	4.11	5.01	5.89	7.04	8.63	9.93	11.1
14	4.07	4.66	5.63	6.57	7.79	9.47	10.8	12.1
15	4.60	5.23	6.26	7.26	8.55	10.3	11.7	13.0
16	5.14	5.81	6.91	7.96	9.31	11.2	12.6	14.0
17	5.70	6.41	7.56	8.67	10.1	12.0	13.5	14.9
18	6.26	7.01	8.23	9.39	10.9	12.9	14.4	15.9
19	6.84	7.63	8.91	10.1	11.7	13.7	15.4	16.9
20	7.43	8.26	9.59	10.9	12.4	14.6	16.3	17.8
21	8.03	8.90	10.3	11.6	13.2	15.4	17.2	18.8
22	8.64	9.54	11.0	12.3	14.0	16.3	18.1	19.7
23	9.26	10.2	11.7	13.1	14.8	17.2	19.0	20.7
24	9.89	10.9	12.4	13.8	15.7	18.1	19.9	21.7
25	10.5	11.5	13.1	14.6	16.5	18.9	20.9	22.6
26	11.2	12.2	13.8	15.4	17.3	19.8	21.8	23.6
27	11.8	12.9	14.6	16.2	18.1	20.7	22.7	24.5
28	12.5	13.6	15.3	16.9	18.9	21.6	23.6	25.5
29	13.1	14.3	16.0	17.7	19.8	22.5	24.6	26.5
30	13.8	15.0	16.8	18.5	20.6	23.4	25.5	27.4
35	17.2	18.5	20.6	22.5	24.8	27.8	30.2	32.3
40	20.7	22.2	24.4	26.5	29.1	32.3	34.9	37.1
45	24.3	25.9	28.4	30.6	33.4	36.9	39.6	42.0
50	28.0	29.7	32.4	34.8	37.7	41.4	44.3	46.9
75	47.2	49.5	52.9	56.1	59.8	64.5	68.1	71.3
100	67.3	70.1	74.2	77.9	82.4	87.9	92.1	95.8

TABLE 3 (*continued*)[a]

q ν	50	60	70	80	90	95	97.5	99	99.5	99.9
1	0.455	0.708	1.07	1.64	2.71	3.84	5.02	6.63	7.88	10.8
2	1.39	1.83	2.41	3.22	4.61	5.99	7.38	9.21	10.6	13.8
3	2.37	2.95	3.67	4.64	6.25	7.81	9.35	11.3	12.8	16.3
4	3.36	4.04	4.88	5.99	7.78	9.49	11.1	13.3	14.9	18.5
5	4.35	5.13	6.06	7.29	9.24	11.1	12.8	15.1	16.7	20.5
6	5.35	6.21	7.23	8.56	10.6	12.6	14.4	16.8	18.5	22.5
7	6.35	7.28	8.38	9.80	12.0	14.1	16.0	18.5	20.3	24.3
8	7.34	8.35	9.52	11.0	13.4	15.5	17.5	20.1	22.0	26.1
9	8.34	9.41	10.7	12.2	14.7	16.9	19.0	21.7	23.6	27.9
10	9.34	10.5	11.8	13.4	16.0	18.3	20.5	23.2	25.2	29.6
11	10.3	11.5	12.9	14.6	17.3	19.7	21.9	24.7	26.8	31.3
12	11.3	12.6	14.0	15.8	18.5	21.0	23.3	26.2	28.3	32.9
13	12.3	13.6	15.1	17.0	19.8	22.4	24.7	27.7	29.8	34.5
14	13.3	14.7	16.2	18.2	21.1	23.7	26.1	29.1	31.3	36.1
15	14.3	15.7	17.3	19.3	22.3	25.0	27.5	30.6	32.8	37.7
16	15.3	16.8	18.4	20.5	23.5	26.3	28.8	32.0	34.3	39.3
17	16.3	17.8	19.5	21.6	24.8	27.6	30.2	33.4	35.7	40.8
18	17.3	18.9	20.6	22.8	26.0	28.9	31.5	34.8	37.2	42.3
19	18.3	19.9	21.7	23.9	27.2	30.1	32.9	36.2	38.6	43.8
20	19.3	21.0	22.8	25.0	28.4	31.4	34.2	37.6	40.0	45.3
21	20.3	22.0	23.9	26.9	29.6	32.7	35.5	38.9	41.4	46.8
22	21.3	23.0	24.9	27.3	30.8	33.9	36.8	40.3	42.8	48.3
23	22.3	24.1	26.0	28.4	32.0	35.2	38.1	41.6	44.2	49.7
24	23.3	25.1	27.1	29.6	33.2	36.4	39.4	43.0	45.6	51.2
25	24.3	26.1	28.2	30.7	34.4	37.7	40.6	44.3	46.9	52.6
26	25.3	27.2	29.2	31.8	35.6	38.9	41.9	45.6	48.3	54.1
27	26.3	28.2	30.3	32.9	36.7	40.1	43.2	47.0	49.6	55.5
28	27.3	29.2	31.4	34.0	37.9	41.3	44.5	48.3	51.0	56.9
29	28.3	30.3	32.5	35.1	39.1	42.6	45.7	49.6	52.3	58.3
30	29.3	31.3	33.5	36.3	40.3	43.8	47.0	50.9	53.7	59.7
35	34.3	36.5	38.9	41.8	46.1	49.8	53.2	57.3	60.3	66.6
40	39.3	41.6	44.2	47.3	51.8	55.8	59.3	63.7	66.8	73.4
45	44.3	46.8	49.5	52.7	57.5	61.7	65.4	70.0	73.2	80.1
50	49.3	51.9	54.7	58.2	63.2	67.5	71.4	76.2	79.5	86.7
75	74.3	77.5	80.9	85.1	91.1	96.2	100.8	106.4	110.3	118.6
100	99.3	102.9	106.9	111.7	118.5	124.3	129.6	135.6	140.2	149.4

[a] Abridged from Table V of A. Hald, "Statistical Tables and Formulas," 1952, Wiley, New York.

TABLE 4. *Critical Values for the Kolmogorov–Smirnov Test of Goodness-of-Fit (Section 2.2.2)*[a]

Sample Size (n)	Significance Level				
	.20	.15	.10	.05	.01
1	.900	.925	.950	.975	.995
2	.684	.726	.776	.842	.929
3	.565	.597	.642	.708	.829
4	.494	.525	.564	.624	.734
5	.446	.474	.510	.563	.669
6	.410	.436	.470	.521	.618
7	.381	.405	.438	.486	.577
8	.358	.381	.411	.457	.543
9	.339	.360	.388	.432	.514
10	.322	.342	.368	.409	.486
11	.307	.326	.352	.391	.468
12	.295	.313	.338	.375	.450
13	.284	.302	.325	.361	.433
14	.274	.292	.314	.349	.418
15	.266	.283	.304	.338	.404
16	.258	.274	.295	.328	.391
17	.250	.266	.286	.318	.380
18.	.244	.259	.278	.309	.370
19	.237	.252	.272	.301	.361
20	.231	.246	.264	.294	.352
25	.21	.22	.24	.264	.32
30	.19	.20	.22	.242	.29
35	.18	.19	.21	.23	.27
40				.21	.25
50				.19	.23
60				.17	.21
70				.16	.19
80				.15	.18
90				.14	
100				.14	
Asymptotic Formula:	$\dfrac{1.07}{\sqrt{n}}$	$\dfrac{1.14}{\sqrt{n}}$	$\dfrac{1.22}{\sqrt{n}}$	$\dfrac{1.36}{\sqrt{n}}$	$\dfrac{1.63}{\sqrt{n}}$

[a] Reproduced from F. J. Massey, Jr. (1951), The Kolmogorov–Smirnov Test for Goodness-of-Fit, *JASA* **46**, 68–71, and Z. W. Birnbaum (1952), Numerical Tabulation of the Distribution of Kolmogorov's Statistic for Finite Sample Size, *JASA* **47**, 425–441, with the kind permission of the authors and the publisher.

TABLE 5. *Percentiles of the Student's t Distribution (Section I.2.7)*[a]

q / ν	60	75	90	95	97.5	99	99.5	99.95
1	.325	1.000	3.078	6.314	12.706	31.821	63.657	636.619
2	.289	.816	1.886	2.920	4.303	6.965	9.925	31.598
3	.277	.765	1.638	2.353	3.182	4.541	5.841	12.941
4	.271	.741	1.533	2.132	2.776	3.747	4.604	8.610
5	.267	.727	1.476	2.015	2.571	3.365	4.032	6.859
6	.265	.718	1.440	1.943	2.447	3.143	3.707	5.959
7	.263	.711	1.415	1.895	2.365	2.998	3.499	5.405
8	.262	.706	2.397	1.860	2.306	2.896	3.355	5.041
9	.261	.703	1.383	1.833	2.262	2.821	3.250	4.781
10	.260	.700	1.372	1.812	2.228	2.764	3.169	4.587
11	.260	.697	1.363	1.796	2.201	2.718	3.106	4.437
12	.259	.695	1.356	1.782	2.179	2.681	3.055	4.318
13	.259	.694	1.350	1.771	2.160	2.650	3.012	4.221
14	.258	.692	1.345	1.761	2.145	2.624	2.977	4.140
15	.258	.691	1.341	1.753	2.131	2.602	2.947	4.073
16	.258	.690	1.337	1.746	2.120	2.583	2.921	4.015
17	.257	.689	1.333	1.740	2.110	2.567	2.898	3.965
18	.257	.688	1.330	1.734	2.101	2.552	2.878	3.922
19	.257	.688	1.328	1.729	2.093	2.539	2.861	3.883
20	.257	.687	1.325	1.725	2.086	2.528	2.845	3.850
21	.257	.686	1.323	1.721	2.080	2.518	2.831	3.819
22	.256	.686	1.321	1.717	2.074	2.508	2.819	3.792
23	.256	.685	1.319	1.714	2.069	2.500	2.807	3.767
24	.256	.685	1.318	1.711	2.064	2.492	2.797	3.745
25	.256	.684	1.316	1.708	2.060	2.485	2.787	3.725
26	.256	.684	1.315	1.706	2.056	2.479	2.779	3.707
27	.256	.684	1.314	1.703	2.052	2.473	2.771	3.690
28	.256	.683	1.313	1.701	2.048	2.467	2.763	3.674
29	.256	.683	1.311	1.699	2.045	2.462	2.756	3.659
30	.256	.683	1.310	1.697	2.042	2.457	2.750	3.646
40	.255	.681	1.303	1.684	2.021	2.423	2.704	3.551
60	.254	.679	1.296	1.671	2.000	2.390	2.660	3.460
120	.254	.677	1.289	1.658	1.980	2.358	2.617	3.373
∞	.253	.674	1.282	1.645	1.960	2.326	2.576	3.291

[a] Table 5 is taken from Table III of R. A. Fisher and F. Yates (1963): "Statistical Tables for Biological, Agricultural and Medical Research," published by Oliver and Boyd, Edinburgh, and used by permission of the authors and publishers.

TABLE 6. *Percentiles of the F Distribution (Section I.2.8)*[a]

90th Percentile

ν_2 \ ν_1	1	2	3	4	5	6	7	8	9	10	12	15	20	24	30	40	60	120	∞
1	39.86	49.50	53.59	55.83	57.24	58.20	58.91	59.44	59.86	60.19	60.71	61.22	61.74	62.00	62.26	62.53	62.79	63.06	63.33
2	8.53	9.00	9.16	9.24	9.29	9.33	9.35	9.37	9.38	9.39	9.41	9.42	9.44	9.45	9.46	9.47	9.47	9.48	9.49
3	5.54	5.46	5.39	5.34	5.31	5.28	5.27	5.25	5.24	5.23	5.22	5.20	5.18	5.18	5.17	5.16	5.15	5.14	5.13
4	4.54	4.32	4.19	4.11	4.05	4.01	3.98	3.95	3.94	3.92	3.90	3.87	3.84	3.83	3.82	3.80	3.79	3.78	3.76
5	4.06	3.78	3.62	3.52	3.45	3.40	3.37	3.34	3.32	3.30	3.27	3.24	3.21	3.19	3.17	3.16	3.14	3.12	3.10
6	3.78	3.46	3.29	3.18	3.11	3.05	3.01	2.98	2.96	2.94	2.90	2.87	2.84	2.82	2.80	2.78	2.76	2.74	2.72
7	3.59	3.26	3.07	2.96	2.88	2.83	2.78	2.75	2.72	2.70	2.67	2.63	2.59	2.58	2.56	2.54	2.51	2.49	2.47
8	3.46	3.11	2.92	2.81	2.73	2.67	2.62	2.59	2.56	2.54	2.50	2.46	2.42	2.40	2.38	2.36	2.34	2.32	2.29
9	3.36	3.01	2.81	2.69	2.61	2.55	2.51	2.47	2.44	2.42	2.38	2.34	2.30	2.28	2.25	2.23	2.21	2.18	2.16
10	3.29	2.92	2.73	2.61	2.52	2.46	2.41	2.38	2.35	2.32	2.28	2.24	2.20	2.18	2.16	2.13	2.11	2.08	2.06
11	3.23	2.86	2.66	2.54	2.45	2.39	2.34	2.30	2.27	2.25	2.21	2.17	2.12	2.10	2.08	2.05	2.03	2.00	1.97
12	3.18	2.81	2.61	2.48	2.39	2.33	2.28	2.24	2.21	2.19	2.15	2.10	2.06	2.04	2.01	1.99	1.96	1.93	1.90
13	3.14	2.76	2.56	2.43	2.35	2.28	2.23	2.20	2.16	2.14	2.10	2.05	2.01	1.98	1.96	1.93	1.90	1.88	1.85
14	3.10	2.73	2.52	2.39	2.31	2.24	2.19	2.15	2.12	2.10	2.05	2.01	1.96	1.94	1.91	1.89	1.86	1.83	1.80
15	3.07	2.70	2.49	2.36	2.27	2.21	2.16	2.12	2.09	2.06	2.02	1.97	1.92	1.90	1.87	1.85	1.82	1.79	1.76
16	3.05	2.67	2.46	2.33	2.24	2.18	2.13	2.09	2.06	2.03	1.99	1.94	1.89	1.87	1.84	1.81	1.78	1.75	1.72
17	3.03	2.64	2.44	2.31	2.22	2.15	2.10	2.06	2.03	2.00	1.96	1.91	1.86	1.84	1.81	1.78	1.75	1.72	1.69
18	3.01	2.62	2.42	2.29	2.20	2.13	2.08	2.04	2.00	1.98	1.93	1.89	1.84	1.81	1.78	1.75	1.72	1.69	1.66
19	2.99	2.61	2.40	2.27	2.18	2.11	2.06	2.02	1.98	1.96	1.91	1.86	1.81	1.79	1.76	1.73	1.70	1.67	1.63
20	2.97	2.59	2.38	2.25	2.16	2.09	2.04	2.00	1.96	1.94	1.89	1.84	1.79	1.77	1.74	1.71	1.68	1.64	1.61
21	2.96	2.57	2.36	2.23	2.14	2.08	2.02	1.98	1.95	1.92	1.87	1.83	1.78	1.75	1.72	1.69	1.66	1.62	1.59
22	2.95	2.56	2.35	2.22	2.13	2.06	2.01	1.97	1.93	1.90	1.86	1.81	1.76	1.73	1.70	1.67	1.64	1.60	1.57
23	2.94	2.55	2.34	2.21	2.11	2.05	1.99	1.95	1.92	1.89	1.84	1.80	1.74	1.72	1.69	1.66	1.62	1.59	1.55
24	2.93	2.54	2.33	2.19	2.10	2.04	1.98	1.94	1.91	1.88	1.83	1.78	1.73	1.70	1.67	1.64	1.61	1.57	1.53
25	2.92	2.53	2.32	2.18	2.09	2.02	1.97	1.93	1.89	1.87	1.82	1.77	1.72	1.69	1.66	1.63	1.59	1.56	1.52
26	2.91	2.52	2.31	2.17	2.08	2.01	1.96	1.92	1.88	1.86	1.81	1.76	1.71	1.68	1.65	1.61	1.58	1.54	1.50
27	2.90	2.51	2.30	2.17	2.07	2.00	1.95	1.91	1.87	1.85	1.80	1.75	1.70	1.67	1.64	1.60	1.57	1.53	1.49
28	2.89	2.50	2.29	2.16	2.06	2.00	1.94	1.90	1.87	1.84	1.79	1.74	1.69	1.66	1.63	1.59	1.56	1.52	1.48
29	2.89	2.50	2.28	2.15	2.06	1.99	1.93	1.89	1.86	1.83	1.78	1.73	1.68	1.65	1.62	1.58	1.55	1.51	1.47
30	2.88	2.49	2.28	2.14	2.05	1.98	1.93	1.88	1.85	1.82	1.77	1.72	1.67	1.64	1.61	1.57	1.54	1.50	1.46
40	2.84	2.44	2.23	2.09	2.00	1.93	1.87	1.83	1.79	1.76	1.71	1.66	1.61	1.57	1.54	1.51	1.47	1.42	1.38
60	2.79	2.39	2.18	2.04	1.95	1.87	1.82	1.77	1.74	1.71	1.66	1.60	1.54	1.51	1.48	1.44	1.40	1.35	1.29
120	2.75	2.35	2.13	1.99	1.90	1.82	1.77	1.72	1.68	1.65	1.60	1.55	1.48	1.45	1.41	1.37	1.32	1.26	1.19
∞	2.71	2.30	2.08	1.94	1.85	1.77	1.72	1.67	1.63	1.60	1.55	1.49	1.42	1.38	1.34	1.30	1.24	1.17	1.00

95th Percentile

ν_1 \ ν_2	1	2	3	4	5	6	7	8	9	10	12	15	20	24	30	40	60	120	∞
1	161.4	199.5	215.7	224.6	230.2	234.0	236.8	238.9	240.5	241.9	243.9	245.9	248.0	249.1	250.1	251.1	252.2	253.3	254.3
2	18.51	19.00	19.16	19.25	19.30	19.33	19.35	19.37	19.38	19.40	19.41	19.43	19.45	19.45	19.46	19.47	19.48	19.49	19.50
3	10.13	9.55	9.28	9.12	9.01	8.94	8.89	8.85	8.81	8.79	8.74	8.70	8.66	8.64	8.62	8.59	8.57	8.55	8.53
4	7.71	6.94	6.59	6.39	6.26	6.16	6.09	6.04	6.00	5.96	5.91	5.86	5.80	5.77	5.75	5.72	5.69	5.66	5.63
5	6.61	5.79	5.41	5.19	5.05	4.95	4.88	4.82	4.77	4.74	4.68	4.62	4.50	4.53	4.50	4.46	4.43	4.40	4.36
6	5.99	5.14	4.76	4.53	4.39	4.28	4.21	4.15	4.10	4.06	4.00	3.94	3.87	3.84	3.81	3.77	3.74	3.70	3.67
7	5.59	4.74	4.35	4.12	3.97	3.87	3.79	3.73	3.68	3.64	3.57	3.51	3.44	3.41	3.38	3.34	3.30	3.27	3.23
8	5.32	4.46	4.07	3.84	3.69	3.58	3.50	3.44	3.39	3.35	3.28	3.22	3.15	3.12	3.08	3.04	3.01	2.97	2.93
9	5.12	4.26	3.86	3.63	3.48	3.37	3.29	3.23	3.18	3.14	3.07	3.01	2.94	2.90	2.86	2.83	2.79	2.75	2.71
10	4.96	4.10	3.71	3.48	3.33	3.22	3.14	3.07	3.02	2.98	2.91	2.85	2.77	2.74	2.70	2.66	2.62	2.58	2.54
11	4.84	3.98	3.59	3.36	3.20	3.09	3.01	2.95	2.90	2.85	2.79	2.72	2.65	2.61	2.57	2.53	2.49	2.45	2.40
12	4.75	3.89	3.49	3.26	3.11	3.00	2.91	2.85	2.80	2.75	2.69	2.62	2.54	2.51	2.47	2.43	2.38	2.34	2.30
13	4.67	3.81	3.41	3.18	3.03	2.92	2.83	2.77	2.71	2.67	2.60	2.53	2.46	2.42	2.38	2.34	2.30	2.25	2.21
14	4.60	3.74	3.34	3.11	2.96	2.85	2.76	2.70	2.65	2.60	2.53	2.46	2.39	2.35	2.31	2.27	2.22	2.18	2.13
15	4.54	3.68	3.29	3.06	2.90	2.79	2.71	2.64	2.59	2.54	2.48	2.40	2.33	2.29	2.25	2.20	2.16	2.11	2.07
16	4.49	3.63	3.24	3.01	2.85	2.74	2.66	2.59	2.54	2.49	2.42	2.35	2.28	2.24	2.19	2.15	2.11	2.06	2.01
17	4.45	3.59	3.20	2.96	2.81	2.70	2.61	2.55	2.49	2.45	2.38	2.31	2.23	2.19	2.15	2.10	2.06	2.01	1.96
18	4.41	3.55	3.16	2.93	2.77	2.66	2.58	2.51	2.46	2.41	2.34	2.27	2.19	2.15	2.11	2.06	2.02	1.97	1.92
19	4.38	3.52	3.13	2.90	2.74	2.63	2.54	2.48	2.42	2.38	2.31	2.23	2.16	2.11	2.07	2.03	1.98	1.93	1.88
20	4.35	3.49	3.10	2.87	2.71	2.60	2.51	2.45	2.39	2.35	2.28	2.20	2.12	2.08	2.04	1.99	1.95	1.90	1.84
21	4.32	3.47	3.07	2.84	2.68	2.57	2.49	2.42	2.37	2.32	2.25	2.18	2.10	2.05	2.01	1.96	1.92	1.87	1.81
22	4.30	3.44	3.05	2.82	2.66	2.55	2.46	2.40	2.34	2.30	2.23	2.15	2.07	2.03	1.98	1.94	1.89	1.84	1.78
23	4.28	3.42	3.03	2.80	2.64	2.53	2.44	2.37	2.32	2.27	2.20	2.13	2.05	2.01	1.96	1.91	1.86	1.81	1.76
24	4.26	3.40	3.01	2.78	2.62	2.51	2.42	2.36	2.30	2.25	2.18	2.11	2.03	1.98	1.94	1.89	1.84	1.79	1.73
25	4.24	3.39	2.99	2.76	2.60	2.49	2.40	2.34	2.28	2.24	2.16	2.09	2.01	1.96	1.92	1.87	1.82	1.77	1.71
26	4.23	3.37	2.98	2.74	2.59	2.47	2.39	2.32	2.27	2.22	2.15	2.07	1.99	1.95	1.90	1.85	1.80	1.75	1.69
27	4.21	3.35	2.96	2.73	2.57	2.46	2.37	2.31	2.25	2.20	2.13	2.06	1.97	1.93	1.88	1.84	1.79	1.73	1.67
28	4.20	3.34	2.95	2.71	2.56	2.45	2.36	2.29	2.24	2.19	2.12	2.04	1.96	1.91	1.87	1.82	1.77	1.71	1.65
29	4.18	3.33	2.93	2.70	2.55	2.43	2.35	2.28	2.22	2.18	2.10	2.03	1.94	1.90	1.85	1.81	1.75	1.70	1.64
30	4.17	3.32	2.92	2.69	2.53	2.42	2.33	2.27	2.21	2.16	2.09	2.01	1.93	1.89	1.84	1.79	1.74	1.68	1.62
40	4.08	3.23	2.84	2.61	2.45	2.34	2.25	2.18	2.12	2.08	2.00	1.92	1.84	1.79	1.74	1.69	1.64	1.58	1.51
60	4.00	3.15	2.76	2.53	2.37	2.25	2.17	2.10	2.04	1.99	1.92	1.84	1.75	1.70	1.65	1.59	1.53	1.47	1.39
120	3.92	3.07	2.68	2.45	2.29	2.17	2.09	2.02	1.96	1.91	1.83	1.75	1.66	1.61	1.55	1.50	1.43	1.35	1.25
∞	3.84	3.00	2.60	2.37	2.21	2.10	2.01	1.94	1.88	1.83	1.75	1.67	1.57	1.52	1.46	1.39	1.32	1.22	1.00

TABLE 6 (*continued*)[a]

97.5th Percentile

ν_2 \ ν_1	1	2	3	4	5	6	7	8	9	10	12	15	20	24	30	40	60	120	∞
1	647.8	799.5	864.2	899.6	921.8	937.1	948.2	956.7	963.3	968.6	976.7	984.9	993.1	997.2	1001	1006	1010	1014	1018
2	38.51	39.00	39.17	39.25	39.30	39.33	39.36	39.37	39.39	39.40	39.41	39.43	39.45	39.46	39.46	39.47	39.48	39.49	39.50
3	17.44	16.04	15.44	15.10	14.88	14.73	14.62	14.54	14.47	14.42	14.34	14.25	14.17	14.12	14.08	14.04	13.99	13.95	13.90
4	12.22	10.65	9.98	9.60	9.36	9.20	9.07	8.98	8.90	8.84	8.75	8.66	8.56	8.51	8.46	8.41	8.36	8.31	8.26
5	10.01	8.43	7.76	7.39	7.15	6.98	6.85	6.76	6.68	6.62	6.52	6.43	6.33	6.28	6.23	6.18	6.12	6.07	6.02
6	8.81	7.26	6.60	6.23	5.99	5.82	5.70	5.60	5.52	5.46	5.37	5.27	5.17	5.12	5.07	5.01	4.96	4.90	4.85
7	8.07	6.54	5.89	5.52	5.29	5.12	4.99	4.90	4.82	4.76	4.67	4.57	4.47	4.42	4.36	4.31	4.25	4.20	4.14
8	7.57	6.06	5.42	5.05	4.82	4.65	4.53	4.43	4.36	4.30	4.20	4.10	4.00	3.95	3.89	3.84	3.78	3.73	3.67
9	7.21	5.71	5.08	4.72	4.48	4.32	4.20	4.10	4.03	3.96	3.87	3.77	3.67	3.61	3.56	3.51	3.45	3.39	3.33
10	6.94	5.46	4.83	4.47	4.24	4.07	3.95	3.85	3.78	3.72	3.62	3.52	3.42	3.37	3.31	3.26	3.20	3.14	3.08
11	6.72	5.26	4.63	4.28	4.04	3.88	3.76	3.66	3.59	3.53	3.43	3.33	3.23	3.17	3.12	3.06	3.00	2.94	2.88
12	6.55	5.10	4.47	4.12	3.89	3.73	3.61	3.51	3.44	3.37	3.28	3.18	3.07	3.02	2.96	2.91	2.85	2.79	2.72
13	6.41	4.97	4.35	4.00	3.77	3.60	3.48	3.39	3.31	3.25	3.15	3.05	2.95	2.89	2.84	2.78	2.72	2.66	2.60
14	6.30	4.86	4.24	3.89	3.66	3.50	3.38	3.29	3.21	3.15	3.05	2.95	2.84	2.79	2.73	2.67	2.61	2.55	2.49
15	6.20	4.77	4.15	3.80	3.58	3.41	3.29	3.20	3.12	3.06	2.96	2.86	2.76	2.70	2.64	2.59	2.52	2.46	2.40
16	6.12	4.69	4.08	3.73	3.50	3.34	3.22	3.12	3.05	2.99	2.89	2.79	2.68	2.63	2.57	2.51	2.45	2.38	2.32
17	6.04	4.62	4.01	3.66	3.44	3.28	3.16	3.06	2.98	2.92	2.82	2.72	2.62	2.56	2.50	2.44	2.38	2.32	2.25
18	5.98	4.56	3.95	3.61	3.38	3.22	3.10	3.01	2.93	2.87	2.77	2.67	2.56	2.50	2.44	2.38	2.32	2.26	2.19
19	5.92	4.51	3.90	3.56	3.33	3.17	3.05	2.96	2.88	2.82	2.72	2.62	2.51	2.45	2.39	2.33	2.27	2.20	2.13
20	5.87	4.46	3.86	3.51	3.29	3.13	3.01	2.91	2.84	2.77	2.68	2.57	2.46	2.41	2.35	2.29	2.22	2.16	2.09
21	5.83	4.42	3.82	3.48	3.25	3.09	2.97	2.87	2.80	2.73	2.64	2.53	2.42	2.37	2.31	2.25	2.18	2.11	2.04
22	5.79	4.38	3.78	3.44	3.22	3.05	2.93	2.84	2.76	2.70	2.60	2.50	2.39	2.33	2.27	2.21	2.14	2.08	2.00
23	5.75	4.35	3.75	3.41	3.18	3.02	2.90	2.81	2.73	2.67	2.57	2.47	2.36	2.30	2.24	2.18	2.11	2.04	1.97
24	5.72	4.32	3.72	3.38	3.15	2.99	2.87	2.78	2.70	2.64	2.54	2.44	2.33	2.27	2.21	2.15	2.08	2.01	1.94
25	5.69	4.29	3.69	3.35	3.13	2.97	2.85	2.75	2.68	2.61	2.51	2.41	2.30	2.24	2.18	2.12	2.05	1.98	1.91
26	5.66	4.27	3.67	3.33	3.10	2.94	2.82	2.73	2.65	2.59	2.49	2.39	2.28	2.22	2.16	2.09	2.03	1.95	1.88
27	5.63	4.24	3.65	3.31	3.08	2.92	2.80	2.71	2.63	2.57	2.47	2.36	2.25	2.19	2.13	2.07	2.00	1.93	1.85
28	5.61	4.22	3.63	3.29	3.06	2.90	2.78	2.69	2.61	2.55	2.45	2.34	2.23	2.17	2.11	2.05	1.98	1.91	1.83
29	5.59	4.20	3.61	3.27	3.04	2.88	2.76	2.67	2.59	2.53	2.43	2.32	2.21	2.15	2.09	2.03	1.96	1.89	1.81
30	5.57	4.18	3.59	3.25	3.03	2.87	2.75	2.65	2.57	2.51	2.41	2.31	2.20	2.14	2.07	2.01	1.94	1.87	1.79
40	5.42	4.05	3.46	3.13	2.90	2.74	2.62	2.53	2.45	2.39	2.29	2.18	2.07	2.01	1.94	1.88	1.80	1.72	1.64
60	5.29	3.93	3.34	3.01	2.79	2.63	2.51	2.41	2.33	2.27	2.17	2.06	1.94	1.88	1.82	1.74	1.67	1.58	1.48
120	5.15	3.80	3.23	2.89	2.67	2.52	2.39	2.30	2.22	2.16	2.05	1.94	1.82	1.76	1.69	1.61	1.53	1.43	1.31
∞	5.02	3.69	3.12	2.79	2.57	2.41	2.29	2.19	2.11	2.05	1.94	1.83	1.71	1.64	1.57	1.48	1.39	1.27	1.00

99th Percentile

v_2 \ v_1	1	2	3	4	5	6	7	8	9	10	12	15	20	24	30	40	60	120	∞
1	4052	4999.5	5403	5625	5764	5859	5928	5981	6022	6056	6106	6157	6209	6235	6261	6287	6313	6339	6366
2	98.50	99.00	99.17	99.25	99.30	99.33	99.36	99.37	99.39	99.40	99.42	99.43	99.45	99.46	99.47	99.47	99.48	99.49	99.50
3	34.12	30.82	29.46	28.71	28.24	27.91	27.67	27.49	27.35	27.23	27.05	26.87	26.69	26.60	26.50	26.41	26.32	26.22	26.13
4	21.20	18.00	16.69	15.98	15.52	15.21	14.98	14.80	14.66	14.55	14.37	14.20	14.02	13.93	13.84	13.75	13.65	13.56	13.46
5	16.26	13.27	12.06	11.39	10.97	10.67	10.46	10.29	10.16	10.05	9.89	9.72	9.55	9.47	9.38	9.29	9.20	9.11	9.02
6	13.75	10.92	9.78	9.15	8.75	8.47	8.26	8.10	7.98	7.87	7.72	7.56	7.40	7.31	7.23	7.14	7.06	6.97	6.88
7	12.25	9.55	8.45	7.85	7.46	7.19	6.99	6.84	6.72	6.62	6.47	6.31	6.16	6.07	5.99	5.91	5.82	5.74	5.65
8	11.26	8.65	7.59	7.01	6.63	6.37	6.18	6.03	5.91	5.81	5.67	5.52	5.36	5.28	5.20	5.12	5.03	4.95	4.86
9	10.56	8.02	6.99	6.42	6.06	5.80	5.61	5.47	5.35	5.26	5.11	4.96	4.81	4.73	4.65	4.57	4.48	4.40	4.31
10	10.04	7.56	6.55	5.99	5.64	5.39	5.20	5.06	4.94	4.85	4.71	4.56	4.41	4.33	4.25	4.17	4.08	4.00	3.91
11	9.65	7.21	6.22	5.67	5.32	5.07	4.89	4.74	4.63	4.54	4.40	4.25	4.10	4.02	3.94	3.86	3.78	3.69	3.60
12	9.33	6.93	5.95	5.41	5.06	4.82	4.64	4.50	4.39	4.30	4.16	4.01	3.86	3.78	3.70	3.62	3.54	3.45	3.36
13	9.07	6.70	5.74	5.21	4.86	4.62	4.44	4.30	4.19	4.10	3.96	3.82	3.66	3.59	3.51	3.43	3.34	3.25	3.17
14	8.86	6.51	5.56	5.04	4.69	4.46	4.28	4.14	4.03	3.94	3.80	3.66	3.51	3.43	3.35	3.27	3.18	3.09	3.00
15	8.68	6.36	5.42	4.89	4.56	4.32	4.14	4.00	3.89	3.80	3.67	3.52	3.37	3.29	3.21	3.13	3.05	2.96	2.87
16	8.53	6.23	5.29	4.77	4.44	4.20	4.03	3.89	3.78	3.69	3.55	3.41	3.26	3.18	3.10	3.02	2.93	2.84	2.75
17	8.40	6.11	5.18	4.67	4.34	4.10	3.93	3.79	3.68	3.59	3.46	3.31	3.16	3.08	3.00	2.92	2.83	2.75	2.65
18	8.29	6.01	5.09	4.58	4.25	4.01	3.84	3.71	3.60	3.51	3.37	3.23	3.08	3.00	2.92	2.84	2.75	2.66	2.57
19	8.18	5.93	5.01	4.50	4.17	3.94	3.77	3.63	3.52	3.43	3.30	3.15	3.00	2.92	2.84	2.76	2.67	2.58	2.49
20	8.10	5.85	4.94	4.43	4.10	3.87	3.70	3.56	3.46	3.37	3.23	3.09	2.94	2.86	2.78	2.69	2.61	2.52	2.42
21	8.02	5.78	4.87	4.37	4.04	3.81	3.64	3.51	3.40	3.31	3.17	3.03	2.88	2.80	2.72	2.64	2.55	2.46	2.36
22	7.95	5.72	4.82	4.31	3.99	3.76	3.59	3.45	3.35	3.26	3.12	2.98	2.83	2.75	2.67	2.58	2.50	2.40	2.31
23	7.88	5.66	4.76	4.26	3.94	3.71	3.54	3.41	3.30	3.21	3.07	2.93	2.78	2.70	2.62	2.54	2.45	2.35	2.26
24	7.82	5.61	4.72	4.22	3.90	3.67	3.50	3.36	3.26	3.17	3.03	2.89	2.74	2.66	2.58	2.49	2.40	2.31	2.21
25	7.77	5.57	4.68	4.18	3.85	3.63	3.46	3.32	3.22	3.13	2.99	2.85	2.70	2.62	2.54	2.45	2.36	2.27	2.17
26	7.72	5.53	4.64	4.14	3.82	3.59	3.42	3.29	3.18	3.09	2.96	2.81	2.66	2.58	2.50	2.42	2.33	2.23	2.13
27	7.68	5.49	4.60	4.11	3.78	3.56	3.39	3.26	3.15	3.06	2.93	2.78	2.63	2.55	2.47	2.38	2.29	2.20	2.10
28	7.64	5.45	4.57	4.07	3.75	3.53	3.36	3.23	3.12	3.03	2.90	2.75	2.60	2.52	2.44	2.35	2.26	2.17	2.06
29	7.60	5.42	4.54	4.04	3.73	3.50	3.33	3.20	3.09	3.00	2.87	2.73	2.57	2.49	2.41	2.33	2.23	2.14	2.03
30	7.56	5.39	4.51	4.02	3.70	3.47	3.30	3.17	3.07	2.98	2.84	2.70	2.55	2.47	2.39	2.30	2.21	2.11	2.01
40	7.31	5.18	4.31	3.83	3.51	3.29	3.12	2.99	2.89	2.80	2.66	2.52	2.37	2.29	2.20	2.11	2.02	1.92	1.80
60	7.08	4.98	4.13	3.65	3.34	3.12	2.95	2.82	2.72	2.63	2.50	2.35	2.20	2.12	2.03	1.94	1.84	1.73	1.60
120	6.85	4.79	3.95	3.48	3.17	2.96	2.79	2.66	2.56	2.47	2.34	2.19	2.03	1.95	1.86	1.76	1.66	1.53	1.38
∞	6.63	4.61	3.78	3.32	3.02	2.80	2.64	2.51	2.41	2.32	2.18	2.04	1.88	1.79	1.70	1.59	1.47	1.32	1.00

TABLE 6 *(continued)*[a]

99.5th Percentile

ν_2 \ ν_1	1	2	3	4	5	6	7	8	9	10	12	15	20	24	30	40	60	120	∞
1	16211	20000	21615	22500	23056	23437	23715	23925	24091	24224	24426	24630	24836	24940	25044	25148	25253	25359	25465
2	198·5	199·0	199·2	199·2	199·3	199·3	199·4	199·4	199·4	199·4	199·4	199·4	199·4	199·5	199·5	199·5	199·5	199·5	199·5
3	55·55	49·80	47·47	46·19	45·39	44·84	44·43	44·13	43·88	43·69	43·39	43·08	42·78	42·62	42·47	42·31	42·15	42·15	41·83
4	31·33	26·28	24·26	23·15	22·46	21·97	21·62	21·35	21·14	20·97	20·70	20·44	20·17	20·03	19·89	19·75	19·61	19·47	19·32
5	22·78	18·31	16·53	15·56	14·94	14·51	14·20	13·96	13·77	13·62	13·38	13·15	12·90	12·78	12·66	12·53	12·40	12·27	12·14
6	18·63	14·54	12·92	12·03	11·46	11·07	10·79	10·57	10·39	10·25	10·03	9·81	9·59	9·47	9·36	9·24	9·12	9·00	8·88
7	16·24	12·40	10·88	10·05	9·52	9·16	8·89	8·68	8·51	8·38	8·18	7·97	7·75	7·65	7·53	7·42	7·31	7·19	7·08
8	14·69	11·04	9·60	8·81	8·30	7·95	7·69	7·50	7·34	7·21	7·01	6·81	6·61	6·50	6·40	6·29	6·18	6·06	5·95
9	13·61	10·11	8·72	7·96	7·47	7·13	6·88	6·69	6·54	6·42	6·23	6·03	5·83	5·73	5·62	5·52	5·41	5·30	5·19
10	12·83	9·43	8·08	7·34	6·87	6·54	6·30	6·12	5·97	5·85	5·66	5·47	5·27	5·17	5·07	4·97	4·86	4·75	4·64
11	12·23	8·91	7·60	6·88	6·42	6·10	5·86	5·68	5·54	5·42	5·24	5·05	4·86	4·76	4·65	4·55	4·44	4·34	4·23
12	11·75	8·51	7·23	6·52	6·07	5·76	5·52	5·35	5·20	5·09	4·91	4·72	4·53	4·43	4·33	4·23	4·12	4·01	3·90
13	11·37	8·19	6·93	6·23	5·79	5·48	5·25	5·08	4·94	4·82	4·64	4·46	4·27	4·17	4·07	3·97	3·87	3·76	3·65
14	11·06	7·92	6·68	6·00	5·56	5·26	5·03	4·86	4·72	4·60	4·43	4·25	4·06	3·96	3·86	3·76	3·66	3·55	3·44
15	10·80	7·70	6·48	5·80	5·37	5·07	4·85	4·67	4·54	4·42	4·25	4·07	3·88	3·79	3·69	3·58	3·48	3·37	3·26
16	10·58	7·51	6·30	5·64	5·21	4·91	4·69	4·52	4·38	4·27	4·10	3·92	3·73	3·64	3·54	3·44	3·33	3·22	3·11
17	10·38	7·35	6·16	5·50	5·07	4·78	4·56	4·39	4·25	4·14	3·97	3·79	3·61	3·51	3·41	3·31	3·21	3·10	2·98
18	10·22	7·21	6·03	5·37	4·96	4·66	4·44	4·28	4·14	4·03	3·86	3·68	3·50	3·40	3·30	3·20	3·10	2·99	2·87
19	10·07	7·09	5·92	5·27	4·85	4·56	4·34	4·18	4·04	3·93	3·76	3·59	3·40	3·31	3·21	3·11	3·00	2·89	2·78
20	9·94	6·99	5·82	5·17	4·76	4·47	4·26	4·09	3·96	3·85	3·68	3·50	3·32	3·22	3·12	3·02	2·92	2·81	2·69
21	9·83	6·89	5·73	5·09	4·68	4·39	4·18	4·01	3·88	3·77	3·60	3·43	3·24	3·15	3·05	2·95	2·84	2·73	2·61
22	9·73	6·81	5·65	5·02	4·61	4·32	4·11	3·94	3·81	3·70	3·54	3·36	3·18	3·08	2·98	2·88	2·77	2·66	2·55
23	9·63	6·73	5·58	4·95	4·54	4·26	4·05	3·88	3·75	3·64	3·47	3·30	3·12	3·02	2·92	2·82	2·71	2·60	2·48
24	9·55	6·66	5·52	4·89	4·49	4·20	3·99	3·83	3·69	3·59	3·42	3·25	3·06	2·97	2·87	2·77	2·66	2·55	2·43
25	9·48	6·60	5·46	4·84	4·43	4·15	3·94	3·78	3·64	3·54	3·37	3·20	3·01	2·92	2·82	2·72	2·61	2·50	2·38
26	9·41	6·54	5·41	4·79	4·38	4·10	3·89	3·73	3·60	3·49	3·33	3·15	2·97	2·87	2·77	2·67	2·56	2·45	2·33
27	9·34	6·49	5·36	4·74	4·34	4·06	3·85	3·69	3·56	3·45	3·28	3·11	2·93	2·83	2·73	2·63	2·52	2·41	2·29
28	9·28	6·44	5·32	4·70	4·30	4·02	3·81	3·65	3·52	3·41	3·25	3·07	2·89	2·79	2·69	2·59	2·48	2·37	2·25
29	9·23	6·40	5·28	4·66	4·26	3·98	3·77	3·61	3·48	3·38	3·21	3·04	2·86	2·76	2·66	2·56	2·45	2·33	2·21
30	9·18	6·35	5·24	4·62	4·23	3·95	3·74	3·58	3·45	3·34	3·18	3·01	2·82	2·73	2·63	2·52	2·42	2·30	2·18
40	8·83	6·07	4·98	4·37	3·99	3·71	3·51	3·35	3·22	3·12	2·95	2·78	2·60	2·50	2·40	2·30	2·18	2·06	1·93
60	8·49	5·79	4·73	4·14	3·76	3·49	3·29	3·13	3·01	2·90	2·74	2·57	2·39	2·29	2·19	2·08	1·96	1·83	1·69
120	8·18	5·54	4·50	3·92	3·55	3·28	3·09	2·93	2·81	2·71	2·54	2·37	2·19	2·09	1·98	1·87	1·75	1·61	1·43
∞	7·88	5·30	4·28	3·72	3·35	3·09	2·90	2·74	2·62	2·52	2·36	2·19	2·00	1·90	1·79	1·67	1·53	1·36	1·00

99.9th Percentile

ν_2 \ ν_1	∞	120	60	40	30	24	20	15	12	10	9	8	7	6	5	4	3	2	1
1	6366*	6340*	6313*	6287*	6261*	6235*	6209*	6158*	6107*	6056*	6023*	5981*	5929*	5859*	5764*	5625*	5404*	5000*	4053*
2	999.5	999.5	999.5	999.5	999.5	999.5	999.4	999.4	999.4	999.4	999.4	999.4	999.4	999.3	999.3	999.2	999.2	999.0	998.5
3	123.5	124.0	124.5	125.0	125.4	125.9	126.4	127.4	128.3	129.2	129.9	130.6	131.6	132.8	134.6	137.1	141.1	148.5	167.0
4	44.05	44.40	44.75	45.09	45.43	45.77	46.10	46.76	47.41	48.05	48.47	49.00	49.66	50.53	51.71	53.44	56.18	61.25	74.14
5	23.79	24.06	24.33	24.60	24.87	25.14	25.39	25.91	26.42	26.92	27.24	27.64	28.16	28.84	29.75	31.09	33.20	37.12	47.18
6	15.75	15.99	16.21	16.44	16.67	16.89	17.12	17.56	17.99	18.41	18.69	19.03	19.46	20.03	20.81	21.92	23.70	27.00	35.51
7	11.70	11.91	12.12	12.33	12.53	12.73	12.93	13.32	13.71	14.08	14.33	14.63	15.02	15.52	16.21	17.19	18.77	21.69	29.25
8	9.33	9.53	9.73	9.92	10.11	10.30	10.48	10.84	11.19	11.54	11.77	12.04	12.40	12.86	13.49	14.39	15.83	18.49	25.42
9	7.81	8.00	8.19	8.37	8.55	8.72	8.90	9.24	9.57	9.89	10.11	10.37	10.70	11.13	11.71	12.56	13.90	16.39	22.86
10	6.76	6.94	7.12	7.30	7.47	7.64	7.80	8.13	8.45	8.75	8.96	9.20	9.52	9.92	10.48	11.28	12.55	14.91	21.04
11	6.00	6.17	6.35	6.52	6.68	6.85	7.01	7.32	7.63	7.92	8.12	8.35	8.66	9.05	9.58	10.35	11.56	13.81	19.69
12	5.42	5.59	5.76	5.93	6.09	6.25	6.40	6.71	7.00	7.29	7.48	7.71	8.00	8.38	8.89	9.63	10.80	12.97	18.64
13	4.97	5.14	5.30	5.47	5.63	5.78	5.93	6.23	6.52	6.80	6.98	7.21	7.49	7.86	8.35	9.07	10.21	12.31	17.81
14	4.60	4.77	4.94	5.10	5.25	5.41	5.56	5.85	6.13	6.40	6.58	6.80	7.08	7.43	7.92	8.62	9.73	11.78	17.14
15	4.31	4.47	4.64	4.80	4.95	5.10	5.25	5.54	5.81	6.08	6.26	6.47	6.74	7.09	7.57	8.25	9.34	11.34	16.59
16	4.06	4.23	4.39	4.54	4.70	4.85	4.99	5.27	5.55	5.81	5.98	6.19	6.46	6.81	7.27	7.94	9.00	10.97	16.12
17	3.85	4.02	4.18	4.33	4.48	4.63	4.78	5.05	5.32	5.58	5.75	5.96	6.22	6.56	7.02	7.68	8.73	10.66	15.72
18	3.67	3.84	4.00	4.15	4.30	4.45	4.59	4.87	5.13	5.39	5.56	5.76	6.02	6.35	6.81	7.46	8.49	10.39	15.38
19	3.51	3.68	3.84	3.99	4.14	4.29	4.43	4.70	4.97	5.22	5.39	5.59	5.85	6.18	6.62	7.26	8.28	10.16	15.08
20	3.38	3.54	3.70	3.86	4.00	4.15	4.29	4.56	4.82	5.08	5.24	5.44	5.69	6.02	6.46	7.10	8.10	9.95	14.82
21	3.26	3.42	3.58	3.74	3.88	4.03	4.17	4.44	4.70	4.95	5.11	5.31	5.56	5.88	6.32	6.95	7.94	9.77	14.59
22	3.15	3.32	3.48	3.63	3.78	3.92	4.06	4.33	4.58	4.83	4.99	5.19	5.44	5.76	6.19	6.81	7.80	9.61	14.38
23	3.05	3.22	3.38	3.53	3.68	3.82	3.96	4.23	4.48	4.73	4.89	5.09	5.33	5.65	6.08	6.69	7.67	9.47	14.19
24	2.97	3.14	3.29	3.45	3.59	3.74	3.87	4.14	4.39	4.64	4.80	4.99	5.23	5.55	5.98	6.59	7.55	9.34	14.03
25	2.89	3.06	3.22	3.37	3.52	3.66	3.79	4.06	4.31	4.56	4.71	4.91	5.15	5.46	5.88	6.49	7.45	9.22	13.88
26	2.82	2.99	3.15	3.30	3.44	3.59	3.72	3.99	4.24	4.48	4.64	4.83	5.07	5.38	5.80	6.41	7.36	9.12	13.74
27	2.75	2.92	3.08	3.23	3.38	3.52	3.66	3.92	4.17	4.41	4.57	4.76	5.00	5.31	5.73	6.33	7.27	9.02	13.61
28	2.69	2.86	3.02	3.18	3.32	3.46	3.60	3.86	4.11	4.35	4.50	4.69	4.93	5.24	5.66	6.25	7.19	8.93	13.50
29	2.64	2.81	2.97	3.12	3.27	3.41	3.54	3.80	4.05	4.29	4.45	4.64	4.87	5.18	5.59	6.19	7.12	8.85	13.39
30	2.59	2.76	2.92	3.07	3.22	3.36	3.49	3.75	4.00	4.24	4.39	4.58	4.82	5.12	5.53	6.12	7.05	8.77	13.29
40	2.23	2.41	2.57	2.73	2.87	3.01	3.15	3.40	3.64	3.87	4.02	4.21	4.44	4.73	5.13	5.70	6.60	8.25	12.61
60	1.89	2.08	2.25	2.41	2.55	2.69	2.83	3.08	3.31	3.54	3.69	3.87	4.09	4.37	4.76	5.31	6.17	7.76	11.97
120	1.54	1.76	1.95	2.11	2.26	2.40	2.53	2.78	3.02	3.24	3.38	3.55	3.77	4.04	4.42	4.95	5.79	7.32	11.38
∞	1.00	1.45	1.66	1.84	1.99	2.13	2.27	2.51	2.74	2.96	3.10	3.27	3.47	3.74	4.10	4.62	5.42	6.91	10.83

a Reprinted from Table 18 of E. S. Pearson and H. O. Hartley (1966), "Biometrika Tables for Statisticians," Vol. I, 3rd Ed., Cambridge Univ. Press, London and New York, with the kind permission of E. S. Pearson.

* Multiply these entries by 100.

TABLE 7. Percentiles of the Studentized Range Distribution (Section 2.4.2)[a]

90th Percentile

ν \ p	2	3	4	5	6	7	8	9	10	11	12	13	14	15	16	17	18	19	20
1	8.93	13.44	16.36	18.49	20.15	21.51	22.64	23.62	24.48	25.24	25.92	26.54	27.10	27.62	28.10	28.54	28.96	29.35	29.71
2	4.13	5.73	6.77	7.54	8.14	8.63	9.05	9.41	9.72	10.01	10.26	10.49	10.70	10.89	11.07	11.24	11.39	11.54	11.68
3	3.33	4.47	5.20	5.74	6.16	6.51	6.81	7.06	7.29	7.49	7.67	7.83	7.98	8.12	8.25	8.37	8.48	8.58	8.68
4	3.01	3.98	4.59	5.03	5.39	5.68	5.93	6.14	6.33	6.49	6.65	6.78	6.91	7.02	7.13	7.23	7.33	7.41	7.50
5	2.85	3.72	4.26	4.66	4.98	5.24	5.46	5.65	5.82	5.97	6.10	6.22	6.34	6.44	6.54	6.63	6.71	6.79	6.86
6	2.75	3.56	4.07	4.44	4.73	4.97	5.17	5.34	5.50	5.64	5.76	5.87	5.98	6.07	6.16	6.25	6.32	6.40	6.47
7	2.68	3.45	3.93	4.28	4.55	4.78	4.97	5.14	5.28	5.41	5.53	5.64	5.74	5.83	5.91	5.99	6.06	6.13	6.19
8	2.63	3.37	3.83	4.17	4.43	4.65	4.83	4.99	5.13	5.25	5.36	5.46	5.56	5.64	5.72	5.80	5.87	5.93	6.00
9	2.59	3.32	3.76	4.08	4.34	4.54	4.72	4.87	5.01	5.13	5.23	5.33	5.42	5.51	5.58	5.66	5.72	5.79	5.85
10	2.56	3.27	3.70	4.02	4.26	4.47	4.64	4.78	4.91	5.03	5.13	5.23	5.32	5.40	5.47	5.54	5.61	5.67	5.73
11	2.54	3.23	3.66	3.96	4.20	4.40	4.57	4.71	4.84	4.95	5.05	5.15	5.23	5.31	5.38	5.45	5.51	5.57	5.63
12	2.52	3.20	3.62	3.92	4.16	4.35	4.51	4.65	4.78	4.89	4.99	5.08	5.16	5.24	5.31	5.37	5.44	5.49	5.55
13	2.50	3.18	3.59	3.88	4.12	4.30	4.46	4.60	4.72	4.83	4.93	5.02	5.10	5.18	5.25	5.31	5.37	5.43	5.48
14	2.49	3.16	3.56	3.85	4.08	4.27	4.42	4.56	4.68	4.79	4.88	4.97	5.05	5.12	5.19	5.26	5.32	5.37	5.43
15	2.48	3.14	3.54	3.83	4.05	4.23	4.39	4.52	4.64	4.75	4.84	4.93	5.01	5.08	5.15	5.21	5.27	5.32	5.38
16	2.47	3.12	3.52	3.80	4.03	4.21	4.36	4.49	4.61	4.71	4.81	4.89	4.97	5.04	5.11	5.17	5.23	5.28	5.33
17	2.46	3.11	3.50	3.78	4.00	4.18	4.33	4.46	4.58	4.68	4.77	4.86	4.93	5.01	5.07	5.13	5.19	5.24	5.30
18	2.45	3.10	3.49	3.77	3.98	4.16	4.31	4.44	4.55	4.65	4.75	4.83	4.90	4.98	5.04	5.10	5.16	5.21	5.26
19	2.45	3.09	3.47	3.75	3.97	4.14	4.29	4.42	4.53	4.63	4.72	4.80	4.88	4.95	5.01	5.07	5.13	5.18	5.23
20	2.44	3.08	3.46	3.74	3.95	4.12	4.27	4.40	4.51	4.61	4.70	4.78	4.85	4.92	4.99	5.05	5.10	5.16	5.20
24	2.42	3.05	3.42	3.69	3.90	4.07	4.21	4.34	4.44	4.54	4.63	4.71	4.78	4.85	4.91	4.97	5.02	5.07	5.12
30	2.40	3.02	3.39	3.65	3.85	4.02	4.16	4.28	4.38	4.47	4.56	4.64	4.71	4.77	4.83	4.89	4.94	4.99	5.03
40	2.38	2.99	3.35	3.60	3.80	3.96	4.10	4.21	4.32	4.41	4.49	4.56	4.63	4.69	4.75	4.81	4.86	4.90	4.95
60	2.36	2.96	3.31	3.56	3.75	3.91	4.04	4.16	4.25	4.34	4.42	4.49	4.56	4.62	4.67	4.73	4.78	4.82	4.86
120	2.34	2.93	3.28	3.52	3.71	3.86	3.99	4.10	4.19	4.28	4.35	4.42	4.48	4.54	4.60	4.65	4.69	4.74	4.78
∞	2.33	2.90	3.24	3.48	3.66	3.81	3.93	4.04	4.13	4.21	4.28	4.35	4.41	4.47	4.52	4.57	4.61	4.65	4.69

95th Percentile

ρ / ν	2	3	4	5	6	7	8	9	10	11	12	13	14	15	16	17	18	19	20
1	17·97	26·98	32·82	37·08	40·41	43·12	45·40	47·36	49·07	50·59	51·96	53·20	54·33	55·36	56·32	57·22	58·04	58·83	59·56
2	6·08	8·33	9·80	10·88	11·74	12·44	13·03	13·54	13·99	14·39	14·75	15·08	15·38	15·65	15·91	16·14	16·37	16·57	16·77
3	4·50	5·91	6·82	7·50	8·04	8·48	8·85	9·18	9·46	9·72	9·95	10·15	10·35	10·52	10·69	10·84	10·98	11·11	11·24
4	3·93	5·04	5·76	6·29	6·71	7·05	7·35	7·60	7·83	8·03	8·21	8·37	8·52	8·66	8·79	8·91	9·03	9·13	9·23
5	3·64	4·60	5·22	5·67	6·03	6·33	6·58	6·80	6·99	7·17	7·32	7·47	7·60	7·72	7·83	7·93	8·03	8·12	8·21
6	3·46	4·34	4·90	5·30	5·63	5·90	6·12	6·32	6·49	6·65	6·79	6·92	7·03	7·14	7·24	7·34	7·43	7·51	7·59
7	3·34	4·16	4·68	5·06	5·36	5·61	5·82	6·00	6·16	6·30	6·43	6·55	6·66	6·76	6·85	6·94	7·02	7·10	7·17
8	3·26	4·04	4·53	4·89	5·17	5·40	5·60	5·77	5·92	6·05	6·18	6·29	6·39	6·48	6·57	6·65	6·73	6·80	6·87
9	3·20	3·95	4·41	4·76	5·02	5·24	5·43	5·59	5·74	5·87	5·98	6·09	6·19	6·28	6·36	6·44	6·51	6·58	6·64
10	3·15	3·88	4·33	4·65	4·91	5·12	5·30	5·46	5·60	5·72	5·83	5·93	6·03	6·11	6·19	6·27	6·34	6·40	6·47
11	3·11	3·82	4·26	4·57	4·82	5·03	5·20	5·35	5·49	5·61	5·71	5·81	5·90	5·98	6·06	6·13	6·20	6·27	6·33
12	3·08	3·77	4·20	4·51	4·75	4·95	5·12	5·27	5·39	5·51	5·61	5·71	5·80	5·88	5·95	6·02	6·09	6·15	6·21
13	3·06	3·73	4·15	4·45	4·69	4·88	5·05	5·19	5·32	5·43	5·53	5·63	5·71	5·79	5·86	5·93	5·99	6·05	6·11
14	3·03	3·70	4·11	4·41	4·64	4·83	4·99	5·13	5·25	5·36	5·46	5·55	5·64	5·71	5·79	5·85	5·91	5·97	6·03
15	3·01	3·67	4·08	4·37	4·59	4·78	4·94	5·08	5·20	5·31	5·40	5·49	5·57	5·65	5·72	5·78	5·85	5·90	5·96
16	3·00	3·65	4·05	4·33	4·56	4·74	4·90	5·03	5·15	5·26	5·35	5·44	5·52	5·59	5·66	5·73	5·79	5·84	5·90
17	2·98	3·63	4·02	4·30	4·52	4·70	4·86	4·99	5·11	5·21	5·31	5·39	5·47	5·54	5·61	5·67	5·73	5·79	5·84
18	2·97	3·61	4·00	4·28	4·49	4·67	4·82	4·96	5·07	5·17	5·27	5·35	5·43	5·50	5·57	5·63	5·69	5·74	5·79
19	2·96	3·59	3·98	4·25	4·47	4·65	4·79	4·92	5·04	5·14	5·23	5·31	5·39	5·46	5·53	5·59	5·65	5·70	5·75
20	2·95	3·58	3·96	4·23	4·45	4·62	4·77	4·90	5·01	5·11	5·20	5·28	5·36	5·43	5·49	5·55	5·61	5·66	5·71
24	2·92	3·53	3·90	4·17	4·37	4·54	4·68	4·81	4·92	5·01	5·10	5·18	5·25	5·32	5·38	5·44	5·49	5·55	5·59
30	2·89	3·49	3·85	4·10	4·30	4·46	4·60	4·72	4·82	4·92	5·00	5·08	5·15	5·21	5·27	5·33	5·38	5·43	5·47
40	2·86	3·44	3·79	4·04	4·23	4·39	4·52	4·63	4·73	4·82	4·90	4·98	5·04	5·11	5·16	5·22	5·27	5·31	5·36
60	2·83	3·40	3·74	3·98	4·16	4·31	4·44	4·55	4·65	4·73	4·81	4·88	4·94	5·00	5·06	5·11	5·15	5·20	5·24
120	2·80	3·36	3·68	3·92	4·10	4·24	4·36	4·47	4·56	4·64	4·71	4·78	4·84	4·90	4·95	5·00	5·04	5·09	5·13
∞	2·77	3·31	3·63	3·86	4·03	4·17	4·29	4·39	4·47	4·55	4·62	4·68	4·74	4·80	4·85	4·89	4·93	4·97	5·01

TABLE 7 (*continued*)[a]

99th Percentile

ρ \ ν	2	3	4	5	6	7	8	9	10	11	12	13	14	15	16	17	18	19	20
1	90·03	135·0	164·3	185·6	202·2	215·8	227·2	237·0	245·6	253·2	260·0	266·2	271·8	277·0	281·8	286·3	290·4	294·3	298·0
2	14·04	19·02	22·29	24·72	26·63	28·20	29·53	30·68	31·69	32·59	33·40	34·13	34·81	35·43	36·00	36·53	37·03	37·50	37·95
3	8·26	10·62	12·17	13·33	14·24	15·00	15·64	16·20	16·69	17·13	17·53	17·89	18·22	18·52	18·81	19·07	19·32	19·55	19·77
4	6·51	8·12	9·17	9·96	10·58	11·10	11·55	11·93	12·27	12·57	12·84	13·09	13·32	13·53	13·73	13·91	14·08	14·24	14·40
5	5·70	6·98	7·80	8·42	8·91	9·32	9·67	9·97	10·24	10·48	10·70	10·89	11·08	11·24	11·40	11·55	11·68	11·81	11·93
6	5·24	6·33	7·03	7·56	7·97	8·32	8·61	8·87	9·10	9·30	9·48	9·65	9·81	9·95	10·08	10·21	10·32	10·43	10·54
7	4·95	5·92	6·54	7·01	7·37	7·68	7·94	8·17	8·37	8·55	8·71	8·86	9·00	9·12	9·24	9·35	9·46	9·55	9·65
8	4·75	5·64	6·20	6·62	6·96	7·24	7·47	7·68	7·86	8·03	8·18	8·31	8·44	8·55	8·66	8·76	8·85	8·94	9·03
9	4·60	5·43	5·96	6·35	6·66	6·91	7·13	7·33	7·49	7·65	7·78	7·91	8·03	8·13	8·23	8·33	8·41	8·49	8·57
10	4·48	5·27	5·77	6·14	6·43	6·67	6·87	7·05	7·21	7·36	7·49	7·60	7·71	7·81	7·91	7·99	8·08	8·15	8·23
11	4·39	5·15	5·62	5·97	6·25	6·48	6·67	6·84	6·99	7·13	7·25	7·36	7·46	7·56	7·65	7·73	7·81	7·88	7·95
12	4·32	5·05	5·50	5·84	6·10	6·32	6·51	6·67	6·81	6·94	7·06	7·17	7·26	7·36	7·44	7·52	7·59	7·66	7·73
13	4·26	4·96	5·40	5·73	5·98	6·19	6·37	6·53	6·67	6·79	6·90	7·01	7·10	7·19	7·27	7·35	7·42	7·48	7·55
14	4·21	4·89	5·32	5·63	5·88	6·08	6·26	6·41	6·54	6·66	6·77	6·87	6·96	7·05	7·13	7·20	7·27	7·33	7·39
15	4·17	4·84	5·25	5·56	5·80	5·99	6·16	6·31	6·44	6·55	6·66	6·76	6·84	6·93	7·00	7·07	7·14	7·20	7·26
16	4·13	4·79	5·19	5·49	5·72	5·92	6·08	6·22	6·35	6·46	6·56	6·66	6·74	6·82	6·90	6·97	7·03	7·09	7·15
17	4·10	4·74	5·14	5·43	5·66	5·85	6·01	6·15	6·27	6·38	6·48	6·57	6·66	6·73	6·81	6·87	6·94	7·00	7·05
18	4·07	4·70	5·09	5·38	5·60	5·79	5·94	6·08	6·20	6·31	6·41	6·50	6·58	6·65	6·73	6·79	6·85	6·91	6·97
19	4·05	4·67	5·05	5·33	5·55	5·73	5·89	6·02	6·14	6·25	6·34	6·43	6·51	6·58	6·65	6·72	6·78	6·84	6·89
20	4·02	4·64	5·02	5·29	5·51	5·69	5·84	5·97	6·09	6·19	6·28	6·37	6·45	6·52	6·59	6·65	6·71	6·77	6·82
24	3·96	4·55	4·91	5·17	5·37	5·54	5·69	5·81	5·92	6·02	6·11	6·19	6·26	6·33	6·39	6·45	6·51	6·56	6·61
30	3·89	4·45	4·80	5·05	5·24	5·40	5·54	5·65	5·76	5·85	5·93	6·01	6·08	6·14	6·20	6·26	6·31	6·36	6·41
40	3·82	4·37	4·70	4·93	5·11	5·26	5·39	5·50	5·60	5·69	5·76	5·83	5·90	5·96	6·02	6·07	6·12	6·16	6·21
60	3·76	4·28	4·59	4·82	4·99	5·13	5·25	5·36	5·45	5·53	5·60	5·67	5·73	5·78	5·84	5·89	5·93	5·97	6·01
120	3·70	4·20	4·50	4·71	4·87	5·01	5·12	5·21	5·30	5·37	5·44	5·50	5·56	5·61	5·66	5·71	5·75	5·79	5·83
∞	3·64	4·12	4·40	4·60	4·76	4·88	4·99	5·08	5·16	5·23	5·29	5·35	5·40	5·45	5·49	5·54	5·57	5·61	5·65

[a] Reprinted from Table 29 of E. S. Pearson and H. O. Hartley (1966), "Biometrika Tables for Statisticians," Vol. I, 3rd Ed., Cambridge Univ. Press, London and New York, with the kind permission of E. S. Pearson.

TABLE 8. *Values of Fisher's z (Section 3.1.4)*[a]

r	.00	.01	.02	.03	.04	.05	.06	.07	.08	.09
.0	.00000	.01000	.02000	.03001	.04002	.05004	.06007	.07012	.08017	.09024
.1	.10034	.11045	.12058	.13074	.14093	.15114	.16139	.17167	.18198	.19234
.2	.20273	.21317	.22366	.23419	.24477	.25541	.26611	.27686	.28768	.29857
.3	.30952	.32055	.33165	.34283	.35409	.36544	.37689	.38842	.40006	.41180
.4	.42365	.43561	.44769	.45990	.47223	.48470	.49731	.51007	.52298	.53606
.5	.54931	.56273	.57634	.59014	.60415	.61838	.63283	.64752	.66246	.67767
.6	.69315	.70892	.72500	.74142	.75817	.77530	.79281	.81074	.82911	.84795
.7	.86730	.88718	.90764	.92873	.95048	.97295	.99621	1.02033	1.04537	1.07143
.8	1.09861	1.12703	1.15682	1.18813	1.22117	1.25615	1.29334	1.33308	1.37577	1.42192
.9	1.47222	1.52752	1.58902	1.65839	1.73805	1.83178	1.94591	2.09229	2.29756	2.64665

[a] Negative *r* produces negative *z*.

TABLE 9. *Charts for Confidence Intervals for Correlation Coefficients (Section 3.1.4)*[a]

Confidence level = 0.95

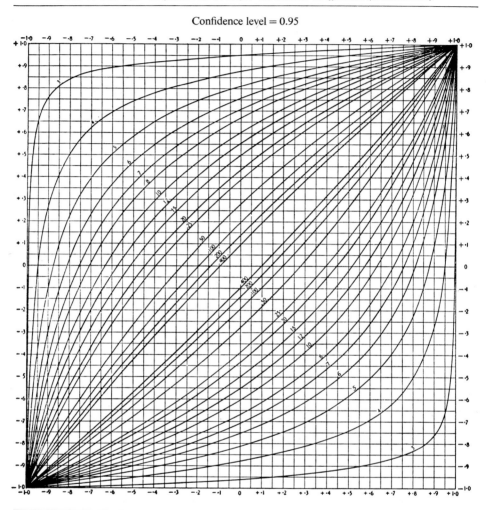

TABLE 9 (*continued*)[a]

Confidence level = 0.99

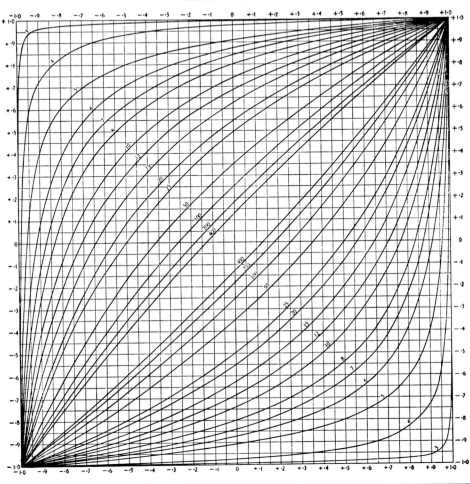

[a] Reprinted from Table 15 of E. S. Pearson and H. O. Hartley (1966), of "Biometrika Tables for Statisticians," Vol. 1, 3rd Ed., Cambridge Univ. Press, London and New York, with the kind permission of E. S. Pearson and F. N. David. The numbers on the curves indicate sample size. Ordinate, scale of ρ (population correlation coefficient); abscissa, scale of r (sample correlation coefficient).

REFERENCES

The letters to the left of each entry indicate the nature of the reference. *p:* paper; *g:* general audience, very little mathematics background required; *r:* restricted, calculus required; *x:* only those with the equivalent of an M.S. in statistics are admitted.

p AFIFI, A. A., and ELASHOFF, R. (1969). Multivariate Two Sample Tests with Dichotomous and Continuous Variables I: The Location Model, *Annals of Mathematical Statistics* **40**, 290–298.

p AFIFI, A. A., RAND, W. M., PALLEY, N. A., SHUBIN, H., and WEIL, M. H. (1971). A Method for Evaluating Changes in Sets of Computer Monitored Physiological Variables, *Computers and Biomedical Research* **4**, 329–339.

p AFIFI, A. A., SACKS, S. T., LIU, V. Y., WEIL, M. H. and SHUBIN, H. (1971). Accumulative Prognostic Index for Patients with Barbiturate, Gluetethemide and Meprobamate Intoxication, *New England Journal of Medicine* **285**, 1497–1502.

x ANDERSON, T. W. (1958). "An Introduction to Multivariate Statistical Analysis," Wiley, New York.

p ANSCOMBE, F. J. (1961). Examination of Residuals, *Proceedings of the Fourth Berkeley Symposium on Mathematical Statistics and Probability*, pp. 1–36. Univ. of California Press, Berkeley.

p ANSCOME, F. J., and TUKEY, J. W. (1963). The Examination and Analysis of Residuals, *Technometrics* **5**, 141–160.

p AZEN, S. P. (1969). Classification of Time-Dependent Observations, Rand Corp. Report, R-471-PR, Rand Corp,. Santa Monica, California.

p AZEN, S. P., and DERR, J. I. (1968). On the Distribution of the Most Significant Hexadecimal Digit, Rand RM-5496-PR, Rand Corp., Santa Monica, California.

p BARTLETT, M. S. (1937). Analysis of Covariance to Missing Values, *Journal of the Royal Statistical Society, Suppl.* **4**, 151.

g BENNETT, C. A., and FRANKLIN, N. L. (1954). "Statistical Analysis in Chemistry and the Chemistry Industry," Wiley, New York.

g BEYER, W. H. (Ed.) (1968). "Handbook of Tables for Probability and Statistics," (2nd ed.), Chemical Rubber Co., Cleveland, Ohio.

p BIRNBAUM, Z. W. (1952). Numerical Tabulation of the Distribution of Kolmogorov's Statistic for Finite Sample Size, *Journal of the American Statistical Association* **47**, 425–441.

g BLISS, C. W. (1967). "Statistics in Biology," Vol. 1, McGraw-Hill, New York.

p Box, G. E. P., and MUELLER, M. E. (1958). A Note on the Generation of Random Normal Deviates, *Annals of Mathematical Statistics* **29**, 610–611.

p Box, G. E. P., and WATSON, G. S. (1962). Robustness to Non-Normality of Regression Tests, *Biometrika* **49**, 99–106.

r BROWNLEE, K. A. (1965). "Statistical Theory and Methodology in Science and Engineering," Wiley, New York.

g BURINGTON, R. S. (1965). "Handbook of Mathematical Tables and Formulas," 4th Edition, McGraw-Hill, New York.

g BURINGTON, R. S. (1970). "Handbook of Probability and Statistics with Tables," (2nd ed.), McGraw-Hill, New York.

p Chen, E. H. (1971). A Random Normal Number Generator for 32-Bit-Word Computers, *Journal of the American Statistical Association* **66**, 400–403.

g CHURCHMAN, C. W., and RATOOSH, P. (1959). "Measurement Definitions and Theory," Wiley, New York.

g COCHRAN, W. G. (1953). "Sampling Techniques," Wiley, New York.

p COCHRAN, W. G. (1954). Some Methods of Strengthening the Common χ^2 Tests, *Biometrics* **10**, 417–451.

p CORNFIELD, J., and TUKEY, J. W. (1956). Average Values of Mean Squares in Factorials, *Annals of Mathematical Statistics* **27**, 907–949.

r Cox, D. R., and LEWIS, P. A. W. (1966). "The Statistical Analysis of Series of Events," Methuen, London.

x CRAMÉR, H. (1946). "Mathematical Methods of Statistics," Princeton Univ. Press, Princeton, New Jersey.

g DAVID, F. N. (1938). "Table of the Correlation Coefficient," Biometrika Office, University College, London.

r DAVIES, O. L. (Ed.) (1954). "Design and Analysis of Industrial Experiments," Oliver & Boyd, Edinburgh.

x DEMPSTER, A. P. (1969). "Elements of Continuous Multivariate Analysis," Addison-Wesley, Reading, Massachusetts.

g DIXON, W. J., and MASSEY, F. J. (1969). "Introduction to Statistical Analysis," (3rd ed.), McGraw-Hill, New York.

r DORN, W. S., and GREENBERG, S. N. (1967). "Mathematics and Computing," Wiley, New York.

r DRAPER, N. R., and SMITH, H. (1968). "Applied Regression Analysis," Wiley, New York.

g DUNN, O. J. (1967). "Basic Statistics: A Primer for the Biomedical Sciences," Wiley, New York.

x FELLER, W. (1966). "An Introduction to Probability Theory and Its Applications," Vol. II, Wiley, New York.

r FELLER, W. (1968). "An Introduction to Probability Theory and Its Applications," Vol. I (3rd ed.), Wiley, New York.

r FERGUSON, T. S. (1967). "Mathematical Statistics—A Decision Theoretic Approach," Academic Press, New York.

p FISHER, R. A. (1918). The Correlation between Relatives on the Supposition of Mendelian Inheritance, *Transactions of the Royal Society of Edinburgh* **52**, 399–433.

r FISHER, R. A. (1925). "Statistical Methods for Research Workers," 1st Edition, Oliver & Boyd, Edinburgh.

r FISHER, R. A. (1935). "The Design of Experiments," Oliver & Boyd, Edinburgh.

p FISHER, R. A. (1936). The Use of Multiple Measurements in Taxonomic Problems, *Annals of Eugenics* **7**, 179–188.

g FISHER, R. A., and YATES, F. (1963). "Statistical Tables for Biological, Agricultural, and Medical Research," (6th ed.), Oliver & Boyd, Edinburgh.

r FISZ, M. (1963). "Probability Theory and Mathematical Statistics," (3rd ed.), Wiley, New York.

p FIX, E., and HODGES, J. L. (1951). Non-Parametric Discrimination: Consistency Properties, USAF School of Aviation Medicine, Project Report 21-49-004, No. 4, Randolph AFB, San Antonio, Texas.

p FIX, E., and HODGES, J. L. (1952). Non-Parametric Discrimination: Small Sample Performance, USAF School of Aviation Medicine, Project Report 21-49-004, No. 11, Randolph AFB, San Antonio, Texas.

r GIBBONS, J. D. (1971). "Nonparametric Statistical Inference," McGraw-Hill, New York.

p GOODMAN, L. A., and KRUSKAL, W. H. (1954). Measures of Association for Cross Classifications, *Journal of the American Statistical Association* **49**, 732–764.

p GOODMAN, L. A., and KRUSKAL, W. H. (1959). Measures of Association for Cross Classifications, II: Further Discussion and Reference, *Journal of the American Statistical Association* **54**, 123–163.

p GOODMAN, L. A., and KRUSKAL, W. H. (1963). Measures of Association for Cross Classifications, III: Approximate Sampling Theory, *Journal of the American Statistical Association* **58**, 310–364.

p GRIZZLE, J. E. (1967), Continuity Correction in the χ^2-Test for 2×2 Tables, *American Statistician* **21**, 28–32.

g HALD, A. (1952). "Statistical Tables and Formulas," Wiley, New York.

r HAMMERSLEY, J. M., and HANDSCOMB, D. C. (1964). "Monte Carlo Methods," Methuen, London.

g "Handbook of Mathematical Tables" (1952). Chemical Rubber Publishing Co., Cleveland, Ohio.

r HARMAN, H. H. (1967). "Modern Factor Analysis," (2nd ed.), Univ. of Chicago Press, Chicago.

p HARTLEY, A. O. (1961). Modified Gauss–Newton Method for Fitting on Nonlinear Regression Functions, *Technometrics* **3**, 269–280.

r HASTINGS, C., Jr. (1955). "Approximations for Digital Computers," Princeton Univ. Press, Princeton, New Jersey.

g HILLS, M. (1966). Allocation Rules and Their Error Rates, *Journal of the Royal Statistical Society, Series B* **28**, 1–20.

r HOEL, P. G. (1963). "Introduction to Mathematical Statistics," Wiley, New York.

r HOGG, R. V., and CRAIG, A. T. (1965). "Introduction to Mathematical Statistics," (2nd ed.), Macmillan, New York.

g HORST, P. (1965). "Factor Analysis of Data Matrices," Holt, New York.

p HOTELLING, H. (1931). The Generalization of Student's Ratio, *Annals of Mathematical Statistics* **2**, 360–378.

p JENNRICH, R. I., and SAMPSON, P. F. (1966). Rotation for Simple Loadings, *Psychometrika* **31**, 313–323.

p KAC, M., KIEFER, J., and WOLFOWITZ, J. (1955). On Tests of Normality and Other Tests of Fit Based on Distance Methods, *American Mathematical Society* **25**, 189–198.

p KAISER, H. F. (1958). The Varimax Criterion for Analytic Rotation in Factor Analysis, *Psychometrika* **23**, 187–200.

r KENDALL, M. G., and STUART, A. (1967). "The Advanced Theory of Statistics, Vol. II: Inference and Relationship," Hafner, New York.

r KENDALL, M. G., and STUART, A. (1968). "The Advanced Theory of Statistics, Vol. III: Design and Analysis, and Time Series," Hafner, New York.

r KENDALL, M. G., and STUART, A. (1969). "The Advanced Theory of Statistics, Vol. I (3rd ed.): Distribution Theory," Hafner, New York.

p LACHENBRUCH, P. A. (1967). An Almost Unbiased Method of Obtaining Confidence Intervals for the Probability of Misclassification in Discriminant Analysis, *Biometrics* **23**, 639–646.

p LACHENBRUCH, P. A., and MICKEY, M. R. (1968). Estimation of Error Rates in Discriminant Analysis, *Technometrics* **10**, 1–11.

r LEHMANN, E. L. (1959). "Testing Statistical Hypotheses," Wiley, New York.

p LEWIS, P. A. W., GOODMAN, A. S., and MILLER, J. M. (1969). A Pseudo-Random Number Generator for the System/360, *IBM Systems Journal* **8**, 136–146.

p LILLIEFORS, H. W. (1967). On the Kolmogorov–Smirnov Test for Normality with Mean and Variance Unknown, *Journal of the American Statistical Association* **62**, 399–402.

r LINDGREN, B. W. (1968). "Statistical Theory," (2nd ed.), Macmillan, New York.

x LOÈVE, M. (1963). "Probability Theory," (3rd ed.), Van Nostrand, Princeton, New Jersey.

p LONGLEY, J. W. (1967). An Appraisal of Least Squares Programs for the Electronic Computer from the Point of View of the User, *Journal of the American Statistical Association* **62**, 819–829.

r McCRACKEN, D. M., and DORN, W. S. (1964). "Numerical Methods and Fortran Programming," Wiley, New York.

p MAHALANOBIS, P. C. (1936). On the Generalized Distance in Statistics, *Proceedings of the National Institute of Sciences of India* **12**, 49–55.

p MARQUARDT, D. W. (1963). An Algorithm for Least-Squares Estimation of Nonlinear Parameters, *Journal of the Society for Industrial and Applied Mathematics* **2**, 431–441.

g MARTIN, F. F. (1968). "Computer Modeling and Simulation," Wiley, New York.

p MASSEY, F. J. (1951). The Kolmogorov–Smirnov Test for Goodness-of-Fit, *Journal of the American Statistical Association* **46**, 68–78.

r MAXWELL, A. E. (1961). "Analysing Qualitative Data," Methuen, London.

r MOOD, A. M., and GRAYBILL, F. A. (1963). "Introduction to the Theory of Statistics," (2nd ed.), McGraw-Hill, New York.

r MORRISON, D. F. (1967). "Multivariate Statistical Methods," McGraw-Hill, New York.

p NEMENYI, P. (1969). Variances: An Elementary Proof and a Nearly Distribution-Free-Test, *American Statistician* **23**, 35–37.

r NOETHER, G. E. (1967). "Elements of Nonparametric Statistics," Wiley, New York.

p PALLEY, N. A., ERBECK, D. H., and TROTTER, J. A., Jr. (1970). Programming in Medical Real Time Environment, *AFIPS Conf. Proc.* **37**, 589–598.

p PALMERSHEIM, J. J. (1970). Nearest Neighbor Classification Rules: Small Sample Performance and Comparison with Linear Discriminant Function and Optimum Rule, Ph.D. Dissertation, Univ. of California, Los Angeles.

r PARZEN, E. (1960). "Modern Probability Theory and Its Applications," Wiley, New York.

g PEARSON, E. S., and HARTLEY, H. O. (1966). "Biometrika Tables for Statisticians," Vol. 1 (3rd ed.), Cambridge Univ. Press, Cambridge.

p PEARSON, K. (1901). On Lines and Planes of Closest Fit to Systems of Points in Space, *Philosophical Magazine*, Series 6 **2**, 559–572.

r RALSTON, A., and WILF, H. S. (1960). "Mathematical Methods for Digital Computers," Wiley, New York.

x RAO, C. R. (1965). "Linear Statistical Inference and Its Application," Wiley, New York.

r RUDIN, W. (1964). "Principles of Mathematical Analysis," (2nd ed.), McGraw-Hill, New York.

p SCHEFFÉ, H. (1953). A Method for Judging All Contrasts in the Analysis of Variance, *Biometrika* **40**, 87–104.

p SCHEFFÉ, H. (1956). Alternative Models for the Analysis of Variance, *Annals of Mathematical Statistics* **27**, 251–271.

r SCHEFFÉ, H. (1959). "The Analysis of Variance," Wiley, New York.

p SHUBIN, H., AFIFI, A. A., RAND, W. M., and WEIL, M. H. (1968). Objective Index of Haemodynamic Status for Quantitation of Severity and Prognosis of Shock Complicating Myocardial Infarction, *Cardiovascular Research* **2**, 329–337.

g SIEGEL, S. (1956). "Non-Parametric Statistics for the Behavioral Sciences," McGraw-Hill, New York.

r SNEDECOR, G. M., and COCHRAN, W. G. (1967). "Statistical Methods," Iowa State Univ. Press, Ames, Iowa.

p STEWART, D. H., ERBECK, D. H., and SHUBIN, H. (1968). Computer System for Real Time Monitoring and Management of the Critically Ill, *AFIPS Conf. Proc.* **33**, 797–807.

g THURSTONE, L. L. (1945). "Multiple Factor Analysis," Univ. of Chicago Press, Chicago.

r TORGERSON, W. S. (1958). "Theory and Methods of Scaling," Wiley, New York.

p TUKEY, J. W. (1949). One Degree of Freedom for Non-Additivity, *Biometrics* **5**, 232–242.

p TUKEY, J. W. (1949). Comparing Individual Means in Analysis of Variance, *Biometrics* **5**, 99.

p TUKEY, J. W. (1962). The Future of Data Analysis, *Annals of Mathematical Statistics* **33**, 1–67.

g WALSH, J. E. (1965). "Handbook of Nonparametric Statistics," Van Nostrand, Princeton, New Jersey.

g WEIL, M. H., and SHUBIN, H. (1967). "The Diagnosis and Treatment of Shock," Williams & Wilkins, Baltimore, Maryland.

p WEIL, M. H., and AFIFI, A. A. (1970). Experimental and Clinical Studies on Lactate and Pyruvate as Indicators of the Severity of Acute Circulatory Failure (Shock), *Circulation* **XLI**, 989–1001.

p WELCH, B. C. (1937). The Significance of the Difference between Two Means when the Population Variances Are Unequal, *Biometrika* **29**, 350–362.

AUTHOR INDEX

SUBJECT INDEX

A

Algorithms, 30
Analysis
  of covariance, 212ff
  of variance, 143ff
    components of variance model, 159
    factorial, 181ff
    fixed effects model, 144, 156ff, 181
    hierarchical, 177
    Latin square, 189
    mixed model, 161, 181
    model, 143
    Model I, 155, 181
    Model II, 155, 181
    in multiple linear regression, 109
    nested model, 177
    one-way, 71, 154ff
    partially nested design, 204
    random effects model, 144, 156, 159ff, 181
    randomized blocks model, 174ff, 186
    in simple linear regression, 94, 99
    split plot design, 187ff
    two-way, 162ff
Anova, *see* Analysis of variance
Average, 293

B

Bayes theorem, 238

B

Binomial distribution, *see also* Distribution
  normal approximation to, 38
Biomedical Computer Programs, 8, 31, 61, 139, 200, 204, 230, 242, 264, 271
Blunders, 36, 43
BMD, *see* Biomedical Computer Programs

C

Cdf, *see* Cumulative distribution function
Central limit theorem, 318
Chi-square distribution, *see also* Distribution
  random generation of, 34
Chi-square test
  for contingency tables, 81
  for goodness-of-fit, 49
  for mean vectors, 231
  for outliers, 228
  for proportions, 41, 81
  for single variance, 55
Class interval, 43
Classification, 235ff
  Bayes procedure for, 238
  in binomial populations, 251
  in $k \geqslant 2$ populations, 246
  table, 248
  in two populations, 235ff
Cochran's theorem, 151
Coding, 11ff
  sheet, 11